MAKING MODERN SCIENCE

D0830603

MAKING

PETER J. BOWLER
AND IWAN RHYS MORUS

MODERN SCIENCE

A HISTORICAL SURVEY

THE UNIVERSITY OF CHICAGO PRESS
Chicago and London

PETER J. BOWLER is professor of the history of science in the School of Anthropological Studies at Queen's University, Belfast, and the author of *Reconciling Science and Religion: The Debate in Early-Twentieth-Century Britain.* IWAN RHYS MORUS is lecturer in the Department of History and Welsh History at the University of Wales, Aberystwyth, and the author of *When Physics Became King.*

The University of Chicago Press, Chicago 60637
The University of Chicago Press, Ltd., London
© 2005 by The University of Chicago
All rights reserved. Published 2005
Printed in the United States of America

14 13 12 11 10 09 08 07 06 05 1 2 3 4 5

ISBN: 0-226-06860-9 (cloth)
ISBN: 0-226-06861-7 (paper)

Library of Congress Cataloging-in-Publication Data
Bowler, Peter J.
Making modern science : a historical survey / Peter J. Bowler and Iwan R. Morus.
p. cm.
Includes bibliographical references and index.
ISBN 0-226-06860-9 (cloth : alk. paper) — 0-226-06861-7 (pbk. : alk. paper)
1. Science—History—Textbooks. 2. Science—History—20th century—Textbooks. I. Title.
II. Morus, Iwan Rhys, 1964–

Q125 .B678 2005
509′.04—dc22 2004019553

⊗The paper used in this publication meets the minimum requirements of the American National Standard for Information Sciences—Permanence of Paper for Printed Library Materials, ANSI Z39.48-1992.

CONTENTS

PREFACE

THIS BOOK AROSE FROM OUR DESPERATE SEARCH for a suitable textbook to accompany newly organized survey courses in the history of science for first-year undergraduates. We soon became convinced that no such text exists, and we suspected that we were by no means the only lecturers crying out for one to be written. At the same time, we realized that the lack of a suitable textbook meant that there was no reliable introduction to the field available to general readers. This book is meant to fill this gap, and we feel that we are in an ideal situation to provide a survey that will be both useful to other lecturers and of interest to general readers (including scientists) looking for an introduction to the way our field now operates. We are both experienced historians, and our interests are complementary, allowing us to provide a survey of the physical, the life, and the earth sciences. Equally important, we are also experienced teachers and writers, and the first drafts of many of the chapters were subjected to the test of experience through circulation to our own students over two academic years. The feedback they have supplied has helped to ensure that what we have written is accessible to this kind of student and (we hope) to the general reader.

Although the book has its origins in our search for a textbook, we have resisted the temptation to go too far down the road of turning it into a conventional textbook, with all the teaching apparatus that implies. This is because we do want the book to serve the additional purpose of providing a survey that will be attractive to the general reader. The interests of the student and the general reader may be significantly different. As explained below, lecturers will seldom use the whole of this book to teach their courses on the history of science—they will pick out a selection of chapters relevant to how they teach. This means that the chapters have to be to some extent self-standing because students will not necessarily be reading them consecutively. The general reader—who may want something more closely resembling a conventional historical narrative—may find this a little disconcerting. At the same time, even some general readers may prefer to sample the book, beginning with their own areas of interest, rather than reading it from cover to cover. Those who are looking for a more coherent narrative should remember that the history of science is a complex and often controversial field, so that any introduction that does justice to the whole will necessarily have to present a wide range of topics and themes.

A major problem confronting anyone planning a survey intended as a textbook is the vast range of different approaches that can be used to teach the history of science, depending on the interests of individual lecturers and the diverse backgrounds of the students to be taught (some will be science students, others will know little or no science). In our own teaching we have adopted two strategies, and this is reflected in the book we have written. One of our courses focusses on particular episodes in the history of science, the other discusses broader themes that may apply to various sciences and historical periods. By writing chapters in both these formats, we think we have generated a text that can be used by instructors adopting a wide range of teaching strategies. Obviously, no text can hope to cover every area in the development of modern science from the Copernican Revolution onward, but we hope that the package of topics we have chosen will be attractive to a large number of both teachers and general readers. It includes some topics that have been standard fare for historians of science for a generation or more, and some that reflect newer trends and interests.

The book is divided into two parts, one on episodes, the other on themes. Cross-references are provided so that students can be assigned clearly defined packages of reading even if this involves reference to chapters in both sections. Thus several of the episodic chapters will raise issues concerned with the interaction of science and religion, so at relevant points in these chapters students are directed to the appropriate thematic chapter for further reading. Should the instructor prefer to teach through themes, the thematic chapters will be the primary reading, and again cross-references direct students to appropriate episodes for further information on the examples used. The cross-references will also help the general reader to fit the material together to give a comprehensive overview of the history of science. Each chapter has a substantial list of references so that anyone wishing to take the topic further is provided with a guide to more specialist reading.

UCTION:
E, SOCIETY, AND HISTORY

Tell someone that you are reading about the history of science and their first reaction will probably be to ask: "What's that?" We instinctively associate science with the modern world, not with the past. Yet a moment's thought resolves the paradox—like any human activity, science has a history, and most people can recall at least a few "great names" associated with key discoveries that have shaped our modern way of thought. Scientists themselves think about the past along similar lines, though they may have a more esoteric list of names at their disposal linked to the major discoveries in their own area. For the scientist, pinpointing a sequence of great advances in our knowledge of the world creates an image of modern science as the continuation of a progressive struggle to drive back the boundaries of ignorance and superstition. But some of the great names familiar to the public evoke images that suggest that the advance of science has not been a smooth process of fact gathering. Almost everyone has heard the story of Galileo's trial by the Inquisition for teaching that the earth goes round the sun, and the controversy sparked by Darwin's theory of evolution remains active still today. As science has come to play an ever-increasing role in our lives the potential for controversy expands so that it now includes our ability to interfere with the most fundamental aspects of our biological and psychological character and even the biosphere of the earth itself. It would be surprising indeed if the history of these areas of science turned out not to be controversial.

The scientists themselves are relatively comfortable with the fact that some of the great discoveries had consequences that forced everyone to rethink their religious, moral, or philosophical values. Science textbooks often tell stories about the great discoveries that present them as steps in a cumulative process by which our understanding of the natural world has expanded. If the new knowledge challenged existing beliefs, then people simply had to learn to live with it. The history of science certainly gains some of its popular audience by exploring the impact of science on the wider world. But it also likes to evaluate the traditional stories that the scientists tell about the past, and in some cases the results are welcomed less eagerly by the scientists. All too often, it turns out that the conventional stories are vastly oversimplified—they are myths that "tidy up" the messy process of controversy surrounding any new innovation (Waller 2002). These myths present a clear-cut image of heroes (who discover or promote the new theory) and villains (who oppose it, usually because their objectivity is subverted by their existing beliefs). Historians often refer to the stories of the great discoveries as a form of "Whig history," a term borrowed from those British historians of the Whig or liberal party who retold the nation's history in terms of the inevitable triumph of their own political values. Nowadays, any history that treats the past as a series of steppingstones toward the present—and assumes that the present is superior to the past—is called Whig history. The conventional stories of the past that appear in the introductory chapters of science textbooks are certainly a form of Whiggism. Historians take great delight in exposing the artificially constructed nature of these stories, and some scientists find the results uncomfortable.

In principle, though, there is no reason why scientists (of all people) should shrink from exposing their ideas to scrutiny, even if the evidence used is based on old books and papers, rather than laboratory tests. If the results paint a more complex and realistic picture of how science works, anyone engaged in modern scientific research ought to recognize the value of portraying past developments in same terms as the present. Instead of cardboard cut-out figures, they can have real heroes, warts and all.

The scientists are understandably less happy when detailed studies of past or present controversies lead people to challenge the actual process by which science claims to advance our knowledge of the world. The modern "science wars"—in which scientists have responded bitterly when the objectivity of science itself has been challenged by sociological critics—illustrate that there is more at stake here than a simple conflict between scientific fact and subjective values. Those who do not like the consequences of science are increasingly inclined to argue that a process that generates potentially dangerous techniques cannot be seen as the mere acquisition of factual knowledge. The history of science has inevitably been sucked into the science wars since some of the ammunition used by those who attack science comes from the reevaluation of key areas where science has generated controversy in the past. The critics argue that the very foundations of scientific "knowledge" are contaminated by values. Science constructs a view of the world that sees it through tinted glasses—so we should hardly be surprised when it turns out that what is offered to us as knowledge tends to reinforce the value system of the military-industrial complex that funds it. Scientists respond with fury when confronted with this line of argument. If science is just another value system no more privileged than anyone else's, why does it work so well when we apply it to manipulate the world via technology and medicine? Those who pay are at least paying for results, not fairy stories. There is a genuine tension here, and the history of science is sucked into the debate as one of the prime sources of information about how science actually works.

Anyone turning to this survey of the history of modern science expecting an uncontroversial list of great discoveries is thus in for a shock. Virtually all the topics and themes we discuss are the subject of intense debate, often sustained by differing perspectives derived from historians' attitudes toward modern science as a whole or toward particular theories and their applications. Teaching as we do in Northern Ireland, we are used to the idea that history can become a battleground on which people with rival opinions seek to validate their beliefs. Irish history can be told from two very different perspectives, depending on whether you approach it from a Nationalist or a Unionist perspective. Was Oliver Cromwell a hero who made British civilization safe in Ireland, or the villain who massacred the inhabitants of Drogheda? It depends on your point of view—each side has constructed its myths of the past, and each may be discomfited when the academic historian uses hard evidence to probe those myths. The history of science certainly challenges many of the myths created by those who present science as a disembodied search for the truth—but does it necessarily support those who claim that science is no more than the expression of a particular value system? Perhaps a middle way is possible, presenting a vision of science as a human activity, albeit one that has more concrete achievements to its credit than most others. In a sense, the very dangers the critics warn about arise from the fact that science performs work, in the sense that it can be applied to change the world we live in.

What we hope you will learn from this book is a willingness to see history as something more than a list of names and dates—it is something that people argue about because the evidence can be interpreted in different ways and they care passionately about the interpretation they support. You will see how historians use evidence to challenge myths, but you should also be cautious and critical in your evaluation of any alternative stories they offer (including our own). It may be hard work, but it will force you to confront important issues—and it will be a lot more fun than learning names and dates.

The rest of this introduction will put flesh on the bare bones of the conflicts outlined above, beginning with a brief survey of how the history of science became the professional field of study it is today. This is important, because many of the older books listed in the readings below—still used because they are classics in their field—were written when the discipline worked very differently from the way it does now. We then outline the more recent developments that have created the modern approach to the subject, including the more sociological techniques that generate the controversies mentioned above. Knowing something about the history of the history of science will help you to understand why the issues discussed in the rest of this book are often so controversial.

The Origins of the History of Science

Something like a history of science in the modern tradition began to emerge in the eighteenth century. This was the Age of Enlightenment, when radical thinkers proclaimed the power of human reason to throw off ancient superstition and provide a better foundation for society. Many of these Enlightenment thinkers were hostile to the Church, which they saw as an agent for the old social hierarchy derived from feudal times. The medieval period was portrayed as one of stagnation, imposed by the Church's rigid endorsement of the traditional worldview. The radicals saw the New Science of the previous century as the first manifestation of a renewed flowering of rational thought and hailed the chief contributors to the modern worldview, including Galileo and Newton, as its heroes. The fact that Galileo had gotten into trouble with the Church for proclaiming Copernican astronomy merely fueled their suspicion of that institution. They carefully suppressed any hint that Newton had dabbled in magic and alchemy. From the Enlightenment's view of its own immediate past we have inherited the assumption that the Scientific Revolution of the seventeenth century was a turning point in the progress of Western thought, and a pantheon of heroes identified with the key steps in the foundation of modern cosmology and physical science.

In 1837 the British scientist and philosopher William Whewell published a massive *History of the Inductive Sciences.* It was Whewell who actually coined the term "scientist," and he had a very specific agenda that in some respects modified the Enlightenment program. He certainly agreed that science was a progressive force, but he had a new vision of how it should set about building an understanding of nature, derived from the German philosopher Immanuel Kant. For Kant and Whewell, knowledge was not simply derived passively from the observation of nature—it was imposed by the human mind via the theories we use to describe the world. The scientific approach rested on the rigorous testing of new hypotheses by observation and experimentation. Whewell subsequently published a *Philosophy of the Inductive Sciences* in which it became clear that his purpose was to use history as a means of illustrating how his vision of the methodology of science was applied in practice. In this respect he contributed to what would become a principal motivation for the creation of the modern discipline of the history of science.

Whewell was more conservative than the Enlightenment thinkers in that he defended the possibility that the scientist might find phenomena that could only be explained as the result of divine intervention. Later on he would refuse to allow a copy of Darwin's *Origin of Species* into the library at Trinity College, Cambridge, because it replaced divine miracle with natural evolution. But to a new generation of radical thinkers in the late nineteenth century, Darwinism confirmed that science was continuing its assault on ancient superstitions, renewing the campaign begun by Galileo. A new generation of histories emerged stressing the inevitability of a "war" between science and religion, a war that science would inevitably win. J. W. Draper's *History of the Conflict between Science and Religion* of 1875 was a pioneering effort in this revival of the Enlightenment program. The metaphor of conflict continues to dominate popular discussion of the relationship, although it has been extensively challenged by later historians.

To those who (like Whewell) retained the hope that science and religion could work in harmony, the materialist program of the Enlightenment was a positive danger to science. It encouraged scientists to abandon their objectivity in favor of the arrogant claim that the laws of nature could explain everything. Alfred North Whitehead's *Science and the Modern World* (1926) urged the scientific community to turn its back on this materialist program and return to an earlier vision in which nature was studied on the assumption that it would reveal evidence of divine purpose. This model of science's history dismisses episodes such as the trial of Galileo as aberrations and portrays the Scientific Revolution as founded on the hope that nature could be seen as the handiwork of a rational and benevolent Creator. For Whitehead and others of his generation, evolution itself could be seen as the unfolding of a divine purpose. This debate between two rival views of science—and hence of its history—is still active today.

In the early twentieth century, the legacy of the rationalist program was transformed in the work of Marxists such J. D. Bernal. Bernal, an eminent crystallographer, berated the scientific community for selling out to the industrialists. In his *Social Function of Science* (1939) he called for a renewed commitment to use science for the good of all. His 1954 *Science in History* was a monumental attempt to depict science as a potential force for good (as in the Enlightenment program) that had been perverted by its absorption into the military-industrial complex. In one important respect, then, the Marxists challenged the assumption that the rise of science represented the progress of human rationality. For them, science had emerged as a by-product of the search for technical mastery over nature, not a disinterested search for knowledge, and the information it accumulated tended to reflect the interests of the society within which the scientist functioned. The aim of the Marxists was not to create a purely objective science but to reshape society so that the science that was done would benefit everyone, not just the capitalists. They dismissed the program advocated by Whitehead as a smokescreen for covering up science's involvement in the rise of capitalism. Similarly, many intellectual historians reacted furiously to what they regarded as the denigration of science implicit in works such as the Soviet historian Boris Hessen's "The Social and Economic Roots of Newton's 'Principia'" from 1931. The outbreak of World War II highlighted two conflicting visions of science's history, both of which linked it to the dangers revealed in Nazi Germany. The optimistic vision of the Enlightenment had vanished along with the idea of inevitable progress in the calamities that the Western world had now experienced. Science must either turn its back on materialism and renew its links with religion or turn its back on capitalism and begin fighting for the common good.

It was at this time that the history of science began to achieve recognition as a distinct academic specialization. There had been earlier efforts, but these had enjoyed limited success. The Belgian scholar George Sarton founded the journal *Isis* in 1912—it continues today as the organ of the History of Science Society—but on moving to America he found it impossible to persuade Harvard University to create a history of science department at that time. The first specialist departments only began to flourish after World War II, reflecting a concern that the technological consequences of science were now so powerful that broader analysis of its history was essential to understand how it had come to play this dominant role in society. But with the outbreak of the Cold War against Soviet Russia, it was inevitable that Bernal's Marxist outlook would be marginalized. Despite the obvious links with technology, the image of science as a by-product of social and economic forces was unacceptable. The alternative was a return to the idea that science represented an important intellectual force in Western culture, paving the way for progress not by its subservience to industry but by its independence and innovation, which had given us a better understanding of nature at a theoretical level. It was the practical applications of this new knowledge that were the by-product—the Marxists had got it the wrong way round. Those applications could be studied quite separately from the development of pure science, which now became, in effect a part of Western culture to be studied by the techniques of intellectual history or the history of ideas. What counted was theoretical innovation at the conceptual level and the process by which theories were tested against the evidence.

This approach to historiography followed the Enlightenment program to the extent that it saw the emergence of the scientific method, and the main steps in the creation of the modern worldview, as major contributions to human progress. Much attention thus focused on the Scientific Revolution of the seventeenth century and the associated developments in astronomy and physics. Later steps were also highlighted and used to define the main line of advance in scientific thought. The advent of Darwinism was seen as a key step forward, and developments in associated sciences such as geology were defined as good or bad depending on whether they seemed to promote the search for natural processes of change. To some extent, the field thus continued and extended the Whiggish approach favored by the scientists themselves, because progress was defined in terms of steps toward what were perceived to be the main components of our modern worldview. In another respect, however, the new historiography of science did go beyond Whiggism: it was willing to admit that scientists were deeply involved with philosophical and religious concerns and often shaped their theories in accordance with their views on these wider questions. A leading influence here was the Russian émigré Alexandre Koyré, working in France and America, who used close textual analysis of classic works in science to demonstrate this wider dimension. Koyré (1978) argued that Galileo was deeply influenced by the Greek philosopher Plato, who had taught that the world of appearances hides an underlying reality structured along mathematical lines. Newton, too, turned out to be a far more complex figure than the old Enlightenment hero, deeply concerned with religious and philosophical issues (Koyré 1965).

The one area of influence that was not considered relevant was the social and economic. Marx's suggestion that Darwin's theory of natural selection reflected the competitive values of the capitalist system was not on the agenda, nor was the association of science with technology and industry. No one doubted that science did have important consequences for society at large, either by influencing religious or political debates or by providing practical information that could be applied through technology or medicine. But these practical applications always came after the science was finished—they had no influence on how the actual research was done. There was supposed to be a clear distinction between the "internal" history of science, which studied the intellectual factors involved in the development of theories, and "external" history, which looked at the wider implications of what was discovered. The postwar generation of historians had a clear preference for internal history—they wanted a history of science firmly situated within the history of ideas, with the external applications left for the separate disciplines of the history of technology and the history of medicine. A good example of this generation's work is Charles C. Gillispie's *Edge of Objectivity* (1960); its most enduring legacy is the monumental *Dictionary of Scientific Biography* (Gillispie 1970–80).

Because of its focus on how new theories were developed, this approach to the history of science revived the program sketched out by Whewell. History was to be used as a source of examples to illustrate the correct application of the scientific method. The history of science and the analysis of the scientific method were supposed to go hand in hand, and several universities now founded departments of the history and philosophy of science. This was, in any case, a period when work in the philosophy of science was extremely active. The old idea of science as a process of fact gathering had been replaced by the "hypothetico-deductive method" in which the scientist proposed hypotheses, deduced testable consequences, and then allowed experimental tests to determine whether the hypothesis should be rejected (Hempel 1966). This emphasis on the scientists' willingness to test and, if necessary, refute hypotheses was carried even further by Karl Popper in his *Logic of Scientific Discovery* (1959). Popper's starting point was the need to establish a line of demarcation separating science from all other intellectual activities such as theology and philosophy. The defining character of science was its reliance on "falsifiability": a scientific hypothesis is always framed in such a way as to maximize its exposure to experimental testing and potential refutation. According to Popper, religious believers, philosophers, and social analysts all evade this requirement by making their propositions so vague that they can explain almost anything and thus can never be refuted. Science thus provides a unique form of knowledge about the world because its theories have all survived rigorous testing.

There was, however, an uncomfortable consequence of the hypothetico-deductive method as far as scientists were concerned. As Popper stressed, no hypothesis can ever be proved to be true because no matter how many positive tests it survives, there is still the possibility that the next one may refute it. The history of science is full of examples showing that a theory can be successful for decades or even centuries and then be exposed as false—think of Einstein's undermining of the conceptual foundations of Newtonian physics. This means that our current theories, too, will eventually be refuted; they can be accepted only provisionally, as the best guides we have available at the moment. Scientists reluctantly accepted this implication of the new philosophy of science, giving up their claim to be providing absolutely true knowledge of the real world. They were willing to do this because Popper offered them a different defense of their objectivity through his criterion for distinguishing science from all other forms of knowledge. Science was objective in the sense that it exposed the weaknesses of its claims as quickly as possible and went on to devise something better.

There was, however, another problem lying at the heart of Popper's methodology that made historians of science instinctively suspicious. For Popper, the good scientist actively seeks to refute the current hypothesis—it is tested in the hope of exposing its weaknesses as quickly as possible. This delineation of what constitutes good science does not fit very well with the observed behavior of scientists, past or present. On the contrary, they get very attached to a successful theory, especially if they have built their careers on it, and are often reluctant to consider, if not actively hostile to, any suggestion that it should be replaced. Here was the point at which the history and philosophy of science began to part company. It seemed to many historians that the more they studied the actual behavior of scientists, the less it fit the idealized picture of the scientific method that the philosophers were devising. The philosophy of science was becoming an armchair discipline, creating ever more elaborate ideas about what scientists ought to do that were increasingly out of touch with how science really worked. The way was opening up for a challenge that would take the history of science in a new direction, creating a sociological model that would study the actual functioning of the scientific community.

Science and Society

The challenge came in the form of Thomas S. Kuhn's *The Structure of Scientific Revolutions* (1962), which sparked immense debate and has since become a classic. Kuhn argued that the replacement of theories is a much more complex affair than the orthodox or Popperian philosophies of science imply (on the resulting debate, see Lakatos and Musgrave [1970]). Kuhn used history to show that successful theories establish themselves as the "paradigm" for scientific activity in the field: they define not only acceptable techniques for tackling problems but also which problems are to be considered relevant for investigation. Not surprisingly, the cards are stacked in the theory's favor because the chance of falsification is minimized by working in "safe" areas. Science done under the influence of a dominant paradigm is what Kuhn calls "normal science": it is real research, but it is more concerned with filling in minor details than probing the foundations. Scientific education involves brainwashing the students so they accept the paradigm uncritically. Even when anomalies (experiments or observations that give unexpected results) begin to appear, the scientific community has become so loyal to the paradigm that older scientists refuse to admit that it has been falsified and continue as though it were still functioning smoothly. Only when the number of anomalies becomes unbearable will a "crisis state" emerge, when younger and more radical scientists begin to look around for a new theory. When a new theory is found that deals with the outstanding problems, it soon establishes itself as the new paradigm and another period of unadventurous normal science begins.

Kuhn's approach stresses that each paradigm represents a new conceptual scheme, incompatible with any other. But it also treats science as a social activity: scientists develop professional loyalties to the paradigm they were educated into that also restrict their ability to challenge the status quo. If this interpretation is valid, there are episodes in which science is anything but objective. On the contrary, scientists will use any trick in the book to defend the theory on which so many careers were founded. Objectivity may seem to be restored at the time of a revolution, but this is soon lost. And although the new paradigm seems to expand our range of knowledge by dealing with facts that could not be incorporated into the old theory, Kuhn notes that there are cases where successful lines of investigation under the old paradigm were abandoned under the new. Not surprisingly, scientists were deeply unhappy about Kuhn's analysis, but historians—while critical of his actual model of revolutions—found his approach a refreshing alternative, one that seemed to offer a more realistic model of how science is actually done.

Sociologists of science such as Robert K. Merton and his followers had also started casting an eye on the sociological conditions that made science possible. While Merton assumed that scientific knowledge was the straightforward result of applying scientific methodology, he argued that particular social conditions, or "norms," needed to be established in order for the scientific community to be able to flourish and apply the scientific method properly (Merton 1973).Without these norms—or generally understood rules of behavior—science would be distorted in various ways by ideological contamination. Merton identified four norms: universalism (scientific claims would be assessed impartially, with no reference to the individual scientists making them), communism (scientific knowledge belonged to the scientific community rather than to individual scientists), disinterestedness (scientists would not develop an emotional or other attachment to their work), and organized skepticism (scientists would systematically subject scientific claims to rigorous checking). Merton's norms were meant to provide a way of distinguishing science from other kinds of activities as well as defining the social circumstances under which science could flourish. Unlike Kuhn, so long as the norms were in operation, Merton did not believe that social circumstances could affect the development of scientific knowledge. Only in societies where the norms could not operate—such as Nazi Germany—did science become contaminated by ideological factors.

Subsequent work has expanded on insights contained explicitly or implicitly in Kuhn's work, sometimes in directions he would not have approved. His book is now seen by some as a pioneering contribution to the mode of analysis called postmodernism, although the main source of this movement derives from French philosophers such as Michel Foucault (1970; see Gutting 1989) or Jacques Derrida. For some, at least, within the postmodernist academic community, science has no privileged position as a source of knowledge because scientific literature forms just one among many rival bodies of texts seeking to gain control of our thoughts and activities. Science's success rests not on any truth value in its propositions but on the power of its proponents to enforce their own interpretations and "readings" of those texts on others. On the model of the history of thought provided by Foucault, Kuhn was quite right to claim that successive paradigms represent different patterns of analysis that cannot be compared objectively with one another. It is like a gestalt switch in psychology: what seems obvious from one perspective simply cannot be seen or understood from the other. The whole idea of science offering cumulative factual knowledge of the world thus goes out the window—leading to howls of outrage from scientists who perceive the "academic left" that endorses this relativist view of knowledge as a major threat to their position (Gross and Levitt 1994; Brown 2001). The resulting controversies, which became known as the "science wars," saw scientists defending their role as experts offering factual information about the world against sociologists who insisted that no one version of knowledge should be accorded such privileged status. Few historians would go so far as some postmodernists in their portrayal of science as a collection of free-floating texts with no reference to the material world. But the ideas of Kuhn and Foucault have forced us to think far more carefully about the literature of earlier periods, driving home the need to avoid reading modern ideas into older texts and alerting us to the possibilities that concepts and distinctions that we take for granted today may have been literally unthinkable for earlier generations of scientists.

The protests against the academic left have also been launched against another major development that has influenced the history of science: the intensification of interest in the way the scientific community functions. Kuhn drew attention to the power that prominent scientists have to shape the way their students and colleagues respond to new hypotheses. Only the most original would be willing to "rock the boat" by suggesting a totally new approach, and this tactic would only succeed when almost everyone had reluctantly begun to admit that the current paradigm was facing difficulties. Historians and sociologists of science then saw that it was often not enough to have good ideas or good evidence to back them up—the successful scientist has to persuade his or her colleagues to take new ideas seriously, often in competition with a host of rival proposals. While it might be nice to imagine that the winner will always be the one with the best evidence, things are rarely so straightforward. It is rare indeed for new evidence to be so unambiguous that it commands immediate assent. Success or failure often hinges on "nonscientific" factors as well, such as access to good research funding, new jobs, or the editorial committees of important journals. The emergence of the modern form of scientific community, with its societies, meetings, and journals, thus becomes a crucial factor in the creation of science as we understand it today. And studying a "revolution" involves showing how the new theory made its way within the political maneuvers that determined who had influence in the community as much as it involves studying conceptual changes and innovations in practice (Golinski 1998).

Investigation of such factors has now gone far beyond the Kuhnian model, however, because it is clear that as the scientific community has grown in size, it has become ever more specialized and fragmented. Theories can often become dominant only within a single narrow community of specialists, and the most innovative work will require the founding of a "splinter group" that establishes itself as a separate research tradition. The processes of professionalization and disciplinary specialization are now seen as crucial to the way science advances, to the extent that some historians no longer concentrate on broad theoretical perspective such as evolutionism in biology. Unless a theory is used to establish a distinct research tradition, it becomes marginalized in this new historiography—leaving some historians to wonder if such a sociological approach may have thrown the baby out with the bathwater. In some cases, theories have gained recognition precisely because they have served as bridges between specializations.

One consequence of this new approach is a recognition that science is a practical activity in which the devising of new techniques is as crucial as conceptual innovation. New specialisms often involve not only new theories but also new forms of apparatus requiring skilled operation to get meaningful results out of them. A now classic study by Steven Shapin and Simon Schaffer (1985) showed how seventeenth-century debates about the nature of the air depended crucially on who had access to the very few air pumps then available, along with the practical skill needed to make these primitive machines work properly. But this focus on the need to see science as a body of practice as well as theory goes far beyond laboratory apparatus. Developments in natural history depended on the founding of museums in which specimens could be used for comparison. Geologists had to develop techniques for mapping strata and representing their order of succession, and as Martin Rudwick (1985) has shown, there was an intense period of negotiation among specialists to agree on which techniques to use. The creation of modern genetics was to a large extent dependent on identifying and learning to control a suitable research organism, most notably the fruit fly *Drosophila melanogaster* (Kohler 1994). More seriously threatening to the old internal-external division is the growing evidence that scientists' choice of research areas and the techniques needed to investigate them often depended on their links with industrialists hoping to exploit the new knowledge. Nineteenth-century physicists such as William Thomson (Lord Kelvin) may have been brilliant theoreticians—but they worked hand in glove with the manufacturers of steam engines and the companies laying telegraph cables, and their work shows clear evidence of their involvement in the resulting practical problems.

Modern scientists have become used to the need for vast amounts of financial support, and few would deny that practical concerns often shape the priorities of researchers, determining which problems get investigated and which do not. But the suggestion that science can be driven by practical concerns points us toward the more controversial claim that what is presented as scientific "knowledge" may itself reflect the interests of those who do the research. Here we enter the domain of the "sociology of knowledge," which insists that science should be studied like any other form of knowledge—by looking at how it expresses and maintains the interests and values of those who construct it. The supposed "objective truth" of scientific theories can play no part in explaining their origins or why their supporters defend them. The parallels between this and the postmodernist view described above are obvious: if each scientific theory must be treated as a conceptual system that cannot be judged by the standards of any other, then no theory can claim to be closer to the truth. The sociology of science movement links the existence of alternative visions of reality to the interests of the groups that promote them. The original exponents of this sociological perspective are often called the Edinburgh school—since many of them originally taught at the Science Studies Unit of the University of Edinburgh (Barnes and Shapin 1979; Barnes, Bloor, and Henry 1996). They argue that science is a social activity like anything else and must be analyzed by sociological methods. The knowledge claims made by scientists should be treated in just the same way as those made by religious thinkers or political leaders. Just as religions and political systems are expressions of the interests of particular groups in society (usually the rulers), so scientific knowledge expresses the values of those who create it. Scientific theories are not collections of facts; they are models of the world that are to some extent capable of being tested by the facts. But those facts do not determine the structure of the theories absolutely, and the theories may thus be shaped by images of the world dictated by social values. As the study by Shapin and Schaffer (1985) showed, those interests may be philosophical or political, as well as economic, or they may reflect professional rivalries. The point is that to understand what is really going on in any piece of scientific research, we cannot simply assume that the whole thing is being determined by the structure of a "real world" that will be accurately represented by any successful model.

Critics of the Edinburgh school argue that their image of science is unrealistic. Science must offer knowledge of the real world or it will not help us to control that world via technology. If social values alone determine what counts as scientific knowledge, scientists would be free to make up any theory they chose and simply manipulate the testing to make it look like the theory was working. The theory would be accepted uncritically by everyone who shared the same social values. It would be rejected by those who had different values, and science could never come to a consensus of which theory was the best. That the community often does come pretty close to a consensus however, clearly cannot rule out the possibility that social factors shaped the origins of the successful theory (Darwin's theory of natural selection is a case in point). Sociologists insist in response that they do not claim that scientists "make it up as they go along." On the contrary, they are particularly interested in the ways in which scientists use the results of their experiments, their instruments, and their measurements to convince others of the superiority of their research programs (Collins 1985; Latour 1987). They point out, however, that in any situation there will be more than one way of pushing research ahead, and more than one way of devising a workable model. Which area of research, and which model, is actually chosen will depend on the interests of the particular group of scientists concerned. The supporters of one model may eventually able to convince the whole community that it offers the best solution, but the fact that even physics has experienced conceptual revolutions suggests that successful theories do not offer "correct" representations of the real world in any absolute sense.

In a complex and value-laden area such as the biology of human nature, it is possible to construct rival models that will each appear to work as the basis for scientific research, and the possibility of convincing everyone that a particular theory is correct are more limited. In part this is because more than one area of science can claim the right to offer theories relevant to the main questions. Biologists will naturally prefer models of human nature that stress the determining role of biological factors, since this allows them to insist that their expertise must be taken into account. Social scientists want to rule out biology so that *they* appear as the only relevant experts. Even more seriously, political values will determine what counts as acceptable theorizing—yet everyone assumes that ideas consistent with their own values are more likely to generate good, uncontaminated science (see chap. 18, "Biology and Ideology"). Political conservatives may try to argue that certain kinds of human behavior, or certain limitations of human ability, are built in by our biology—they are "natural" and hence inevitable, imposing constraints on social structures that we ignore at our peril. Liberals may want to deny the role of such factors so they can claim that improved conditions will indeed be able to promote a better society.

Each side will try to exploit the alleged objectivity of science to its advantage. It will try to discredit its opponents' position as "bad" or distorted science. The good guys always do hard, objective science, the bad guys let themselves be led astray by their political, religious, or philosophical preferences. The fact that some debates seem hard to resolve, however, suggests that neither side's claim for complete objectivity is valid. Each allows its criteria for what makes "good" science to be determined by its preconceptions. The sociologists of science argue that both sides are equally wrong—it is politics that forces people into polarized positions in which one side or the other is dismissed as trivial or irrelevant for practical purposes. Since the rival positions reflect deeply entrenched social and political values, it is hardly surprising that neither side seems able to score a permanent victory in the debate, even though each claims to be doing good science.

The controversies that have raged (and still rage) in some areas of biology suggest that we cannot ignore the sociologists' challenge to the objectivity of science. Physical scientists may claim that their knowledge is "harder" because it is backed up more easily by experimental tests, but the sociologists will have none of the distinction between hard and soft sciences. And history certainly offers examples where the search for knowledge in physics has reflected the scientists' wider beliefs and values. In the end, though, we do not want to present the history of science in a way that forces us to take a position on either side of the science wars. Both the history and the sociology of science provide ample evidence that science is a human activity, not an automated process that could be done equally well by a giant computer. Philosophical commitments, religious beliefs, political values, and professional interests have all helped to shape the way scientists have constructed and promoted their models of the world. At best only a few radical postmodernists have claimed that science is just make-believe. Sociologists of scientific knowledge like the Edinburgh school and the historians of science who have adopted their insights know that to make a research program stick, its proponents have to produce measurable results, and in this case "knowledge"—in the sense of our ability to describe and control nature—expands. In this respect, some of the spokespersons for science in the science wars seem to be aiming at the wrong target. Whether this link to practice satisfies the philosophers' criterion of objectivity isn't really the point: if the scientists were happy with Popper's warning that they could provide only provisionally valid information, they ought to be able to accept the more realistic model of science provided by sociologically inclined historians. In the end, scientists, too, have something to gain from a model of scientific development that accepts that it does indeed provide far more sophisticated knowledge of how the world works but refuses to see it as constructing a totally disinterested and immutably true model of nature. We live in an age where the general public often sees scientists having to take sides on controversial issues related to public health or the environment. They need to know that scientific research is a complex process in which it is not impossible for two perfectly legitimate projects to suggest opposing positions on some controversial question. Anything that helps people to understand why new research cannot offer instant answers to every complex problem will be a bonus rather than a danger to those trying to defend science's integrity and authority.

Why Modern Science?

This book offers a history of modern science, and we conclude with a few words explaining why we focus so strongly on the past few centuries. A previous generation of scholars would have taken it for granted that a survey of the history of science must begin with the natural philosophy of the ancient Greeks, acknowledge the important contributions of Islam, and then deal with the revival of learning in the medieval West, before moving on to tackle the Scientific Revolution of the sixteenth and seventeenth centuries. In taking that revolution as our starting point, we do not intend to suggest that the earlier developments were insignificant and we urge those who wish to know more about the foundations on which modern science has been built to consult David Lindberg's survey *The Beginnings of Western Science* (1992). It is particularly important that we recognize the debt that modern science owes not just to classical antiquity but also to the civilization of Islam, which nurtured and extended the traditions of ancient natural philosophy and provided a vital foundation for later developments in Europe. We should also note that Chinese culture produced many important inventions, including gunpowder and the magnetic compass, along with a philosophy of nature very different to that which eventually emerged in the West. Joseph Needham's monumental survey *Science and Civilisation in China* celebrates this alternative tradition. Needham also tried to answer the vexed question of why China did not build on this foundation to generate a scientific revolution equivalent to that which occurred in Europe (Needham 1969).

By recognizing the contributions made by other cultures we avoid the implication that the Scientific Revolution with which we begin was a genuine revolution in which an entirely new approach to nature appeared from nowhere to put Europe on course for world dominance in the study of nature. One product of the new sociological approach to history is Steven Shapin's account of the "revolution" (1996) that declares openly that there was no such thing because modern science emerged from a complex of changing attitudes and activities that influenced all areas of life and belief at the time. But in the end, a new kind of activity that we call science did emerge, resulting in an explosion of new methods, theories, organizations, and practical applications. The new developments in the history of science described above have tended to focus on the modern period precisely because it is during the past few centuries that the kind of activity that we recognize as science emerged—and the changes become even more striking when we move into the modern era of "big science" driven by industrial and military concerns. Compare the annual *Critical Bibliography* issued by the journal *Isis* for, say, 1975 with one for a recent year, and the change of emphasis is striking. The number of publications on ancient science, Islamic science, medieval science, and Renaissance science has remained more or less static (and has decreased as a proportion of the whole). Publications on the period from the seventeenth to the nineteenth centuries have increased slightly. But studies of twentieth-century science have increased dramatically, making it now by far the biggest category of publications. And a large proportion of those twentieth-century studies focus on American science—because that is where the most history, as well as the most science, is being done.

This change of emphasis is almost certainly a reflection of the modern tendency to see the history of science less in terms of conceptual (theoretical) innovations and more in terms of research schools, practical developments, and the ever-increasing influence of government and industry. When the focus was on the history of scientific ideas (including the idea of the scientific method itself) it seemed obvious that the natural philosophy of the Greeks should form the starting point—to begin with the Scientific Revolution would leave the whole project without a foundation. But when science is defined more in terms of how the modern scientific community operates, then forms of natural knowledge gained under different social environments seem less obviously foundational (although the study of how science functions in those other societies ought to be of interest for comparative purposes). Historians have become more interested in the creation of professional networks defined by scientific societies, journals, and university and government departments and in the interaction of scientists with industry, government, and the general public. These are all institutions and connections that were established in the period from the seventeenth to the twentieth centuries. There has also been massive increase in the actual amount of science being done in the modern period, with more being added all the time (what was new science in 1975 is history now). At the same time, the history of science has gained a new role within science studies departments, and here the focus is almost necessarily on developments that lead straight into the dilemmas of the modern world.

In recognition of this change of emphasis we have chosen to focus on science since the seventeenth century and to include as wide a range of topics within that area as is practical for a single book. Our first part deals more conventionally with developments within science itself, beginning with the Scientific Revolution and then focusing on major themes within individual sciences. Here we have tried to combine the traditional interest in the emergence of new theories with the modern approach based on the emergence of disciplines and research programs, with illustrations of the reassessments made possible by the new methods of study. Part 2 offers a more thematic set of cross-sections through the history of science, including traditional interests such as the links with technology, medicine, and religion, along with newer areas of study such as popular science. Whichever section you begin with, remember that you can always gain a wider perspective by looking for the cross-references that show how all these topics and themes intertwine. We don't pretend that it will be easy to build up an overview, but we hope that in the process you will gain a new respect for science and a better understanding of its importance for our lives.

References

Barnes, Barry, and Steven Shapin, eds. 1979. *Natural Order: Historical Studies of Scientific Culture.* Beverly Hills, CA, and London: Sage Publications.

Barnes, Barry, David Bloor, and John Henry. 1996. *Scientific Knowledge: A Sociological Analysis.* London: Athlone.

Bernal, J. D. 1954. *Science in History.* 3 vols. 3d ed., Cambridge, MA: MIT Press, 1969.

Brown, James Robert. 2001. *Who Rules in Science? An Opinionated Guide to the Wars.* Cambridge, MA: Harvard University Press.

Collins, Harry. 1985. *Changing Order: Replication and Induction in Scientific Practice.* London: Sage.

Foucault, Michel. 1970. *The Order of Things: The Archaeology of the Human Sciences.* New York: Pantheon.

Gillispie, Charles C. 1960 *The Edge of Objectivity: An Essay in the History of Scientific Ideas.* Princeton, NJ: Princeton University Press.

———, ed. 1970–80. *Dictionary of Scientific Biography.* 16 vols. New York: Scribners.

Golinski, Jan. 1998. *Making Natural Knowledge: Constructivism in the History of Science.* Cambridge: Cambridge University Press.

Gross, Paul R., and Norman Levitt. 1994. *Higher Superstition: The Academic Left and Its Quarrel with Science.* Baltimore: Johns Hopkins University Press.

Gutting, Gary. 1989. *Michel Foucault's Archaeology of Scientific Reason.* Cambridge: Cambridge University Press.

Hempel, Karl. 1966. *Philosophy of Natural Science.* Englewood Cliffs, NJ: Prentice Hall.

Kohler. Robert E. 1994. *Lords of the Fly:* Drosophila *Genetics and the Experimental Life.* Chicago: University of Chicago Press.

Koyré, Alexandre, 1965. *Newtonian Studies.* Chicago: University of Chicago Press.

———. 1978. *Galileo Studies.* Atlantic Highlands, NJ: Humanities Press; Hassocks: Harvester.
Kuhn, Thomas S. 1962. *The Structure of Scientific Revolutions.* Chicago: University of Chicago Press.
Lakatos, Imre, and Alan Musgrave, eds. 1979. *Criticism and the Growth of Knowledge.* Cambridge: Cambridge University Press.
Latour, Bruno. 1987. *Science in Action: How to Follow Scientists and Engineers through Society.* Milton Keynes: Open University Press.
Lindberg, David C. 1992. *The Beginnings of Western Science: The European Scientific Tradition in its Philosophical, Religious and Institutional Contexts, 600 B.C. to A.D. 1450.* Chicago: University of Chicago Press.
Merton, Robert K. 1973. *The Sociology of Science: Theoretical and Empirical Investigations.* Chicago: University of Chicago Press.
Needham, Joseph. 1969. *The Grand Titration: Science and Society in East and West.* London: Allen & Unwin.
Popper, Karl. 1959. *The Logic of Scientific Discovery.* London: Hutchinson.
Rudwick, Martin J. S. 1985. *The Great Devonian Controversy: The Shaping of Scientific Knowledge among Gentlemanly Specialists.* Chicago: University of Chicago Press.
Shapin, Steven. 1996. *The Scientific Revolution.* Chicago: University of Chicago Press.
Shapin, Steven, and Simon Schaffer. 1985. *Leviathan and the Air Pump: Hobbes, Boyle and the Experimental Life.* Princeton, NJ: Princeton University Press.
Waller, John. 2002. *Fabulous Science: Fact and Fiction in the History of Scientific Discovery.* Oxford: Oxford University Press.
Whitehead, A. N. 1926. *Science and the Modern World.* Cambridge: Cambridge University Press.

PART I

EPISODES IN THE DEVELOPMENT OF SCIENCE

CHAPTER 2

THE SCIENTIFIC REVOLUTION

WAS THERE REALLY A "SCIENTIFIC REVOLUTION" during the seventeenth century? The traditional historical answer to this question has been a resounding yes. According to this view, the fundamental changes that took place in the ways in which Western culture viewed the universe and the methods used for finding out about the universe during this period were so cataclysmic that they deserve to be described as revolutionary. Not only that, but these changes had an impact on our understanding of the cosmos and our place within it to such an extent that they should be considered unique. In other words, what happened in the seventeenth century was not just any scientific revolution. It was *the* Scientific Revolution. From this perspective, what happened during the Scientific Revolution was nothing less than the birth of modern science. As a result, if this view of history is correct, then the great names that we associate with the Scientific Revolution—people such as Copernicus, Descartes, Galileo, Kepler, and Newton—have a real claim to be considered the parents of modern science. Not only did they make great discoveries and formulate new theories, but they also inaugurated a new method—the scientific method—that could provide us with sure and reliable knowledge about the world around us.

This way of looking at the history of science has a history of its own as well. Many sixteenth- and seventeenth-century protagonists in the philosophical debates and discoveries that shaped the Scientific Revolution quite clearly regarded themselves as being at the forefront of a revolutionary intellectual movement. The English philosopher-courtier Francis Bacon took a very dim view of ancient Greek philosophy, for instance—"a kind of wisdom most adverse to the inquisition of truth"—as compared to the scientific achievements of his own age. The key in his opinion was a willingness to experiment and a recognition that knowledge "must be sought from the light of nature, not fetched back out of the darkness of antiquity." In a similar vein, the Enlightenment writer Voltaire celebrated the achievements of Bacon, Robert Boyle, and Isaac Newton at the expense of Aristotle, Plato, and Pythagoras. At the very least, the events of the seventeenth century were looked back on in the nineteenth and twentieth centuries as a reflowering of the human intellect and its possibilities after the long stagnation of the Middle Ages. The twentieth-century historian Alexandre Koyré maintained that the achievement of the founders of modern science had been "to destroy one world and to replace it by another" (Koyré 1968). His contemporary Herbert Butterfield in his classic *The Origins of Modern Science* asserted of the Scientific Revolution that "it outshines everything since the rise of Christianity and reduces the Renaissance and the Reformation to the rank of mere episodes" (Butterfield 1949).

Historians' views of the Scientific Revolution—and particularly of its unique status—have undergone considerable revision in recent years (Shapin 1996). There are a number of reasons for this. Historians now are far less happy to accept that it makes sense to talk at all about "science" during the seventeenth century. There is a recognition that in actuality seventeenth-century men of science and natural philosophers (as they would describe themselves) engaged in a whole range of activities that may or may not fit comfortably with modern notions of science. We also now know a great deal more about knowledge-making activities during the Middle Ages with the result that many historians can argue that there are significant continuities between medieval ideas and practices and later ones. It is far more difficult, therefore, to assert that what happened during the seventeenth century constituted an unprecedented break with the past after all. More generally, most historians of science are increasingly uncomfortable with the idea that there is just one, unique scientific method. Without that belief in *the* scientific method, it is less and less clear just what *the* Scientific Revolution was about. One good reason for keeping the notion remains, however. As we have just seen, many seventeenth-century commentators certainly thought that they were engaged in a revolutionary process. If we are to take our subjects and their beliefs seriously, then looking to see just what they were doing and why they thought it was so important certainly remains worth doing.

This chapter will provide a necessarily brief and sketchy overview of the Scientific Revolution. We will start by looking at the enormous transformations that took place in astronomy, a science that according to traditional accounts, at least, underwent a literally earth-moving transformation during our period. It is this massive change in perspective from an earth-centered (or geocentric) universe to a heliocentric universe in which the earth is relegated to no more than another planet orbiting the sun that most people have in mind when they think of the Scientific Revolution. We will then survey the mechanical philosophy that many seventeenth-century commentators themselves regarded as being at the very core of the new views of nature that were being advanced during this period. We will also look at the emergence of new ways of knowing during this period, as well as new ideas. Philosophers talked of experiment and of mathematics as providing new tools and even a new language that could be used to understand nature. We will end the chapter with a look at the celebrated Isaac Newton, hailed by many of his contemporaries as the man who single-handedly brought the New Science into being. A look at his achievements will give us our best chance of answering the question with which we started this chapter—was there really a Scientific Revolution?

Relocating the Heavens

Astronomy was certainly one of the hot topics of the Scientific Revolution as it is usually understood. Many of the great names we automatically associate with that great intellectual transformation are those of practicing astronomers. Think of Tycho Brahe, Copernicus, Galileo, Kepler, or even Newton. Strictly speaking, however, astronomy before the seventeenth century was not really a part of natural philosophy at all. Astronomy, like mathematics, was held to deal only with accidents and appearances while it was left to natural philosophy to deal with the real causes of things. This was far more than just a technical distinction. It meant that astronomy occupied a different place from natural philosophy in university curricula, for example. It also meant that its practitioners, like mathematicians, were held to be of lower intellectual and social status than professors of natural philosophy. This was one of the reasons, as we shall see later, that Galileo was so pleased to have persuaded Cosimo de'Medici to engage him as court philosopher rather than court mathematician. Because they were supposed to deal only with the appearances of things rather than the reality, however, astronomers were not expected to produce models of the heavens that were in any sense realistic. Their task was simply to find models that would allow them to describe and predict the apparent motions of heavenly bodies accurately rather than find ways of describing what the structure of the universe was really like. That was a task reserved for natural philosophers.

Broadly speaking, sixteenth-century natural philosophers subscribed to

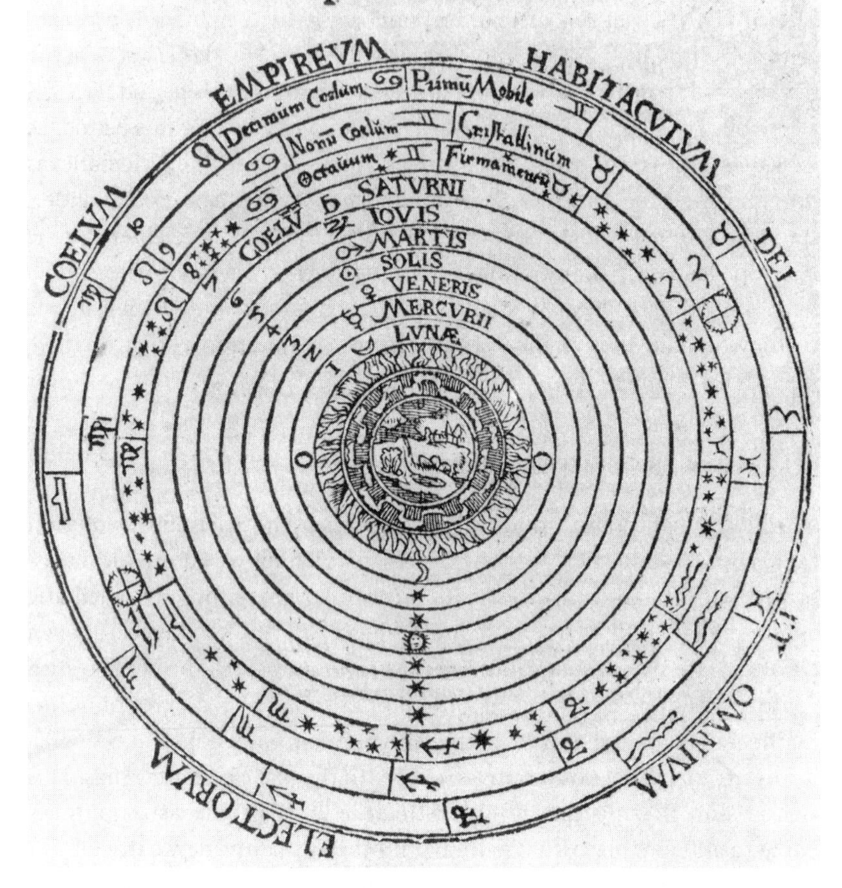

FIGURE 2.1 The Ptolemaic universe as illustrated in Petrus Apianus, *Cosmographia* (1539). The earth is at the center of the universe with the moon, the sun, and the five planets circling around it. The sphere of fixed stars marks the outer boundary of the universe.

an Aristotelian view of the universe. According to this model, the earth lay at the exact center of the cosmos, while the moon, sun, and the planets orbited around it on a number of spheres. The sphere carrying the moon defined the boundary between the corruptible and changing sublunary world and the incorruptible and unchanging heavens above it. Most astronomers at this time adopted a version of the Ptolemaic model of the universe (fig. 2.1), developed by the Alexandrian astronomer Claudius Ptolemy

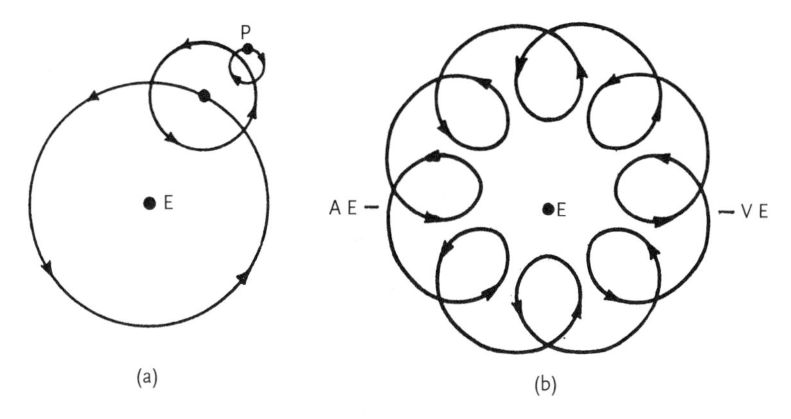

FIGURE 2.2 An example of how Ptolemy used geometrical constructions such as epicycles to generate a more accurate picture of planetary motions. In this instance, the arrangement of epicycles on the left generates the movement pictured on the right.

during the second century A.D. Ptolemy had perfected a number of refinements to the basic Aristotelian model allowing for more accurate description and prediction of the apparent movements of the heavenly bodies. He introduced innovations such as epicycles (whereby the planets were supposed to describe circular orbits around fixed points on their spheres as well as circular orbits around the earth; (fig 2.2) and equants (a complex device whereby the velocity at which a heavenly body circled the earth was calculated such that it was constant with respect to a point other than the orbit's center). Using devices like these, Ptolemy's followers could produce highly accurate charts and tables of the movements of the heavens. Nobody supposed, however, that these epicycles and equants in any way described reality. They were simply geometrical techniques used to "save the appearances." Aristotelian natural philosophy maintained that only perfect circular motion was possible in the incorruptible superlunary sphere (Kuhn 1966; Lloyd 1970, 1973).

When the Polish cleric Nicolaus Copernicus published his *De revolutionibus orbis coelestium* in 1543 it was very easy for his contemporaries to read it in just this way as well. In fact any other way of reading it would have been regarded as a little peculiar. Copernicus argued that it was possible to produce more accurate predictions of the movements of the heavenly bodies—and do away with some of the more aesthetically dubious aspects of the Ptolemaic model such as the equant—by assuming that the sun rather than the earth was placed at the center of the cosmos. To many of his readers this simply looked like another ingenious attempt to "save the appear-

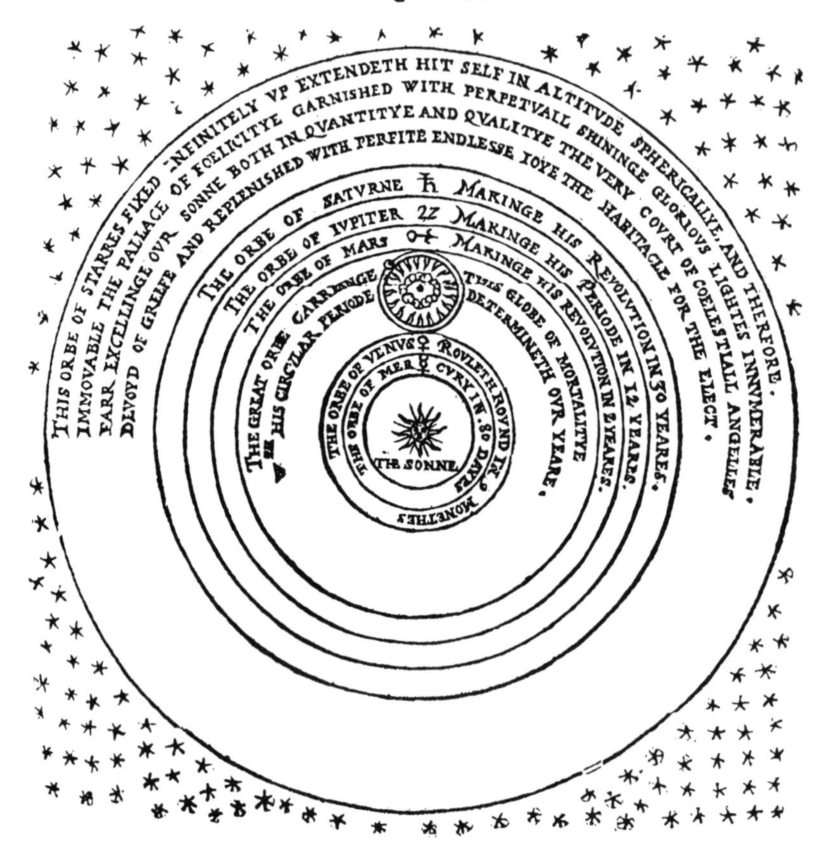

FIGURE 2.3 The Copernican Universe as illustrated in Thomas Digges, *A Perfit Description of the Coelestiall Orbes* (1576). The sun is placed at the center of the universe with the earth and the other planets circling around it, while the moon circles the earth. Notice that the universe is still bound by the sphere of fixed stars.

ances" and produce more accurate star charts and tables. The preface to *De revolutionibus* did, however, make a more startling claim. It suggested that Copernicus's model should be taken to reflect physical reality (fig. 2.3). Copernicus seemed to be arguing that astronomy had a claim to intellectual territory usually firmly occupied by natural philosophy. If Copernicus was right, then his book was truly revolutionary in its implications. Not only did it suggest that astronomers could compete with natural philosophers for intellectual authority and status, it also suggested that the earth and humankind were not, after all, at the center of the universe. The effect was rather spoiled, however, by the insertion in the published text of an unsigned foreword written by Copernicus's friend, the Lutheran clergyman Andreas Osiander, asserting that the suggestion of the heliocentric model's physical reality was, after all, no more than an intellectual conceit. There was no indication that this was not Copernicus's view, and since he died shortly after publication there was no way of checking.

Copernicus had made little or no effort to publicize his potentially earth-moving innovation. For probably his most famous follower, the Italian astronomer, mathematician and natural philosopher Galileo Galilei, however, publicity was the name of the game. In the summer of 1609, Galileo, then a humble professor of mathematics at the University of Padua, turned his newly improved telescope on the heavens and used it to make a number of startling claims and discoveries. In his *Sidereus nuncius* published a year later, Galileo claimed to have seen novelties through the telescope. He had seen countless new stars never before observed or cataloged, and he had seen imperfections on the surface of the allegedly incorruptible moon. Most important, he had discovered four new planets that he claimed orbited the planet Jupiter rather than orbiting the earth as the other planets were supposed to do. Galileo named the new planets the Medicean Stars and dedicated his book to Grand Duke Cosimo de'Medici of Tuscany in an ultimately successful attempt to attract that powerful magnate's patronage (Biagioli 1993). Galileo's reward was a major change in status. He was made professor of philosophy at the University of Pisa and appointed court philosopher and mathematician to Cosimo. This was a change in the status of astronomy as well. Indeed, to sustain his newfound status Galileo had to argue that his astronomical discoveries had profound philosophical consequences, too.

By the time he published his notorious *Dialogue Concerning the Two Chief World Systems* in 1632 Galileo had long made something of a name for himself as a controversialist. In many ways that was part of his job. He was expected to entertain his patrons at the Florentine court with witty disputations. In the *Dialogue* he went a step further, however. He explicitly drew on his telescopic discoveries and other arguments to mount a thinly veiled defense of the physical truth of the Copernican theory. He argued that the evidence of the heavens as uncovered by his telescope favored the Copernican theory and produced a swathe of physical arguments defending the idea of the earth's rotation. The result was a personal disaster. He was summoned before the Holy Office in Rome, forced to recant his Copernican beliefs, and sent into exile. His book was banned. It is important to be clear about just what the arguments on the matter were between Galileo and the Catholic Church (see chap. 15, "Science and Religion"). The Church in the past had shown no objection to discussions of Copernican ideas provided they were couched in hypothetical terms and acknowledged the ultimate authority of Scripture in deciding the truth of the matter. Galileo's sin lay, therefore, not so much in what he said but in the way that he said it. His challenge was against the authority of the Church and its legitimacy as an intellectual arbiter as much as it was against the validity of the Aristotelian theory of the universe (Redondi 1987).

The example of Galileo's career underlines the increasing importance of patronage in sustaining astronomical work during the sixteenth and seventeenth centuries. To make a name for himself, Galileo needed the financial and cultural backing of Cosimo de'Medici. The role of patronage is just as clear in the career of the Danish astronomer Tycho Brahe. Brahe was himself a nobleman, the son of an influential member of the Danish court. He was in the enviable position of being able to finance his own career in astronomy as well as acquiring unprecedented support from the Danish crown. He was even granted an entire island by the king on which to build his own private observatory of Uraniborg (fig. 2.4). His career was not without difficulties, however. Astronomy was hardly a usual occupation for a member of the nobility. Tycho had his work cut out in persuading both his family and his aristocratic peers to allow him to indulge his passion for astronomy and in persuading the astronomical community of scholars to accept him as a serious member. He made a name for himself with a series of detailed observations on the New Star that appeared in the heavens in 1572—what would now be called a supernova. Tycho's observations were particularly interesting because he argued that they showed no evidence of stellar parallax. In other words they showed that the New Star was too far away to be inside the sublunary sphere of Aristotelian physics. Instead it could be treated as evidence of corruption and change in the supposedly incorruptible and unchanging superlunary sphere.

Ensconced at Uraniborg, Tycho Brahe made a name for himself with astronomical observations of unprecedented accuracy. He did not use a telescope for these observations. What he did was design and commission the very best astronomical instruments that his considerable financial resources allowed. He used these to pinpoint the positions of the planets. This kind of observational work had a vital role to play in producing astronomical tables, used among other things for working out the calendar and establishing the proper times of Church festivals (such as Easter). This was

FIGURE 2.4 An impression of Tycho Brahe's observatory at Uraniborg from Tycho Brahe, *Astronomiae instauraiae mechanica* (1587). Notice the instruments and the assistants working in the background.

one of the main uses to which the new Copernican model of the universe was put, and Tycho's observations could be used to make those astronomical tables even more accurate. Brahe was no Copernican himself, however. While sympathetic to Copernican adherents, he doubted the motion of the earth. Instead he came up with his own solution, a system in which the earth remained at the center of the universe with the sun and moon orbiting around it but with the remaining planets orbiting the sun. It was a system that might easily seem to preserve the best of both worlds, maintaining the integrity and plausibility of the geocentric Aristotelian cosmos while adding to it the accuracy and simplicity of the Copernican model.

It was the controversy surrounding the origins of Tycho's system of the universe that brought Johannes Kepler to the Danish astronomer's attention. Tycho was engaged in a virulent exchange with another German, Nicolai Reymers Ursus, over the new system, large elements of which Tycho claimed Ursus had plagiarized. Tycho became Kepler's patron and enlisted him in his campaign to discredit Ursus. Tycho himself had by now moved to Prague to become a client of Rudolph II, the Holy Roman Emperor. He hired Kepler to write a defense of his claims to originality against Ursus and to reduce the vast mass of observational data he had amassed during his career into a form that could be used to prove the superiority of the Tychonic system. Kepler, a student of the German astronomer Michael Maestlin, was already well on the way to making a name for himself in astronomical circles. Following Tycho's death in 1601, he quickly found himself succeeding him as Rudolph's imperial mathematician, inheriting Tycho's precious astronomical instruments and even more precious observational records along the way. It was another example of the importance of royal and aristocratic patronage in sustaining astronomical work and of the vital importance of access to resources.

Kepler was not all that anxious to use Tycho's accumulated mass of observations to defend his former master's celestial system. Like many of his early seventeenth-century contemporaries he was a Platonist, convinced that the universe operated according to harmonic principles. He took the music of the spheres seriously. Unlike most of his contemporaries, though, he was also a committed Copernican. In his *Mysterium cosmographicum* of 1596 he had already articulated a system of the universe in which the distances between the orbits in which the planets circled around the sun were determined by the sequence of the regular Platonic solids (fig. 15.2, p. 352). It took Kepler years of work to reduce Tycho's observations to the simple law that as a convinced Platonist he was positive the planets should follow. He published the results in 1607, showing that both Copernicus and Tycho

Brahe were wrong. The planets did not orbit around the sun in circles. The path that each planet followed instead was an ellipse. Having paid his dues to his master, Kepler returned to his fascination with harmony, publishing his *Harmonice mundi* in 1619, which established his conviction that the universe operated according to the laws of harmony. It was an example of the new status that astronomy had acquired, that a mere astronomer and mathematician (albeit the former imperial mathematician to the Holy Roman Emperor) could make serious contributions to this kind of natural philosophical discussion.

During the course of the century or so following Copernicus's publication of *De revolutionibus,* astronomical opinion gradually shifted toward acceptance of the heliocentric position. As long as astronomy remained subordinate to natural philosophy and limited to the aim of "saving the appearances" nothing much hinged on this gradual acceptance. The Copernican system simply offered a better way of calculating the movements of the planets. Arguably, at least, the really decisive shift was not the move from geocentricism to heliocentricism but the breaking down of the barrier between the sublunary and superlunary spheres and the extension of earthly corruption to embrace the movements of the stars. These shifts were part and parcel of changes in the social and cultural status of astronomers and natural philosophers. Along with the physical barrier between earth and the heavens, the social barrier between astronomy and natural philosophy was breaking down. It was increasingly becoming legitimate for mere astronomers to have views on philosophical matters. The social place of astronomy was shifting too. All the astronomers we have discussed here made names for themselves outside the cloistered world of the universities. In astronomy and, as we shall see, in natural philosophy as well, the place of knowledge was increasingly coming to be the civic forum.

Magic and Mechanism

What kind of world did the new systems of natural philosophy that emerged around this period describe? One common feature of the various new natural philosophical systems that were offered up during the sixteenth and seventeenth centuries is that they were quite self-consciously novel. Authors offered up books with titles like *Novum organum* (Francis Bacon), *Due nuove scienze* (Galileo), or *Phonurgia nova* (Athanasius Kircher). There was no mistaking these authors' ambitions. They wanted to set the study of the natural world on a completely new footing. It is difficult for the historian to generalize too freely when trying to find ways of characterizing these new systems of natural philosophy. We now know that in their specifics, at least, these attempts at creating a new science varied a great deal. There was a great deal of disagreement over what the New Science would look like, what the most secure way of proceeding might be, and what the results of the investigation might deliver. At least some of the avenues that the Scientific Revolution's protagonists followed in their search for knowledge appear distinctly unpromising from a modern perspective. Others fit more comfortably into our conceptions of what science should be like. It is important to remember, however, that these early modern natural philosophers had very different conceptions of the world from ours—and very different views of what science should be able to deliver (Lindberg and Westman 1990).

Magic seemed, to some natural philosophers at least, to be a promising way of investigating nature. Sixteenth- and seventeenth-century magicians traced their traditions back to the mythological figure of Hermes Trismegistus. Magic was regarded as a search for the hidden, "occult" qualities of natural objects and phenomena. Understanding these occult qualities would provide a way of comprehending the hidden operations of nature and the relationships between different kinds of natural objects (Yates 1964). Particular objects—such as magnets, for example—could clearly be seen to influence other objects without apparent contact. Astrology seemed to many a promising avenue of occult enquiry as well. Trying to understand the ways in which the movements of the stars and planets influenced the unfolding of earthly events was a way of coming to grips with the hidden operations of the universe. Similarly, alchemy appeared to offer a way of understanding the ways in which different substances influenced each other and what their essential qualities might be. There was a flourishing tradition of natural magic in the sixteenth and seventeenth centuries as well. Natural magicians like the Elizabethan courtier and mathematician John Dee or the Jesuit scholar and polymath Athenasius Kircher could produce spectacular phenomena at will. Kircher, for example, was famed for his invention of the magic lantern and a clock driven by a sunflower seed that followed the course of the sun from sunrise to sunset just like a sunflower blossom, demonstrating the sun's occult influence over natural objects.

Less controversial—to modern sensibilities at least—than magic as a tool for understanding nature was the mechanical philosophy. This was the view that the best way of understanding the cosmos was to regard it as a huge machine and that the task of natural philosophy was to understand the principles by which the machine operated. In some ways, at least, the mechanical philosophy was the antithesis of the magical tradition since it denied the very existence of the occult qualities that magic tried to investigate. Clockwork was the dominant metaphor of the mechanical philosophy. All the parts of a clock worked in harmony together to produce the final motion. This was how some natural philosophers visualized the workings of the universe, too—all the parts worked in unison to produce the movements of the earth and the planets. The clockwork metaphor had the major advantage of implying the existence of a celestial clockmaker as well: if the universe were a piece of a complex mechanism like a clock, then just as clocks had clockmakers, the universe must have had a creator too. Mechanical philosophy did not apply only to large-scale phenomena like the movements of the planets, though. Mechanical philosophers devoted their ingenuity to finding out mechanisms for all the phenomena of nature. They aimed to banish occult qualities completely from natural philosophy by showing that even the most mysterious forces could be reduced to the operations of simple mechanical principles.

The doyen of the mechanical philosophy in the early seventeenth century was undoubtedly the French natural philosopher and mathematician René Descartes. A Jesuit-trained scholar and former mercenary soldier during the Thirty Years' War, Descartes had famously resolved to reduce all human knowledge to first principles, eventually producing what is probably the most recognizable philosophical dictum in modern history—*cogito ergo sum*. Descartes laid out his plans for a new and ambitious philosophy of nature in his *Discourse on Method* (1637). The picture of the universe he produced was unambiguously mechanical. Descartes conceived of the universe as a plenum, that is to say, full of matter. There was no room for a vacuum in Descartes's cosmology. Since the universe was full of matter, if one part of it moved, then other parts of it had to move as well. The simplest means of achieving this was movement in a circle—hence the circular motion of the planets around the sun. For Descartes, therefore, the universe was made up of an indefinite number of vortices, each swirling around a sun or a star and carrying the planets around with them. The planets were held in stable orbits by the constant pressure of subtle matter continually swirling outward from the central sun. Descartes could even use his theory of vortices to explain the motion of the tides—one of the most intractable problems of seventeenth-century practical mathematics.

Like other mechanical philosophers, Descartes's theories accounted for more than large-scale phenomena such as the motions of the planets and the movements of the tides. Everything in Descartes's universe was made up of particles of matter. Light, for example, consisted in a stream of subtle

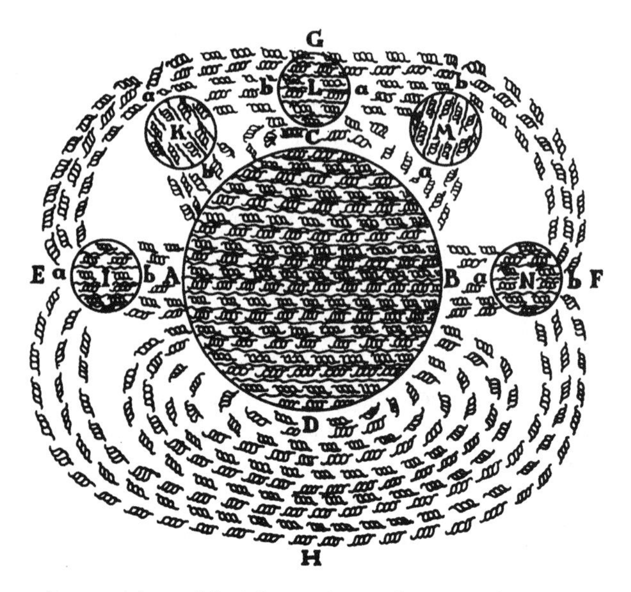

FIGURE 2.5 Descartes's model of the mechanical origins of magnetism. The magnetic body emanates a stream of magnetic particles, shaped like little screws. When those screwlike particles pass through other bodies, they cause them to move either toward or away from the magnetic body, depending on whether the screws are right- or left-handed.

particles flowing outward from the sun. He also tried to account for the phenomenon of magnetism on mechanical, corpuscularian principles (fig. 2.5). Magnetism was one of natural magicians' favorite examples of evidence for the existence of occult qualities. Even William Gilbert, author of *De magnete*—the first comprehensive account of magnetism published in 1600—had compared the actions of the magnet to those of the soul. According to Descartes, magnetism was the result of a stream of corpuscles (or particles) flowing out from the magnetic body. These corpuscles were shaped as either left-handed or right-handed screws, so that depending on their shapes they could cause objects they passed through to move either toward or away from the magnet. Descartes's mechanical philosophy even extended to animals and people. Descartes famously described all animals as no more than complex machines. He took the same view of human bodies as well, except that humans were possessed of an animating soul that controlled their bodies through the medium of the pineal gland. Descartes was convinced that proper management of the body's mechanism through proper diet could lead to the indefinite extension of human life (see chap. 19, "Science and Medicine").

The Anglo-Irish natural philosopher Robert Boyle shared with Descartes the view that all natural phenomena could be accounted for by the mechanical action of minute particles or corpuscles of matter. Boyle held that at the original creation of the universe, the uniform, homogenous matter of which it originally consisted was divided into a whole array of differently shaped and textured moving corpuscles. It was the different sizes, shapes, and textures of these particles of matter, along with the different ways in which they moved, that explained the various visible and tangible properties of matter. Boyle differed from Descartes in being rather more cautious in specifying just what the various shapes and sizes of these invisible particles might actually be. Where Descartes was prepared to specify the precise shapes of the particles causing magnetism, for example, Boyle was satisfied to leave such matters open. What mattered as far as he was concerned was simply that these kinds of mechanical explanations of natural phenomena, in terms of the behavior and form of material corpuscles, were accepted as the most plausible ones available. While Boyle accepted that in general the best explanation of the color or texture of an object, for example, would be in terms of the kinds of particles from which it was made, he also accepted that the issue of just what those particles were like remained a matter for speculation.

The caution with which Boyle approached the issue of providing specific rather than general mechanical explanations of the phenomena is clear in his accounts of his famous air-pump experiments—of which more later in this chapter. During the late 1650s and early 1660s Boyle carried out a number of experiments with a new item of philosophical apparatus—the air pump—designed to investigate the properties of the air. On the basis of these experiments, Boyle argued that the air was made up of springlike particles. It was as a result of this springlike character of the particles from which it was made up that air could resist any force exerted on it and could expand when those forces were removed. In his *New Experiments Physico-Mechanical Touching the Spring of the Air* (1660), Boyle argued that while he could be effectively certain of the truth of the phenomena produced in his air pump—in other words, air did indeed act as he had described it—he could have no such certainty concerning the details of a causal explanation of those phenomena. As a mechanical philosopher he could be sure that the causes of the phenomenon were mechanical in nature but any account of the detailed mechanism could be, at best, only probable. Particles of air might be just like steel springs but, conversely, they might not.

Despite this kind of circumspection, the mechanical philosophy certainly seemed to many of its adherents to be the best way of formulating causal accounts of natural phenomena. The English natural philosopher Robert Hooke, once Boyle's assistant experimenter, even suggested that in due course it might be possible actually to see the basic particles of matter through the medium of the recently invented microscope. Even in the absence of such direct sensory evidence of the existence of these "small Machines of Nature," most natural philosophers were willing to concede that hypothesizing their existence was the best way of proceeding in constructing philosophically respectable accounts of nature. It was certainly a better alternative than falling back on the assumption of occult qualities inherent in different kinds of matter. When Evangelista Torricelli carried out experiments with pumps and liquids in 1644 it was with the aim of showing that the phenomena could be explained mechanically without having to fall back on the notion that "nature abhors a vacuum." Blaise Pascal had the same kind of end in view when he repeated the experiments in 1648 on the slopes of Puy de Dôme in France. One advantage many pointed to in adopting mechanical explanations was that it did away with the temptation of endowing matter with animistic qualities. As the French priest Marin Mersenne argued, making matter active could lead to the dangerous possibility of breaking down the distinction between God and nature. It was far better to follow the mechanists and accept that matter was essentially passive and only differed in the size and shape of its particles.

As we have already seen with Descartes's example, these kinds of mechanistic arguments applied to animal and human bodies as much as they more conventionally did to the inanimate world. The English physician William Harvey's description of the circulation of the blood was widely hailed by contemporaries as a classic example of the mechanical philosophy applied to animated bodies, though Harvey himself was dubious of the mechanical philosophy's merits. His *De motu cordis et sanguinis* (1628) argued that blood circulated through the body, passing through the heart and lungs into the arteries and through them to the extremities of the body before returning through the veins to the heart. Following what they regarded as Harvey's example, self-proclaimed iatromechanists (from the Greek *iatro,* meaning doctor) such as Giovanni Borelli argued that understanding the human body as a complex machine was the key to improving medicine. Hermann Boerhaave argued that all the anatomical components that made up the body could be seen to have their equivalents in various kinds of machinery: "We find some of them resembling *Pillars, Props, Crossbeams, Fences, Coverings;* some like *Axes, Wedges Leavers* and *Pullies;* others like *Cords, Presses,* or *Bellowes;* others again like *Sieves, Strainers, Pipes, Conduits,* and *Receivers;* and the Faculty of performing various motions by these Instruments, is called their Functions; which are all performed by Mechanical Laws, and by them only are intelligible." For Boerhaave, the human body was just like a complex hydraulic machine (see chap. 19, "Science and Medicine").

Proponents of the mechanical philosophy often ranged themselves quite explicitly in opposition to practitioners of magic and those who argued for the existence of occult qualities in nature. To many, it seemed that simply explaining some feature of nature in terms of the inherent qualities of matter was no explanation at all. This was what dramatist Jean-Baptiste Molière lampooned with his description of natural philosophers explaining the sleep-inducing capacities of opium by reference to its possession of "dormative qualities." Recent historians have been rather more careful than their mechanically minded contemporaries in dismissing magical practitioners completely out of hand. Most historians of the Scientific Revolution accept that magic had an important role to play in the intellectual debates of the period. Magicians and mechanical philosophers do appear to have shared a concern to elucidate the properties of matter by examining its hidden qualities, whether those qualities were regarded as being innate or not. They also shared the stance of self-conscious novelty. Most natural philosophers during this period shared a perception that they were engaged in a fundamentally novel project, however they might then characterize the details of the project itself.

New Ways of Knowing

When celebrating the novelty of the New Science, its practitioners had more in mind than just what they were finding out about the nature of the universe. Just as important in their view was the question of how that new knowledge had been acquired. Almost all were united in agreeing that the big difference between their brand of knowledge as compared to previous varieties was that it was based on experience rather than authority. The "Schoolmen"—as previous generations of scholars were dismissed—were held to have based their claims to knowledge simply on the authority of ancient texts, primarily Aristotle and his medieval interpreters. Promoters of the New Science, in contrast, claimed that their knowledge was based on actual experience of the world. We have already noted the extent to which seventeenth-century natural philosophers emphasized the novelty of their science. This is what they mainly had in mind with such claims. Their science was new because it was based on an entirely different set of assumptions about how we might best go about acquiring knowledge in the first place. Where previous generations of scholars had searched for knowledge in the books of Aristotle, the new generation prided themselves on their realization that the best approach to knowledge was to read it in the "book of nature."

Increasingly, also, many natural philosophers argued that the language in which the book of nature was written was the language of mathematics. This represented a major shift in the epistemological—and social—status of mathematics. As we have already seen, mathematics had traditionally been regarded as epistemologically inferior to natural philosophy. Natural philosophy was taken to deal with the real nature of things—their essences. Mathematics, in contrast, simply dealt with accidental qualities like numbers. Mathematics was certainly regarded as providing certainty of a particular kind, but natural philosophers argued that what it provided was a very limited kind of certainty. Conclusions derived by means of mathematical reasoning were true only insofar as the premises from which the argument started were taken to be true—and establishing the truth of such premises was held to be beyond the scope of mathematical reasoning. Along with these differences in epistemological status came differences in social status. Mathematics did not occupy as exalted a position in the university curriculum as did natural philosophy. Professors of mathematics, as Galileo, for one, was well aware, earned less than their philosophical counterparts. It was also widely regarded as a far more practical endeavor than natural philosophy.

Mathematics not only embraced those aspects that might now be characterized as "pure" reasoning, such as geometry, for example; it also embraced more practical activities, such as arithmetic. Some commentators argued that mathematics was not, properly speaking, an academic activity at all—it was something that mechanics did, "the business of *Traders, Merchants, Seamen, Carpenters, Surveyors of Lands,* or the like." This is an extreme example, but it does, nevertheless, highlight the sense in which mathematics was considered by some, at least, to be a socially inferior epistemological practice. Practical mathematics was an activity built around the manipulation of different kinds of mathematical instruments, such as sextants, quadrants, or calculating devices like the slide rule (fig. 2.6). During an age that saw increasing maritime travel and exploration as well as the start of the drive toward land enclosures and accurate mapmaking, practical mathematics was, however, undeniably useful. Gentleman landowners (and gentleman adventurers) increasingly found themselves in need of the skills of practical mathematicians and even started acquiring a certain level of mathematical proficiency themselves (see chap. 17,

FIGURE 2.6 The frontispiece of Jonas Moore, *A New System of Mathematicks* (1681). The range of mathematical instruments illustrated here suggests how important practical mathematics was becoming during the seventeenth century.

"Science and Technology"). All of this certainly resulted in a raising of mathematicians' cultural visibility, particularly around the princely courts and aristocratic households toward which the intellectual center of gravity was decisively shifting, away from the Aristotelian-dominated universities.

It was this kind of shift, as we saw earlier, that Galileo took advantage of in his move from professor of mathematics at Padua to philosopher at the Medici court in Florence. In the same way that he did with his astronomy, part of Galileo's strategy in making the move was to insist on the philosophical status of mathematics. As he and others increasingly argued, the book of nature was written in the language of mathematics. Galileo argued that natural philosophy should be expressed in terms of mathematics because nature was mathematical in its structure. The main aim of natural philosophy, therefore, should be to produce mathematically expressed laws of nature, such as Galileo's own mathematical law of falling bodies, according to which all bodies fell to earth at the same rate regardless of their weight. There was even an ancient pedigree for this claim, to match the Aristotelian pedigree of the Schoolmen. Mathematicians turned to the authority of Plato and Pythagoras to establish the mathematical nature of the natural world. This is what Kepler, for example, did with his early argument that the distances between the planets' orbits were defined by the series of five Platonic solids—cube, tetrahedron, dodecahedron, octahedron, and icosahedron.

Just what the status of mathematical descriptions of the natural world was remained the subject of dispute, however. It did not escape critics' attention, for example, that Galileo's law of falling bodies did not hold true in the real world, but only in a mathematically idealized one. To overcome this, Galileo had to argue that it was actually his idealized, frictionless mathematical model rather than messy reality, that somehow properly captured the essence of the phenomena. Natural philosophers worried about just what kind of epistemological status—what degree of certainty—should be accorded to the results of mathematical arguments concerning the operations of the natural world. Just what was the nature of the fit between the mechanical universe made up of particles in motion and mathematical descriptions of it? How could the integrity of that match be guaranteed? Even a mechanical philosopher like Robert Boyle, who was in principle happy to proclaim that the book of nature was "written in mathematical letters," was in practice far more circumspect about writing his own natural philosophy in mathematical language. One of Boyle's problems with mathematics was that, like many of his contemporaries, he was convinced that to sustain its authority—to appeal to as many people's commonly accepted experience of the world as possible—natural philosophy had to be accessible. The problem with mathematics was that it was not.

Boyle, like many others, was keen to emphasize that the New Science was empirical science. Rather than depending on the authority of the ancients, he and his philosophical contemporaries aimed to base their science on the authority of their own senses. Experience was to be the key to constructing new theories about the natural world. This seems relatively unproblematic from a straightforward modern perspective. This appearance is itself a testimony to the success of early modern natural philosophers in establishing this view of the proper basis for inquiry into the workings of nature. Seventeenth-century commentators themselves, however, were acutely aware of the philosophical problems to be encountered in translating everyday experience into secure knowledge. They knew that reasoning from individual experiences to universal generalizations was fraught with difficulty. They knew that ways were needed of judging which kinds of experiences were to be regarded as trustworthy and which not. This was a period when the horizons of human experience in the Western world were expanding massively, as travelers and explorers brought back tales of strange encounters in distant lands as well as exotic specimens of plants and animals. On the one hand, these novel sources of information seemed to justify skepticism concerning the reliability of ancient authority. On the other hand, as contemporaries were uncomfortably aware, they also raised questions concerning just what kinds of experiences should be considered as legitimate sources of knowledge and what sources of evidence could be trusted.

One of the foremost philosophical advocates of empirical knowledge was the English courtier and lawyer Francis Bacon. According to Bacon, there could be no question that well-attested experience rather than ancient authority was the only credible foundation for real knowledge. But, Bacon argued, to be useful experience had to be properly policed. Drawing explicitly on his legal experiences and his background as a state inquisitor, he insisted that experience had to be organized to be useful. "Just as if some kingdom or state were to direct its counsels and affairs not by letters and reports from ambassadors and trustworthy messengers, but by the gossip in the streets," he scoffed, "such exactly is the system of management introduced into philosophy with relation to experience." Bacon's solution was to make the business of empirical fact-finding into a collective, highly regulated system. In his *New Atlantis,* Bacon advocated the institution of Solomon's House—an establishment dedicated to the collaborative, disciplined acquisition of empirical knowledge. Bacon envisaged a hierarchy of researchers, ranging from humble fact gatherers at the bottom to philosophers on top, all engaged in the systematic production of empirical knowledge. Solomon's House was never built, although Bacon's vision certainly played a role in the seventeenth-century establishment of such collaborative scientific institutions as the Royal Society of London or the Paris Academy of Sciences (see chap. 14, "The Organization of Science"). His sense that distilling knowledge from experience required disciplined method and that not just any (or anybody's) experiences could be considered as a reliable basis for knowledge was widely shared, however (Martin 1992).

Disciplined and carefully regulated experience lay at the heart of Robert Boyle's experimental project, as exemplified in his air-pump experiments. Boyle's experiments were widely regarded—in England, at least—as exemplars of proper experimental practice. Boyle used his experiments to establish a number of claims about the constitution and nature of air (see chap. 3, "The Chemical Revolution"). He was well aware, however, that there was nothing straightforward about the procedure. Everything that took place inside the air pump was artificial, for example. It was not self-evident that the way the air behaved under such circumstances accurately reflected its natural behavior. Even given a general acceptance of a homology between what happened inside the air pump and what happened in nature, Boyle still had to work hard to convince his skeptical audience of the validity of his claims. He produced minutely detailed reports of just what he had seen during the course of the experiments. He carried out experiments in public, before witnesses. All of this was essential if he was to persuade others that his testimony about his experiences with the air pump was to be accepted as reliable. This was one reason that he and others like him felt that establishing scientific societies such as the Royal Society was so important. Even so, Boyle remained circumspect about what might be inferred from his experiments. As we have seen, while he regarded his reports about the behavior of the air as having the status of truth, any inferences from that behavior to the real constitution of the air remained hypothetical (Shapin and Schaffer 1985).

As we have suggested, seventeenth-century practitioners' acute awareness of the need to demonstrate the validity of experience was one factor in the rise of scientific societies. Most philosophical commentators agreed that the key to reliable empirical information lay in the trustworthiness of the witnesses. This was why Boyle and many other made experiments in public. The more witnesses—and the higher those witnesses' social status—the more reliable the results of the experiment. In the absence of witnesses, experimenters made every effort to produce sufficiently detailed and technical reports of their experiences that others would be convinced of their veracity. There was also a new vogue for cabinets of curiosities (Findlen 1994). Natural philosophers and their patrons collected and displayed natural (and artificial) curiosities of all kinds as a way of demonstrating the variety of nature—and their own prestige, of course (see chap. 16, "Popular Science"). Many empiricist natural philosophers concurred with Francis Bacon in his conviction that making new knowledge was an essentially collaborative enterprise. This provided one reason why it mattered that they should be able to trust each others' observations, which was, in turn, a reason why experimenters should be gentlemen, too, as opposed to artisans, tradesmen, women, or even foreigners. Gentlemen were conventionally regarded as more trustworthy because they were meant to be economically independent and therefore free from external influence. Many of them also agreed with Bacon that natural philosophy should be a civic matter as it was something that had an important role to play for the good of the commonwealth—another reason why it might be best carried out by gentlemen. Among other things, this suggested that one role for the new experimental natural philosophy should be to produce useful knowledge as well (Shapin 1994).

As we said earlier, this concern with the openness of natural philosophical knowledge was one of the reasons underlying Robert Boyle's and others' suspicions concerning the place of mathematics in the new mechanical philosophy. The key to making the New Science reliable, as far as they were concerned, was to make it as accessible as possible. New knowledge could be passed around, tested, and attested to and thereby slowly build up into a new consensus. It would become part of the common and universal stock of experience. The insistence that the book of nature was written in the language of mathematics was something of an impediment in this respect. Mathematics was far from being an accessible and commonly understood language in the seventeenth century. On the contrary, it was a highly technical practice that only a very few experts had fully mastered. Despite such concerns, however, few if any enthusiasts for the New Science would deny that mathematics was the language of nature. It was certainly increasingly held up as an exemplar of clear reasoning. Models of good ways of reasoning were just what seventeenth-century natural philosophers were looking for, after all. They wanted to be sure that their way of knowing, as well as their knowledge itself, was built on a secure foundation.

Let Newton Be!

Many of his contemporaries and immediate disciples regarded Sir Isaac Newton as having put the finishing touches on the Scientific Revolution. As the poet Alexander Pope rhapsodized:

> Nature, and Nature's Laws lay hid in Night.
> God said, *Let Newton be!* And all was *Light.*

Newton had succeeded in bringing together the disparate and fragmentary elements of the New Science and had forged them into a coherent whole. In many ways, he was the epitome of the natural philosopher as well: acerbic, difficult, and solitary. He was the archetype of the scientific genius for succeeding generations. Born the son of a prosperous Lincolnshire yeoman on Christmas Day 1642 (or 4 January 1643, as far as the rest of Europe was concerned, having already adopted the Gregorian calendar), Newton studied at the local grammar school before entering Trinity College Cambridge. It was as a fellow of Trinity that Newton produced the two books on which his claim to fame rested: the *Principia,* published in 1687, and the *Opticks,* eventually published in 1704 after Newton's elevation to the presidency of the Royal Society and, not coincidentally, following the death of his archnemesis Robert Hooke. By the time of his own death in 1727, he had transformed himself from a reclusive scholar into a powerful and influential public figure, gathering around himself a coterie of self-confessed Newtonians committed to his vision of what natural philosophy should be and how it should be practiced.

It is worth pausing a moment over the title page of Newton's great mathematical work. The *Principia*'s full title was the *Philosophiae naturalis principia mathematica,* or the *Mathematical Principles of Natural Philosophy.* It heralded an ambitious project. Newton was certainly committing himself to the view that mathematics was the language of nature and that the task of natural philosophy was to uncover the hidden mathematical laws that governed the universe's operations (Cunningham 1991). He was also making it clear to his readers that he knew what those mathematical laws were. In effect, the title page of Newton's *Principia* announced to the world that he had uncovered the secrets of the universe. For such an ambitious book, the *Principia* had relatively obscure origins. According to anecdote, the book began as a response to a question from the astronomer Edmund Halley (discoverer of the eponymous comet), who at a meeting with Newton in 1684 had asked him if he could work out what kind of path an object (such as a planet) would follow under the influence of a force that varied as the inverse square of the distance from the center. Newton replied that he had calculated that such a path would be an ellipse—just like the orbits of the planets around the sun—but that he had mislaid the proof. Halley shrugged his shoulders knowingly and returned to London. Newton sat down to recover the proof. The result a few years later was the *Principia.*

Newton started off the *Principia* with a series of definitions of the physical properties of natural bodies—things like mass, momentum, inertia, and force—that he would deal with in the rest of the book. He then followed with a statement of his three fundamental laws of motion: that a body will rigidly maintain its state of uniform motion in a straight line, or its state of rest, unless it is acted on by an impressed force; that any change in the motion of a body is proportional to the motive force impressed; and that for every action there is an equal and opposite reaction. In the following three books of the *Principia,* Newton put these propositions to work. In book 1 he studied the motion of bodies under the actions of different kinds of forces, showing among other things that if a body follows an elliptical path, then the force acting on it must vary as the inverse square of distance from the center. In book 2 he studied the motions of bodies in various resisting media. In book 3, the "System of the World" he applied the general theory he had developed in book 1 specifically to the motions of heavenly bodies, establishing his universal law of gravitation along the way. Having established that the force acting to maintain the moon in its orbit was the same as that causing the acceleration of falling bodies at the earth's surface, he argued that "the economy of nature requires us to make gravity responsible for the orbital force acting on each of the planets." It was—and was widely recognized as—a veritable tour de force.

Newton's *Opticks* was in many ways a very different book. Despite (or perhaps because of) its comparative accessibility compared with the highly technical mathematics of the *Principia,* it was also considerably more controversial. The *Opticks* was set out as an exposition of the theory of colors that Newton had first developed several decades previously in his "New Theory about Light and Colours" published in the Royal Society's *Philosophical Transactions* in 1672. In that paper, Newton had attacked the prevailing idea that colors were the result of modifications of white light and suggested instead that white light was itself the result of the combination of different colors of light. He used his famous prism experiments in which glass prisms were used first to turn white light into separate colors and then to recombine those colors into white light. It is important to be clear just what status Newton accorded to this experiment. As far as he was concerned the experiment proved his theory of colors—it was an *experimentum crucis,* a crucial experiment that established his theory beyond reasonable doubt. This was why Newton reacted so furiously to Robert Hooke's suggestion that the experiment could in fact be interpreted differently. As far as he was concerned this was not just an attack on his interpretation of the experiment, it was an attack on his personal integrity.

There was far more to Newton's *Opticks* than just his theory of colors. He used the book and its succeeding editions to outline his vision of the future course of natural philosophy. In particular, he introduced a number of Queries that included his view on any number of natural philosophical issues such as the nature of light, the causes of newly discovered electrical and magnetic phenomena, and the possible existence of a universal ether filling all space. The first edition of the book contained sixteen of these Queries, a number that had swelled to thirty-one by the final edition. These Queries—as the name suggests—were frankly speculative in nature, despite the famous motto, *hypotheses non fingo* (I do not feign hypotheses), that he added to the 1713 edition of the *Principia.* He asked, for example, "Are not the rays of light very small particles emitted from shining substances?" The thirty-first query was seemingly the most speculative of all. "Is not infinite Space the Sensorium of a Being incorporeal, living, and intelligent," Newton asked, "who sees the things themselves intimately, and thoroughly perceives them, and comprehends them wholly by their immediate presence to himself?" These were dangerous questions to raise. They are also an indication of the degree to which Newton placed his version of the mechanical philosophy in a thoroughly theological perspective.

While developing the work that led to the *Principia,* Newton was also working on another line of inquiry that he considered at least as important. He was trawling through ancient scriptural texts in an effort to recover a pristine and uncorrupted sacred history of creation. Newton was in fact an Arian—a heretic who denied the validity of the Trinity, the central belief of orthodox Catholicism and Protestantism. Newton held that the early Church had deliberately falsified and obscured the meanings of original biblical writings in order to mystify and confuse its followers. In his view, the ancients had known the truth about the mathematical structure of the universe, but the early Church fathers had deliberately conspired to obscure those truths. His scriptural researches were a systematic effort to recover the original meanings of biblical texts and hence to recover the lost wisdom of the ancients. That is just what he regarded his natural philosophy as doing as well. It was a process of rediscovery rather than discovery. Newton was sure that not only Plato and Pythagoras but Moses and the mythological Hermes Trismegistus as well had known about the Copernican system of the universe and the universal law of gravitation. All he was doing was rescuing that knowledge from the obscurity into which the early Church had condemned it.

Alchemy was another line of inquiry that Newton pursued in his efforts to recover lost knowledge. Newton delved enthusiastically into alchemical texts, producing copious notes and commentaries. He also carried out his own alchemical researches in his laboratory at Trinity College. Writings and experiments like this provided another possible avenue through which he might be able to rediscover what ancient philosophers had known about the nature and structure of the world. Newton regarded the arcane language and symbolism in which alchemical texts were presented as deliberate attempts to keep secret knowledge hidden from the eyes of the vulgar. In reading alchemical texts and attempting to reproduce the experimental procedures they described he was engaged in just the same kind of recovery exercise as he was when trying to wrest sense out of ancient scriptural writings or, for that matter, when producing the *Principia.* Unlike many other enthusiasts for the mechanical philosophy, Newton was also more sympathetic toward the idea of occult qualities in nature. Unlike many other mechanists, he was willing to leave open the question of the physical cause of gravity. He also suggested the possibility that matter might be endowed with "active powers." The German mathematician and philosopher Gottfried Wilhelm Leibniz explicitly accused Newton of reintroducing occult principles into natural philosophy on just these grounds.

It was partly in order to defend himself against such attacks that Newton surrounded himself with disciples. His defense against Leibniz—and his claim that Leibniz had stolen the idea of calculus from him—was undertaken by the young Anglican (and like Newton, secret Arian) clergyman Samuel Clarke. Despite charges such as those leveled by Leibniz, however, Newton's reputation in the early years of the eighteenth century could hardly have been higher. In England, he was regarded as the greatest flowering of English natural philosophy. On the Continent, particularly in France, he was regarded as the harbinger of enlightened rationality. The French writer Voltaire was a particular fan. According to Voltaire, a genius like Newton was born only once in a thousand years. Even Voltaire had to admit, however, that few of his disciples had read Newton, particularly the difficult *Principia.* As he reported back to France, few in London had read the great man "because one must be very learned to understand him." One friend of Voltaire's who clearly had read the *Principia* was Emilie du Châtelet, who wrote the first French translation and helped her lover Voltaire with the mathematical sections of his *Eléments de la philosophie de Newton*

(1738). Even while paying lip service to the mathematical bravura of the *Principia,* most of Newton's self-proclaimed eighteenth-century followers were more likely have derived their inspiration from the *Opticks* and its speculative Queries. Experimenters and instrument makers such as Francis Hauksbee or John Desaguliers saw themselves as devising experimental apparatus and techniques that could be used to demonstrate Newton's speculations concerning active powers with spectacular displays of electric or magnetic powers.

Newton's eighteenth-century legacy was in many ways all things to all men. Historians have struggled to define a coherent natural philosophy shared by all those who described themselves as Newton's followers. One strategy has been to divide them into two camps—those who took their Newton from the pages of the *Opticks* and those who imbibed him from the *Principia* instead. Those who read the *Opticks* followed Newton's experimental line of inquiry, studying the phenomena of electricity, heat, magnetism, or light—the active powers that Newton had identified. The *Principia*'s readers devoted themselves to expanding and refining Newton's mathematical treatment and applying it to new problems. There is something rather unsatisfying about this picture, implying, as it does, that the authors of the *Opticks* and the *Principia* had quite different and even unrelated concerns. It might be more promising to recognize that there simply was no coherent "Newtonian" tradition. Different eighteenth-century practitioners borrowed some parts of what they took to be Newton's approach and discarded others. They were all certainly keen to associate themselves with his name, if for no other reason than the tremendous authority it had acquired. Those, like Voltaire, who knew of his unpublished biblical researches found them an embarrassment. Newton had become an icon of the eighteenth-century Enlightenment and its cult of rationality.

Conclusions

And so to return to the question with which we began this chapter, was there really a scientific revolution? It is worth reminding ourselves of just what is involved in the claim that the radical changes that took place in our culture's way of viewing the universe during the seventeenth century or thereabouts constituted no less than a scientific revolution. To begin with, historians have traditionally regarded this a unique event. There may have been several scientific revolutions but only one Scientific Revolution. In other words, the original claim is that the events that took place around the seventeenth century were sufficiently momentous and unprecedented to be considered revolutionary, that this was a unique set of events without parallel elsewhere in history, and that something decisively recognizable as modern science emerged as a result. Until comparatively recently this interpretation would scarcely have appeared worth challenging. All of its elements, after all, appeared to be self-evident. It is a view that would have been shared to some degree or another by historians of science from the eighteenth century until the present day. It is nevertheless worth asking ourselves, in view of the brief sketch presented here, whether the traditional picture really stands up to rigorous scrutiny.

In many ways, it is clear that the traditional account of the Scientific Revolution simply does not add up. Indeed, it fails in all three of its basic assumptions. Historians now typically agree that cataclysmic as the intellectual changes of the Scientific Revolution might arguably be, they are not unique in history. Other changes in worldview have been just as momentous. The term "revolution" itself has been exposed as problematic. Historians have exposed clear continuities between early modern approaches to understanding the natural world and earlier perspectives. There does not seem to be a particular point or event in history to which we can point and say that the Scientific Revolution started there. If this was a revolution, then it was one without a clearly defined beginning and with no decisive ending either. Finally, it is now clear that whatever emerged from the Scientific Revolution, it was not modern science. Certainly there are aspects of Newton's work, for example, that look recognizably modern. This is hardly surprising. At the same time, there are aspects of his work—like his fascination with sacred histories—that seem irrecoverably alien. It simply will not do to bracket off that portion of his work and proclaim the sanitized remainder as the origin of modern science if for no other reason than it would do a gross injustice to Newton's own perception of the enterprise in which he was engaged.

At the same time, despite all this, as we suggested at the beginning of this chapter, many of the protagonists who participated in the Scientific Revolution unquestionably appear to have been convinced in their own minds that something momentous was going on. They demonstrate a rare degree of unanimity (a very rare degree for the period in question) not only that something significant was going on in terms of their understanding of the universe but also regarding just what that something was. On the whole, protagonists agreed that what was special about their approach to knowledge was that it was based on the interrogation of experience rather than authority. Instead of consulting Aristotle they were consulting their own senses. It is a moot point whether this perception was accurate. Modern historians of medieval philosophy take a rather less jaundiced view of its practices than did those who were, after all, explicitly rejecting it. It is nevertheless how they presented their activities. In this light, at least, if we want to take historical participants' own views of what they did at all seriously, then we have to accord some degree of validity to the idea of the Scientific Revolution. It is also true that what they had to say about their activities in this respect does strike a chord with modern perceptions of science. We rather like to think that modern science is based on experience rather than authority as well.

The best answer to our question, in the end, is probably to conclude that it is simply the wrong question to ask. Whether the Scientific Revolution is a useful historical category is largely a matter of perspective. At the very least, such categorizations should be taken with a fairly healthy pinch of salt. They certainly should not be allowed to cloud historical judgment. Categories like the Scientific Revolution are useful, after all, only insofar as they help us understand past science and its place in culture. When defending the category becomes an end in itself, then it is probably time to let it go. What matters about our historical study of the period in question is that we try to understand what happened then and what the various protagonists were trying to achieve on their own terms. Fitting it into a picture that leads from them to us is an important but secondary concern. If we go at it from the other direction—actively looking for precursors of modern science rather than appreciating the full picture—we will almost certainly end up getting hold of the wrong end of the stick.

References and Further Reading

Bennett, J. A. 1986. "The Mechanics' Philosophy and the Mechanical Philosophy." *History of Science* 24:1–28.

Biagioli, Mario. 1993. *Galileo Courtier: The Practice of Science in the Culture of Absolutism.* Chicago: University of Chicago Press.

Burtt, Edwin. 1924. *The Metaphysical Foundations of Modern Physical Science.* New York: Humanities Press.

Butterfield, Herbert. 1949. *The Origins of Modern Science, 1300–1800.* London: G. Bell.

Cunningham, Andrew. 1991. "How the *Principia* Got Its Name; or, Taking Natural Philosophy Seriously." *History of Science* 29:377–92.

Dear, Peter. 1995. *Discipline and Experience: The Mathematical Way in the Scientific Revolution.* Chicago: University of Chicago Press.

Fauvel, John, Raymond Flood, Michael Shortland, and Robin Wilson, eds. 1988. *Let Newton Be!* Oxford: Oxford University Press.

Findlen, Paula. 1994. *Possessing Nature: Museums, Collecting, and Scientific Culture in Early Modern Italy.* London and Berkeley, University of California Press.

Hall, Rupert. 1954. *The Scientific Revolution, 1500–1800.* London: Longmans, Green.
Hessen, Boris. [1931] 1971. "The Social and Economic Roots of Newton's 'Principia,'" In *Science at the Cross-Roads,* edited by N. I. Bukharin et al. Reprint ed., edited by Gary Werskey. London: Frank Cass, 149–212.
Iliffe, Rob. 1992. "In the Warehouse: Privacy, Property and Propriety in the Early Royal Society." *History of Science* 30:29–68.
Koyré, Alexandre. 1953. *From the Closed World to the Infinite Universe.* Baltimore: Johns Hopkins University Press.
———. 1968. *Metaphysics and Measurement: Essays in Scientific Revolution.* Cambridge, MA: Harvard University Press.
Kuhn, Thomas. 1966. *The Copernican Revolution.* Cambridge, MA: Harvard University Press.
Lindberg, David, and Robert Westman, eds. 1990. *Reappraisals of the Scientific Revolution.* Cambridge: Cambridge University Press.
Lloyd, Geoffrey E. R. 1970. *Early Greek Science.* London: Chatto & Windus.
———. 1973. *Greek Science after Aristotle.* London: Chatto & Windus.
Mayr, Otto. 1986. *Authority, Liberty and Automatic Machinery.* Baltimore: Johns Hopkins University Press.
Martin, Julian. 1992. *Francis Bacon, the State, and the Reform of Natural Philosophy.* Cambridge: Cambridge University Press.
Hunter, Michael, and Simon Schaffer, eds. 1989. *Robert Hooke: New Studies.* Woodbridge: Boydell.
Redondi, Pietro. 1987. *Galileo Heretic.* Princeton, NJ: Princeton University Press.
Shapin, Steven. 1994. *A Social History of Truth: Civility and Science in Seventeenth-Century England.* Chicago: University of Chicago Press.
———. 1996. *The Scientific Revolution.* Chicago: University of Chicago Press.
Shapin, Steven, and Simon Schaffer. 1985. *Leviathan and the Air-Pump: Hobbes, Boyle and the Experimental Life.* Princeton, NJ: Princeton University Press.
Thoren, Victor. 1990. *Lord of Uraniborg: A Biography of Tycho Brahe.* Cambridge: Cambridge University Press.
Westfall, Richard. 1971. *The Construction of Modern Science: Mechanisms and Mechanics.* Cambridge: Cambridge University Press.
Westfall, Richard. 1980. *Never at Rest: A Biography of Isaac Newton.* Cambridge: Cambridge University Press.
Whiteside, D. Thomas, ed. 1969. *The Mathematical Papers of Isaac Newton.* Cambridge: Cambridge University Press.
Yates, Frances. 1964. *Giordano Bruno and the Hermetic Tradition.* London: Routledge.

CHAPTER 3

THE CHEMICAL REVOLUTION

Chemistry often gets treated as the poor relation in histories of science. Traditionally, historians of science have had a great deal to say about major developments in the physical sciences during and since the Scientific Revolution. Similarly, a great deal of historical attention has been devoted to the life sciences, particularly in the context of Darwinism, its origins, and its consequences. Developments in the chemical sciences, by contrast, have been regarded as rather less earth shaking in their consequences. There are a number of possible reasons for this comparative neglect. Historically, many of the practices and ideas that we might now characterize as chemical originated in a wide variety of different contexts and places. Alchemists, apothecaries, doctors, dyers, and metalworkers all engaged in activities that we might now think of as having to do with the origins of chemistry. Faced with this wide variety of origins, historians of chemistry have sometimes found it difficult to come up with a unified view of the science's development. Another problem has to do with the perception of chemistry as a practical rather than a theoretical science. Until comparatively recently, historians of science have regarded themselves as historians of ideas. From this perspective, practical sciences such as chemistry have often simply seemed less worthy of historical attention. Physics and biology have had their big philosophical ideas. There seem to be no clear equivalents in the history of chemistry.

The traditional view has been that chemistry did not play a major role in the so-called Scientific Revolution of the sixteenth and seventeenth centuries. On the contrary, according to at least one historian, chemistry was behind the times by almost a century (Butterfield 1949). According to this view it was only at the end of the eighteenth century that the "delayed Scientific Revolution in chemistry" eventually took place. Before the French chemist Antoine-Laurent Lavoisier's systematic reform of chemical ideas and language and the overturning of the phlogiston theory in the closing decades of the eighteenth century, chemistry remained in a kind of scientific Dark Ages. While physics (or, more properly, natural philosophy) had embraced the Newtonian ideal of a rigorously quantitative and experimental methodology, chemistry remained wedded to woefully vague and qualitative approaches. More recent historians recognize that this view of chemistry before Lavoisier begs several questions. As we have seen already, few historians would now agree with the idea that there was a uniquely scientific revolution during the sixteenth and seventeenth centuries, still less that it resulted in a uniquely defined scientific method. In much the same way, historians of chemistry are now far less likely to regard Lavoisier's contributions in themselves as having decisively inaugurated a new era (Ihde 1964).

In that respect we need to think quite carefully about the proposition that there was a chemical revolution at the end of the eighteenth century. As in the case of the Scientific Revolution more generally, it is important to recognize just what is being argued. To accept the case for regarding the changes in chemical theories and practices that took place during this period as constituting a uniquely defined chemical revolution, we would need to accept that the chemistry that emerged from the end of the eighteenth century was in some way recognizably modern in a way that previous chemistry was not. We would also need to accept that this transformation was unique. Historians are now far more aware of the range and complexity of chemical theories and practices before Lavoisier and the important contributions that earlier chemists made. It is also clear that the debates surrounding chemistry at the end of the eighteenth century can no longer be regarded plausibly as a straightforward battle between enlightened supporters of Lavoisier's chemical reforms on the one hand and blinkered rejectionists on the other. In reality the range of positions was far more complex. Neither were Lavoisier's reforms as decisive as they were once considered to be. Many aspects of Lavoisier's theories would appear as peculiar to modern chemists as those of his predecessors and adversaries.

We will commence this chapter with an overview of "unreformed" chemistry during the seventeenth and early eighteenth centuries. It should become clear that regardless of the views of later generations of chemists and historians of chemistry, chemical practitioners such as Robert Boyle, Paracelsus, and Georg Stahl regarded themselves as fully committed to the New Science. We will then look at the development of pneumatic chemistry during the eighteenth century, particularly the work of English chemist and natural philosopher Joseph Priestley. Looking at Priestley's contributions will help to make clear the role chemistry played in eighteenth-century science and culture and the wider ramifications of phlogiston theory. Against this background we will then examine Lavoisier's contribution to chemistry, in particular his rejection of phlogiston theory in favor of his own theory of oxygen and his efforts to establish a new, reformed chemical language. We will see how Lavoisier's chemical innovations can be located within the particular context of late eighteenth-century developments in French chemistry and natural philosophy. Finally we will look at developments in chemistry in the immediate aftermath of Lavoisier's innovations during the opening decades of the nineteenth century. In particular, we will look at John Dalton's development of atomic theory. This will help to make clear the extent to which his immediate chemical successors regarded Lavoisier's innovations as decisive and the extent to which they regarded his theories as providing only one of a number of possible approaches to reforming chemistry.

Chemistry Unreformed?

Many practitioners engaged in activities we might now characterize as "chemical" certainly regarded themselves as being at the forefront of the New Science during the sixteenth and seventeenth centuries. Alchemists such as Michael Sendivogus or even Sir Isaac Newton viewed themselves as the inheritors of a tradition stretching back into antiquity. The aim of their science was to understand the hidden relationship between natural substances and to find the key that would allow the transmutation of one element into another. Apothecaries and doctors were interested in the medicinal properties of substances. Medical reformers like Paracelsus and Joan-Baptista van Helmont wanted to develop new theories of matter that would lead to new understandings of the medical applications of natural substances. Metallurgists such as Vannoccio Biringuccio developed and tabulated new recipes for the improved production of metals as well as other industrial products such as dyes and gunpowder. Early eighteenth-century phlogiston theorists such as Georg Ernst Stahl were trained in this metallurgical tradition. The mechanical philosopher Robert Boyle carried out chemical experiments as a way of trying to understand the fundamental mechanical properties of matter. As we have seen already, far from being regarded as working in an unreformed and antiquated tradition, many of his contemporaries considered Boyle to be the archetypal new natural philosopher. Other chemical practitioners were equally convinced of the novelty and importance of their activities (Debus 1987).

Renaissance and early modern alchemists worked in a tradition that stretched back to the Greeks. Greek alchemists had tried to understand the methods involved in industrial processes such as metalworking and pigment making in terms of ideas about the fundamental elements of matter. Their medieval Islamic inheritors such as (the possibly mythical) Jabir ibn Hayyan and Al-Razi developed these ideas to form an extensive corpus of alchemical writings that were later borrowed in the Latin West. Early modern alchemists such as Michael Sendivogus, alchemist to the Holy Roman Emperor Rudolph II, claimed to be able to transmute elements into one another and to have particular mystic insights into the operations of nature. The Holy Grail of alchemy was the search for the philosopher's stone that was the key to the transmutation of one metal to another. Finding the stone would deliver not only limitless wealth (through the ability to transmute base metals into gold) but ultimate understanding of the secret nature of matter as well. Sendivogus was read by, among others, Sir Isaac Newton who investigated alchemy as part of his grand scheme to recover systematically the lost knowledge of the ancients. Alchemists developed a range of techniques and equipment designed to investigate the properties of different substances. They also developed an arcane language and symbolism to hide their knowledge of such matters from the uninitiated (fig. 3.1).

Alchemical tracts such as (the fictitious) Basil Valentine's *Triumphant Chariot of Antimony* (1604) emphasized the medicinal properties of substances. This was the main concern of apothecaries and doctors investigating the properties of matter. The medical reformer Paracelsus (full name: Theophrastus Phillippus Aureolus Bombastus von Hohenheim—you can see why he changed it) were adamant that a new understanding of the ultimate properties of matter was a prerequisite of a reformed medicine. Like many another proponent of the New Science, Paracelsus had nothing but contempt for his predecessors in medicine such as the Alexandrian medical authority Galen. He chose his new name (Para-Celsus) to symbolize his superiority over the past in the form of the Roman medical writer Celsus. The aim of medicine was to prepare arcana—remedies for disease based on the properties of natural substances. Paracelsus called his new practice iatrochemistry (from the Greek *iatro,* meaning doctor). The task of the iatrochemist was to use the doctrine of signatures—the knowledge of the relationship between earthly bodies and astral essences—in order to identify what substances could be used to cure particular diseases. Substances were

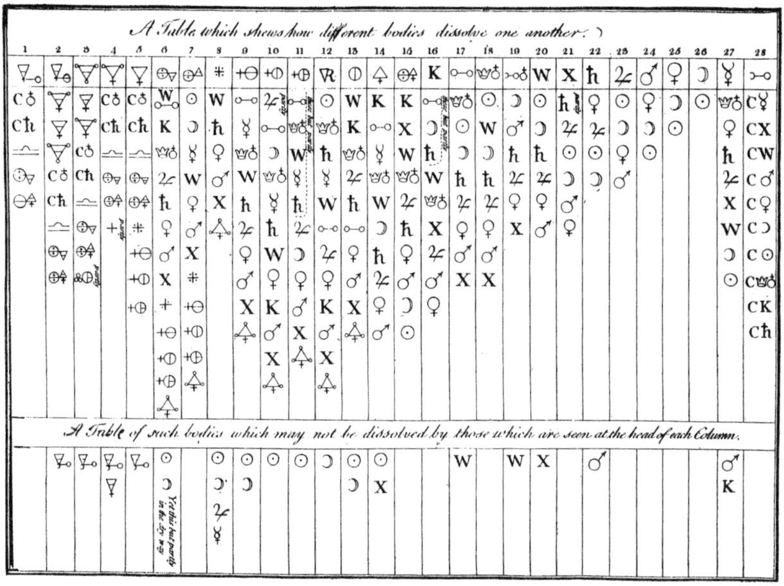

A Table which shews how different bodies dissolve one another.

A Table of such bodies which may not be dissolved by those which are seen at the head of each Column.

FIGURE 3.1 A table of alchemical symbols from G. E. Gellert, *Metallurgic Chemistry* (1776).

made up of the four elements (air, earth, fire, water) combined with the *tria prima* (three principles) of salt, sulfur, and mercury (or body, soul, and spirit). Like alchemists, Paracelsus argued that knowledge like this could only be made available to the initiated adept (Debus 1977).

Some iatrochemists, while applauding their master's insistence that chemistry was the foundation of proper medicine, abandoned some of Paracelsus's broader cosmological principles such as the doctrine of signatures and the *tria prima.* Van Helmont, a Flemish nobleman and follower of Paracelsus, denied the existence of the four elements and the *tria prima.* He held that there was only one element—water—along with the modifying principle of fermentation. Van Helmont demonstrated his claim in a famous experiment in which he grew a willow tree in two hundred pounds of dried earth regularly nourished with distilled rainwater. After five years the tree had grown in weight from five to one hundred and sixty-nine pounds while the weight of the earth remained the same. From this van Helmont concluded that the increase in size of the tree had been entirely due to the added water. Like many iatrochemists Helmont was interested in the chemistry of physiological processes such as digestion, which he interpreted as a fermentation process. His followers such as Franciscus Silvius expanded the theory to explain digestion in terms of conflict between opposite principles of salts and acids. Van Helmont was a pantheist who denied any distinction between matter and spirit. Like Paracelsus he also regarded chemical knowledge as the particular preserve of the initiated few (Pagel 1982).

Helmontianism was popular in England during the first half of the seventeenth century, but following the Civil War and the Commonwealth its mystical aspects and overtones of personal revelation started to bring it under suspicion. A new generation of chemists like Robert Boyle turned to the mechanical philosophy rather than Helmont's or Paracelsus's politically dangerous pantheism as a source of chemical explanation. Boyle's *Sceptical Chymist* (1661) dismissed Aristotelian, Paracelsian, and Helmontian theories of matter in favor of a corpuscular perspective. According to Boyle everything was made up of matter in motion. Rather than trying to explain the particular chemical and physical properties of substances in terms of the innate qualities of the various elements, Boyle argued that they should be seen as resulting from the particular shapes and arrangements of the corpuscles (or particles) making up those substances. One of Boyle's aims in embracing the mechanical philosophy as an explanation for chemical phenomena was to make chemistry itself part of natural philosophy. He wanted to do away with the arcane mysticism of Paracelsian or Helmontian approaches to chemistry with their overtones of charlatanry and make chemistry a practice in which gentlemen could take part without suspicion. He extolled the benefits in terms of medicine and the arts of a properly philosophical approach to chemistry (Kargon 1966; Thackray 1970).

Chemistry was increasingly recognized as a useful source of new knowledge in the development of metallurgical and other industrial processes. In his *Pirotechnica* (1540) the sixteenth-century Italian chemist Vannoccio Biringuccio provided detailed recipes of metallurgical processes and the manufacture of industrially and militarily useful substances such as gunpowder. Chemical knowledge could be put to work in improving the purification of metals from their ores and in the production of alloys. Chemical skills and know-how were needed to improve the production of dyes and pigments for the cloth industries. Johann Becher's chemical investigations into the origins of minerals in the earth were an explicit effort to find new ways of exploiting such resources for economic gain. His *Physica Subterranea* (1667) argued that minerals were made up of three types of earth—*terra fluida* (mercurous earth), *terra pinguis* (fatty earth), and *terra lapidea* (vitreous earth)—that defined their various properties. Becher's work was taken up by Georg Ernst Stahl, professor of medicine at the University of Halle, in the early eighteenth century as he developed his theory of phlogiston as a way of explaining metallurgical processes. He renamed Becher's *terra pinguis* phlogiston and identified it as the principle of combustion in the production of metals from their ores. According to Stahl's theory, pure metals were the result of the combination of metal ores (or calxes) with phlogiston during the heating process (Brock 1992).

There seems little doubt that most if not all of these sixteenth- and seventeenth-century chemical practitioners would have regarded themselves as fully fledged participants in the production of the New Science. Even alchemists working in what they at least regarded as age-old traditions viewed what they were doing as an important contribution to contemporary knowledge. Newton, for example, was interested in alchemy precisely because it offered a path to recovering lost knowledge in just the same way as he regarded his universal theory of gravitation as doing. To seventeenth-century eyes there was no contradiction between investigating ancient systems of knowledge and discovering new ones. Paracelsus and van Helmont, while deeply steeped in alchemical lore, also regarded what they were doing as a radical break from past practice. Like other proponents of the New Science, such as Galileo and Boyle, chemists also promoted the utilitarian aspects of their practices. Chemistry could contribute to improving the manufacturing arts and the wealth of nations. Becher, for example, was a cameralist—an advocate of the systematic intervention of the state to support commerce and manufacturing industry. His research into the theory of mineral production, carried out under the patronage of the Holy Roman Emperor Leopold I was quite straightforwardly part of his efforts to improve mining technology for the benefit of the state. If the defining feature of the Scientific Revolution is taken to be its participants' efforts to reform and reorganize knowledge then by their lights, at least, chemists were active participants in those efforts.

Pneumatic Chemistry

Joseph Wright of Derby's famous painting *An Experiment on a Bird in the Air-Pump* (fig. 3.2) painted in 1768 captures the increasingly important role of chemical investigations in eighteenth-century science and culture very well. In particular it highlights the central role played by investigations into the chemistry of gases—pneumatic chemistry, as it was called. Before the eighteenth century, the air was usually taken to be a single substance, one of the four Aristotelian elements. Eighteenth-century chemists, however, started to discover different kinds of air, with a variety of chemical properties and effects. Wright's painting shows a chemist demonstrating

FIGURE 3.2 Joseph Wright's "Experiment on a Bird in an Airpump" (1768) (courtesy of the National Gallery, London). A chemist demonstrates his experiments to a group of fashionable onlookers. The painting illustrates the increasing cultural importance of chemistry and natural philosophy during the eighteenth century.

the properties of one of these new airs by showing whether a bird could survive breathing it. The chemist is demonstrating the experiment to a group of well-dressed middle-class witnesses. The newly prosperous middle class was an important new audience for science during the eighteenth century. They were attracted by its utility and the lessons that might be learned by studying the order of nature. In the hands of radical natural philosophers and chemists such as Joseph Priestley even the chemistry of gases could be shown to carry important political messages. It was also a source of new technologies and played a key role in transforming the language of chemistry at the end of the century.

Investigating the chemical properties of the air was an eighteenth-century innovation. Seventeenth-century chemists usually assumed that air was chemically inert and therefore played no role in chemical reactions. The English clergyman and natural philosopher Stephen Hales, known for his investigations into the natural philosophy of plants (*Vegetable Staticks*) and animals (*Haemostaticks*), was one of the first to suggest that the air was chemically active. He had started investigating the air on discovering, during the course of experiments on plants, that large quantities of air were "fixed" in solid matter and could be released by heating. The instrument he developed to collect this air—later developed by the English doctor William Brownrigg into the pneumatic trough—was a key tool for chemical investigation throughout the rest of the century. Air produced by heating was washed of impurities by being passed through water before being collected in an inverted jar. Hales's observation that air could combine with other forms of matter focused chemists' attention. His discovery was pursued further by the Scottish chemist Joseph Black, among others. Black found that by heating the substance *magnesia alba* (a form of magnesium carbonate) he could produce a kind of air with distinct properties, which he called "fixed air"—what we would call carbon dioxide. He developed new ways of testing the air and determining its chemical properties through studying its reactions with acids and alkalis (Schofield 1970).

The key figure in eighteenth-century pneumatic chemistry was the English chemist, dissenting minister, natural philosopher, and political radical Joseph Priestley. The breadth of Priestley's activities captures well the broader context of chemistry during this period (Anderson and Lawrence 1987). Born to a religiously nonconformist family in the English midlands, Priestley trained as a minister in a dissenting academy and served as minister to a number of congregations before being appointed as a tutor at Warrington Academy in 1761. While there he established links with leading religious radicals such as the Welshman Richard Price and befriended, among others, the soon-to-be American revolutionary, Benjamin Franklin. He established his reputation as a natural philosopher in 1767 with his *History and Present State of Electricity* and made his name as a chemist with *Experiments and Observations on Different Kinds of Air* in 1774. Picking up on Hales's and Black's observations, Priestley established the existence of a number of distinct kinds of air, each with specific properties. The two best known of these discoveries were nitrous air (now known as nitrous oxide or laughing gas) and dephlogisticated air (oxygen). In 1780 Priestley took up the ministry of the New Meeting House in Birmingham and while there joined the Lunar Society of natural philosophical enthusiasts including the industrialists James Watt and Josiah Wedgwood and the radical doctor and proponent of evolution Erasmus Darwin (Schofield 1963; Uglow 2002).

Priestley used his chemical discoveries as the foundations of a whole new philosophy of nature. To explain the different chemical properties of the different kinds of air he had established he turned to Stahl's theory of phlogiston. Different kinds of air had a range of chemical properties depending on the quantities of phlogiston they contained. Some airs, like

Black's fixed air, contained relatively large amounts of phlogiston, others less. For a time Priestley assumed that normal atmospheric air was the air that contained the least phlogiston until he made a spectacular discovery in 1774. He found that by heating red calx of mercury he could produce an air that seemed to contain little or no phlogiston at all. According to Priestley's view of the "aerial economy"—the role different airs played in the natural order—this new dephlogisticated air was the best kind of air possible. Priestley argued that phlogiston, the principle of combustion (and corruption) was at the heart of the natural economy. Some processes, such as combustion, respiration, and the decomposition of animal bodies, released phlogiston into the atmosphere. Other processes, such as the actions of plants or the movement of water, removed it, thus maintaining a natural equilibrium. The best kinds of airs for human life were those that contained the least phlogiston. The newly discovered dephlogisticated air was therefore the most virtuous (Golinski 1992).

Priestley regarded this aerial economy as proof of divine benevolence. It showed the natural mechanism through which God kept the cosmos in a state of equilibrium. Everything in nature—plants, animals, the movements of wind and water, thunderstorms, earthquakes, and even volcanic eruptions—had a role to play in maintaining the economy of nature by adding to or subtracting from the amount of phlogiston in circulation. For a political and religious radical like Priestley, this view of nature's economy had important political and social consequences. Priestley famously claimed that "the English hierarchy, if there be anything unsound in its constitution, has reason to tremble even at an air-pump or an electrical machine." What he meant by this was that these scientific instruments helped reveal the proper order of nature. Since the social order should be based on this natural order and if, therefore, there was something wrong with the prevailing social order (and Priestley thought there was) scientific instruments could also be political instruments by showing how social injustices were at odds with nature. Priestley, as an outspoken political radical was an ardent supporter of both the American and French Revolutions. As a result of this support, his house and laboratory in Birmingham were burnt by a loyalist, "Church and King" mob in 1791, leading to his emigration to Pennsylvania in 1794 (Schofield 1970).

Priestley's pneumatic chemistry had other connotations as well, however. Some of his supporters, such as the Oxford professor of chemistry Thomas Beddoes—a student of the Scottish chemist Joseph Black—felt that Priestley's discoveries could provide the basis for a new system of med-

FIGURE 3.3 "Scientific Researches!" by James Gillray (NPG D13036; image courtesy of the National Portrait Gallery, London). Pneumatic experiments at the Royal Institution, satirized by James Gillray. The Royal Institution's professor of chemistry, Thomas Garnett, is administering gas to a member of the audience. The man standing behind him wielding a bellows and wearing a satanic grin is Humphry Davy. The gentleman with a large nose on the far right, looking on benevolently, is the institution's founder, Count Rumford.

icine. As well as being an advocate of Priestley's views, Beddoes was a supporter of the medical theories of John Brown, who argued that health could be achieved by maintaining a proper balance of stimulants and sedatives in the body. Beddoes believed that the newly discovered airs could be used in this way. Following his dismissal from Oxford for his radical political views, Beddoes established the Pneumatic Institute in Bristol in order to put his theories about the medical benefits of breathing different airs to practical use. He hired a promising young apothecary-surgeon's apprentice, Humphry Davy, to carry out experiments on the chemical and medicinal properties of the various kinds of air. Davy carried out a systematic program of chemical analysis on the airs, abandoning Priestley's phlogiston theory as he did so in favor of Lavoisier's new system of chemistry. His experiments on the physiological effects of breathing the various airs—particularly nitrous oxide—made him both famous and notorious in late eighteenth-century England (fig. 3.3) and helped him secure the plum position of professor of chemistry at the newly established Royal Institution in 1803 (Fullmer 2000).

Beddoes's and Davy's efforts to put pneumatic chemistry to medical use remind us that there was more to phlogiston than simply a theoretical principle. It was also the basis of a practical chemical technology. Priestley had himself been one of the first to attempt the exploitation of pneumatic chemistry's medical potential when he patented a method of dissolving fixed air in water to produce the world's first artificially prepared fizzy drink. Priestley assumed that his artificial soda water would have the same medicinal qualities as the mineral waters drunk in gallons at spa resorts such as Bath or Malvern. He also developed an instrument that could measure the amount of phlogiston present in different kinds of air and thus assess its capacity to sustain animal and human life. The eudiometer worked by mixing the air to be tested in a glass tube with a quantity of nitrous air. The extent to which the test sample changed in volume as the phlogiston in it combined with the nitrous air, was a measure of the air's virtue. The science of eudiometry was particularly popular in industrial Britain where it was used to assess the quality of air in manufacturing districts and in Italy where the Milanese professor of experimental physics, Marsilio Landriani, devised an eudiometer he could use to demonstrate the effects of *mal aria* on his fellow citizens' health.

Priestley's example in particular shows how chemistry was at the very heart of the eighteenth-century Enlightenment. Far from being an example of a science that had somehow not yet caught up with the advances made in other areas, chemistry was widely recognized by many contemporaries as an example of just how important science could be to eighteenth-century society. Chemists not only demonstrated that they were at the forefront of scientific progress as they saw it—developing powerful new theories and practical technologies—but that their science was making a major contribution to social progress too. It should also alert us to how careful historians of science ought to be in approaching past ideas that may now appear wrong or misguided. Some historians point to phlogiston theory—particularly the French chemist Guyton de Morveau's suggestion that phlogiston might have negative weight (since substances seemed to gain weight as they lost phlogiston during combustion)—as a prime example of how preconceived ideas can hold science back. This kind of "whiggish" approach is guilty of not taking past science seriously on its own terms and those of its practitioners. Phlogiston did not seem at all silly to its promoters, as, for example, Priestley, though as it happens few of them took de Morveau's suggestion seriously either. Most argued that phlogiston was an immaterial principle and as such made no contribution to the weight of a substance one way or another.

Phlogiston versus *Oxygène*

One ongoing dispute in the history of chemistry involves the issue of who should be considered the discoverer of the gas oxygen. The historian and philosopher of science Thomas Kuhn uses the episode as a classic example of the difficulties involved in reconstructing the "historical structure of scientific discovery" (Kuhn 1977). In the case of the discovery of oxygen, we have three candidates for the status of discovery. The first is Carl Scheele, a Swedish chemist who during the early 1770s succeeded in isolating what he called "fire air" through a variety of methods. He did not announce his results publicly until much later, however. The second candidate is Joseph Priestley with his isolation of a new air in 1774 and his identification of it as dephlogisticated air in 1775. The final candidate is Antoine-Laurent Lavoisier, who repeated Priestely's experiments and redesignated the air as oxygen in 1776, using it as the cornerstone of his new system of chemistry. Kuhn wanted to make two points about discoveries using this example. In the first place, he pointed out that discoveries were not simple straightforward events. They have a historical structure. He pointed out, for example, that it took time and several efforts at identification before anyone recognized oxygen for what it really was. Second, he pointed out that discoveries were only possible within the context of a theoretical system. Whether dephlogisticated air or oxygen gas had been discovered depended on whether Priestley's or Lavoisier's systems of chemistry were accepted.

Kuhn saw Lavoisier's new system of chemistry as an example of a scientific revolution. It was his recognition that this new substance was an anomaly that did not fit into established systems that led to his conceptual breakthrough and development of a new way of understanding chemical processes. Lavoisier, by the 1770s, was a highly respected French chemist and a member of the Academie des Sciences. He came from a prosperous middle-class background and had originally been intended for a career in the law before taking up chemical studies at the Collège Mazarin. His teacher there, Guillaume-François Rouelle, was an exponent of Stahl's theory of phlogiston. By the mid-1760s Lavoisier was already making a name for himself in French philosophical circles as an ambitious young chemist. He was appointed to the lowest rank of the Academie des Sciences in 1768 and commenced on a career as a "scientific civil servant," putting his chemical expertise at the service of the French state (Brock 1992; Donovan 1996).

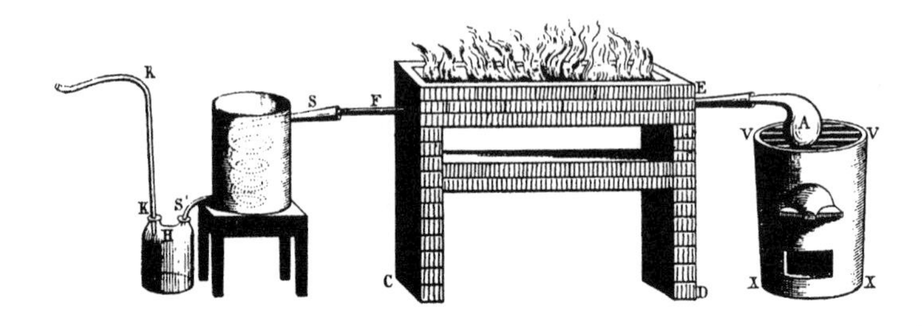

FIGURE 3.4 An eighteenth-century chemical experiment, showing the decomposition of steam by iron.

Lavoisier was independently wealthy after his father made over a large inheritance to him. He used his wealth to buy shares in the Ferme Générale—a company that had acquired the rights to collect taxes on behalf of the state. It was his status as a shareholder in the Ferme that led to his execution by guillotine in 1794 in the aftermath of the French Revolution.

During the late 1760s Lavoisier was particularly interested in the chemistry of the air and the role it played in combustion and the isolation of metals from their ores (calxes). Phlogiston theorists argued that metals were a combination of a calx and phlogiston. During combustion, phlogiston from the fire combined with the calx to produce a metal. By about 1770 Lavoisier was convinced that the air must play some role in the reaction too. In 1772, on the basis of experiments carried out with the Academie des Sciences's great burning lens, he suggested that gaseous air was in fact a combination of aerial matter and phlogiston (fig. 3.4). Heating metal in air thus led to the production of a calx (a combination of metal and aerial matter) and liberated phlogiston in the form of heat. On the basis of these and other experiments he deposited a sealed note at the Academie, laying claim to the hypothesis that the basic process taking place during combustion was the combination of the burning substance (like a metal) with aerial matter and that this accounted for the fact that substances increased in weight on combustion. By 1775, having come across Priestley's account of dephlogisticated air he refined his account further. He now argued that it was this dephlogisticated air, which he called *oxygène* that played the key role in combustion (Guerlac 1961).

In introducing *oxygène,* Lavoisier abandoned the phlogiston theory. In its place he offered a comprehensive new theory based around the new gas. The word "oxygen" came from the Greek, meaning "acid former" since Lavoisier had noticed that the substances formed by the combination of metals or carbon with this new principle were all acids. Oxygen gas, he argued, was composed of oxygen (the principle of acidity) and caloric (heat). During combustion, the principle of acidity combined with the metal to produce an acidic calx while the caloric from the gas was released in the form of heat. Lavoisier wanted his theory to do far more than explain the principles of metallic combustion, however. He wanted it to be the basis of a new and unified chemical system. One problem in this respect was the anomalous production of "inflammable air" when a metal was treated with an acid. This was easy to explain according to the phlogiston theory. The acid combined with the calx in the metal to produce a salt while releasing phlogiston as inflammable air. Lavoisier only solved this problem in the 1780s when the English chemist Henry Cavendish carried out experiments that seemed to show that water was a compound of dephlogisticated and inflammable air. Lavoisier could now argue that when metals combined with acids, the inflammable air came from the water in which the acid was dissolved. He called the gas hydrogen, meaning "water former."

One particularly important feature of Lavoisier's attempt to reform chemistry was the way in which he developed a whole new chemical language using his new theory. In 1782 Lavoisier, along with his French fellow chemists Guyton de Morveau, Claude-Louis Berthollet, and Antoine Fourcroy, published the *Méthode de nomenclature chimique* in which they described a new way of naming chemicals on the basis of the oxygen theory. All substances that could not be decomposed any further (like carbon, iron, or sulfur) were taken to be elements and formed the basis of the naming system. What had been called calxes were now called oxides, since they were the result of combining simple elements with oxygen, giving oxides of carbon, iron, or zinc, for example. Acids were named after their elements according to the amount of oxygen involved in their creation, as in sulfurous or sulfuric acids, respectively. As well as the metals and the bases of various salts along with hydrogen and oxygen, Lavoisier's list of elements also contained one other gas—azote (now called nitrogen). It also contained two other elements—caloric and light. The new system embodied Lavoisier's chemical theories. Simply by using it, chemists were signaling their acceptance of the oxygen theory around which it was based.

Lavoisier's reform of chemistry was widely recognized as radical and controversial. Some supporters of the phlogiston theory, notably Joseph Priestley, never accepted it. Another English chemist who remained convinced of the superiority of the theory of phlogiston was Henry Cavendish, despite the fact that his observations on inflammable air had formed one of the key factors in Lavoisier's reform. A number of English chemists were, however, converted to the oxygen theory within a comparatively short period. The rising star of late eighteenth-century English chemistry, Humphry Davy, was a supporter of Lavoisier's new chemical system, although as we shall see, he was soon to become one of its most committed opponents. In Scotland, by the 1790s the chemist Joseph Black was also teaching the new chemistry, and he along with his successors at Edinburgh introduced oxygen to new generations of medical students. In the German lands, opposition to the oxygen theory remained common until the early years of the nineteenth century. Yet even there, translations of Lavoisier's key works were being published by the early 1790s. In France, acceptance of the new theory was particularly rapid. Even prominent supporters of the phlogiston theory such as Guyton de Morveau were quickly converted—even, as we have seen, collaborating with Lavoisier in spreading the new doctrine.

One reason that Lavoisier's chemical system succeeded so quickly in France was the way in which it fit in with other contemporary developments in French science and philosophy. For a new generation of French natural philosophers, the keys to progress in science were quantification and accurate measurement. Natural philosophers such as the rising star Pierre-Simon Laplace were convinced that this was the only way to make sure that Newton's success in astronomy and mechanics could be repeated in other areas of physics. Lavoisier's emphasis on carefully weighing the ingredients and products of chemical reactions and his insistence that changes in weight provided the crucial evidence for what went on in such reactions fit in well with this concern for quantification. In the same way, his efforts to reform the language of chemistry and his insistence on the need for a comprehensive system of chemistry chimed well with broader French philosophical concerns. Philosophers such as Denis Diderot and Jean le Rond d'Alembert argued that the whole of philosophy needed systematic reform. The philosopher Étienne Bonnot de Condillac argued that reforming language was an essential prerequisite of reforming the ways in which people thought. In many ways, therefore, it seemed to his French contemporaries that Lavoisier's reforms in chemistry were part of a bigger picture. They were part of a larger reordering of the French intellectual world (Holmes 1985).

Past historians of chemistry have regarded Lavoisier's rejection of phlogiston and his reform of the language of chemistry as decisive moments in the chemical revolution. Before Lavoisier, chemistry was stuck in the Dark Ages. After Lavoisier, it was a recognizably modern science. It is worth pausing briefly here to consider how accurate this view is. However familiar many of its central features may appear to us, such as the role of oxygen in combustion and the new nomenclature, many features of Lavoisier's chemistry should also appear quite strange. While he had banished phlogiston from his system, the immaterial principle of heat remained in the form of caloric. Neither was caloric the only immaterial principle that took its place in Lavoisier's table of elements. Lavoisier's identification of oxygen as the principle of acidity—which formed the linchpin of his system—has also long been abandoned by modern chemists. At the same time, there can be little question that the phlogiston theory that Lavoisier abandoned was itself a powerful and versatile theoretical tool. It might sound peculiar from a modern perspective, but in the hands of experienced practitioners such as Joseph Priestley or Henry Cavendish it provided highly sophisticated explanations of known chemical phenomena and of recent discoveries such as the new kinds of airs. In that respect, at least, there was nothing inevitable or self-evident about the success of Lavoisier's theory or about its status as the key to the revolution in chemistry.

Chemistry Reformed?

One way of assessing the significance of Lavoisier's revolution in chemistry is by looking at the state of chemical knowledge in the decades immediately following the introduction of his reforms. Was Lavoisier's new chemistry quickly and universally adopted? How long was it before Lavoisier's reforms were themselves reformed? Kuhn characterizes a scientific revolution as a period of massive intellectual change followed by a period of "normal science" during which the implications of new conceptual frameworks and theories are explored and articulated. Did a period of such "normal science" follow the revolution in chemistry? It seems relatively clear, as we have already seen, that Lavoisier's reforms were adopted comparatively quickly and comprehensively. By the beginning of the nineteenth century, there were very few chemists who were still fully committed to the phlogiston theory. At the same time, there were also comparatively few chemists who were fully committed to Lavoisier's theory either. In that respect, at least, it is difficult to characterize the immediate aftermath of the chemical revolution as "normal science." Earlier supporters of Lavoisier's ideas were, by the 1800s, casting doubts on some of their key assertions. Other chemists, such as the Englishman John Dalton or the Swede Jöns Jacob Berzelius, were coming up with new theoretical frameworks of their own.

The Cornish chemist Humphry Davy had learned the basics of chemistry from William Nicholson's presentation of Lavoisier's ideas to an English audience. By the 1800s, however, following his appointment as professor of chemistry at London's Royal Institution, Davy was starting to cast serious doubt on the adequacy of some of Lavoisier's fundamental ideas. In the first place, Davy's experiments undermined the idea that acidity was due to the presence of oxygen. Davy showed that some acids such as muriatic acid (now called hydrochloric acid) did not contain oxygen. Similarly, he demonstrated that oxymuriatic acid not only contained no oxygen but was, in fact, an element in its own right, which he dubbed chlorine. By 1813 he had succeeded in isolating another similar element, called iodine. Davy made his name primarily by way of spectacular electrical experiments. He used the Royal Institution's powerful and expensive electric batteries to isolate not only chlorine and iodine but sodium and potassium too (Golinski 1992). Davy also argued against the existence of caloric, which played a key role in Lavoisier's chemical system. Heat, according to Davy, was not an immaterial fluid. Instead, he argued that it was a form of motion. If Davy was to be believed, not only was Lavoisier's oxygen misnamed—it was not an acid former—but, furthermore, neither it nor caloric played the critical part in chemical reactions that Lavoisier had assigned to them.

Lavoisier's definition of an element was largely pragmatic. Chemical elements were just those substances that chemists had been unable to break down into simpler constituent parts. In the hands of the English chemist John Dalton, however, the idea of an element developed different connotations. The idea that matter might be composed of indivisible particles or atoms went back to the Greeks. Seventeenth-century chemists such as Robert Boyle embraced the idea of atoms as a central tenet of the new mechanical philosophy. Where Lavoisier considered discussions of the ultimate nature of the elements as being metaphysical and beyond the reach of chemistry, Dalton set out to give the elements a real, physical existence. Dalton was born in northwest England to a Quaker family. At fifteen years old he started a school in Kendall in the Lake District with his brother before later moving to Manchester. During his time in the Lake District, Dalton, who had taught himself the rudiments of Newtonian natural philosophy, developed an interest in meteorology (the study of weather) and kept detailed diaries of local conditions that he published in 1793 as *Meteorological Essays*. The *Essays* helped make Dalton's philosophical reputation, and he employed the same approach toward looking for regularities in large quantities of data to produce his atomic theory of chemical elements (Patterson 1970).

The key difference between Dalton's atomic theory and the corpuscularianism espoused by previous chemical practitioners such as Boyle lay in the fact that Dalton assumed that each element had a unique atom associated with it. Boyle and other eighteenth-century proponents of atomism assumed that all atoms were the same (Thackray 1970). Building on this assumption, Dalton set out to try to define the relative weights of the atoms of the different elements. To do this he had to make a number of assumptions about the ways in which atoms combined together to make different substances. Simply speaking, he argued that elements would always combine together in the simplest possible ways. Since, for example, there was only one known combination of hydrogen and oxygen, Dalton argued that it must be a simple binary compound, with one atom of hydrogen combining with one of oxygen. Where more than one combination was known, more complex combinations (like two to one) were permissible. In the first part of his *New System of Chemical Philosophy* (1808), Dalton used these assumptions to calculate the relative atomic weights of Lavoisier's different elements from the known data concerning the relative quantities of different elements in chemical combinations. Since the relative weight of oxygen to hydrogen in water, for example, was known to be approximately seven to one, Dalton argued that a single atom of oxygen weighed seven times as much as one of hydrogen, the lightest element known (fig. 3.5).

On the basis of his own spectacular electrical experiments, Humphry Davy had concluded that the forces joining chemical elements together into compounds—the forces of chemical affinity as they were known—were electrical in nature. The Swedish chemist Berzelius built on Davy's conclusion along with what he knew of Dalton's atomic theory to come up with an electrochemical view of the way elements combined together. Berzelius classified elements into two kinds—electropositive and electronegative—depending on whether they were released from the positive or negative pole of a galvanic battery when decomposed. The terminology was later reversed to match the conventions introduced by Humphry Davy. The position of any particular element on the scale, with oxygen being the most electronegative and potassium the most electropositive, decided the way that element would combine with others. In atomic terms what this meant was that the individual atoms of the various elements had positive

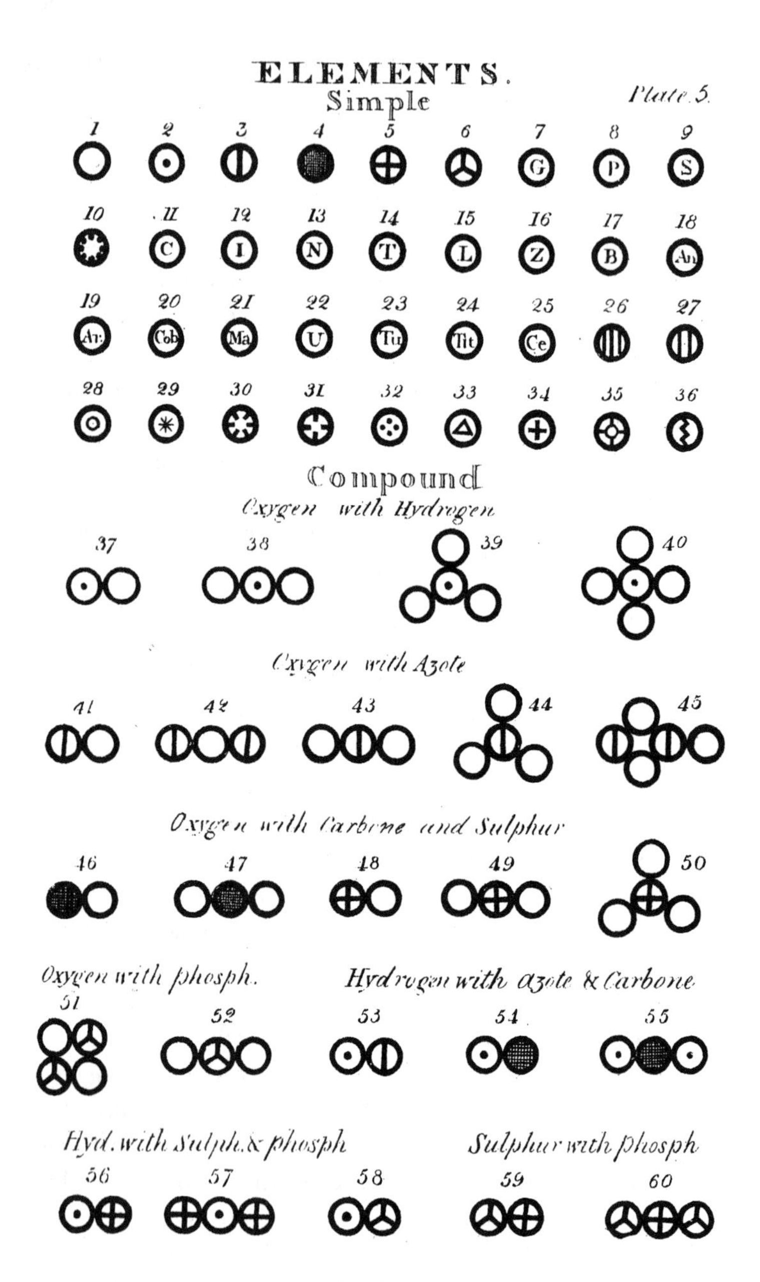

FIGURE 3.5 An example of John Dalton's new chemical notation from his *New System of Chemical Philosophy*. The notation was intended to emphasize the reality of chemical atoms.

or negative electrical charges associated with them that determined the ways in which they could attach themselves to atoms of other elements to form chemical compounds. Berzelius's comprehensive account of his electrochemical atomic theory was published as *Essai sur la théorie des proportions chimiques et sur l'influence chimique de l'électricité* in 1818.

Berzelius had originally studied medicine at the University of Uppsala and as professor of chemistry at Stockholm was responsible for teaching pharmacy to medical students. As a result, he was particularly aware that by the early nineteenth century most pharmaceutical texts were increasingly anachronistic by the standards of new chemical theories. It was in the context of his efforts to bring pharmacy up to date that he introduced a new conventional nomenclature into chemistry, based on his own electrochemical theories. The various elements were represented by a number of letters and abbreviations (like O for oxygen or Fe for iron) and their combinations represented by sequences of these symbols, with the most electropositive element being written first. Numbers of atoms were represented by numerical superscripts (later subscripts). So carbon dioxide, for example, would be represented as CO^2. Berzelius's new convention was just one of many introduced during the early decades of the nineteenth century and itself underwent many modifications. John Dalton in particular never accepted it, being concerned that the use of conventional symbols to represent the various elements tended to undermine acceptance that chemical atoms had a real physical existence. Dalton used his own notation, which he argued emphasized atoms' physical reality.

Dalton's objection to Berzelius's notation underlines one of the key issues surrounding the atomic theory. Were chemical atoms to be accepted as having a physical reality or were they just a convenient way of talking about chemical reactions and the proportions in which elements combined (Thackray 1972; Rocke 1984)? Dalton himself was convinced that chemical atoms were real. In this he was probably in the minority. Certainly by the middle of the nineteenth century, few chemists took the physical reality of atoms seriously. Chemists regarded the atomic theory—along with other generalizations such as the French chemist Joseph Louis Gay-Lussac's observations that volumes of gases combined with each other in simple ratios—as no more than useful empirical tools. It is not clear that even Berzelius took the reality of atoms too seriously. What is clear, however, is that very few if any early nineteenth-century chemists took Lavoisier's revolution of chemistry as definitively establishing a new chemical worldview. On the contrary, it is arguable that little beyond his rejection of the phlogiston theory survived in its original form beyond the new century's opening decades. With the consolidation of thermodynamics and its rejection of heat as an immaterial principle by midcentury, even caloric's pivotal role in chemical reactions was rejected. It seems that early nineteenth-century chemists did not regard their practice as having been reformed decisively. They were still in the process of reforming it.

Conclusions

So what are we to make of the delayed eighteenth-century chemical revolution? It seems that just as we have rejected the traditional account of the Scientific Revolution of the sixteenth and seventeenth centuries we have little choice but to reject the chemical revolution as well—and for many of the same reasons. As we have seen, it is difficult to sustain the notion that chemistry during the sixteenth and seventeenth centuries was somehow left out of the Scientific Revolution. The ideas and practices of a Becher, Boyle, or Paracelsus may seem peculiar to us now, but there is no evidence that they were considered peculiar at the time. Far from it, these practitioners were widely recognized by their contemporaries as important contributors to the New Science. Neither did eighteenth-century natural philosophers regard chemists as being behind the times. Chemists such as Joseph Priestley or Joseph Black were viewed as having made important contributions to natural philosophy as well as chemistry. More generally, contemporaries thought of chemistry as a vital and progressive component of Enlightenment science. Far from being outside the Newtonian synthesis as eighteenth-century practitioners saw it, many chemists were regarded as being in the vanguard (Knight 1978, 1992). Historians increasingly recognize that chemists before Lavoisier made decisive contributions and that their chemistry needs to be understood in the context of their own particular concerns in order to be fully appreciated.

There can be little question, either, that Lavoisier's reforms of chemistry had a major impact. His rejection of the phlogiston theory was in the end decisive and his introduction of quantitative methods and careful measurement set new standards of accuracy in chemical analysis. Again, however, it is also clear that Lavoisier's chemistry cannot be considered as having decisively ushered in the era of modern chemistry. In that sense, at least, his contribution was not revolutionary. As we have seen, very few elements of Lavoisier's chemical system survived the opening decades of the nineteenth century unscathed. Chemists such as Berzelius or Dalton did not regard themselves as working within the confines of a system already established. They were trying to establish their own systems of chemistry. There seems to be something peculiarly arbitrary about the choice of the late eighteenth century and Lavoisier's work as the locus for a unique chemical revolution. More generally, Lavoisier's "chemical revolution" ought maybe to alert us to the problematics of approaching the history of science in terms of a revolutionary perspective at all. On closer inspection, very few revolutions in science turn out to be as coherent or decisive as they might at first have appeared. In that respect, at least, there was nothing peculiar about the revolution in chemistry.

References and Further Reading

Anderson, R., and C. Lawrence, eds. 1987. *Science, Medicine and Dissent.* London: Wellcome Trust.

Brock, William H. 1992. *The Fontana/Norton History of Chemistry.* London: Fontana; New York: Norton.

Butterfield, Herbert. 1949. *The Origins of Modern Science, 1300–1800.* London: G. Bell.

Debus, Alan G. 1977. *The Chemical Philosophy: Paracelsian Science and Medicine in the Sixteenth Century.* New York: Science History Publications.

———. 1987. *Chemistry, Alchemy and the New Philosophy, 1550–1700.* London: Variorum.

Donovan, Arthur. 1996 *Antoine Lavoisier: Science, Administration and Revolution* Cambridge: Cambridge University Press.

Fullmer, J. Z. 2000. *Young Humphry Davy: The Making of an Experimental Chemist* Philadelphia: American Philosophical Society.

Golinski, Jan. 1992. *Science as Public Culture: Chemistry and Enlightenment in Britain, 1760–1820.* Cambridge: Cambridge University Press.

Guerlac, Henry. 1961. *Lavoisier, the Crucial Year.* Ithaca, NY: Cornell University Press.

Holmes, Frederick L. 1985. *Lavoisier and the Chemistry of Life.* Madison: University of Wisconsin Press.

Ihde, A. 1964. *The Development of Modern Chemistry.* New York: Harper & Row.

Kargon, Robert. 1966. *Atomism in England from Hariot to Newton.* Oxford: Oxford University Press.

Knight, David. 1978. *The Transcendental Part of Chemistry.* Folkestone: Dawson.

———. 1992. *Ideas in Chemistry.* New Brunswick NJ: Rutgers University Press.

Kuhn, Thomas S. 1977. "The Historical Structure of Scientific Discovery." In *The Essential Tension: Selected Studies in Scientific Tradition and Change.* Chicago: University of Chicago Press.

Pagel, Walter. 1982. *Johan Baptista van Helmont: Reformer of Science and Medicine.* Cambridge: Cambridge University Press.

Patterson, E. 1970. *John Dalton and the Atomic Theory.* New York: Doubleday.

Rocke, A. 1984. *Chemical Atomism in the Nineteenth Century.* Columbus: Ohio State University Press.

Schofield, Robert. 1963. *The Lunar Society of Birmingham.* Oxford: Oxford University Press.

———. 1970. *Mechanism and Materialism: British Natural Philosophy in an Age of Reason.* Princeton, NJ: Princeton University Press.
Thackray, Arnold. 1970. *Atoms and Powers.* Oxford: Oxford University Press.
———. 1972. *John Dalton.* Cambridge, MA: Harvard University Press.
Uglow, J. 2002. *The Lunar Men.* London: Faber & Faber.

CHAPTER 4

THE CONSERVATION OF ENERGY

IN A FAMOUS PAPER, THE PHILOSOPHER THOMAS KUHN raised what seemed to him a curious question about the discovery of the conservation of energy about halfway through the nineteenth century (Kuhn 1977). Kuhn observed that this was a simultaneous discovery—within a period of about thirty years, between the mid-1820s and the mid-1850s, a number of scientists more or less independently came up with the idea of the conservation of energy. Kuhn suggested that three factors in particular played a key role in that simultaneous discovery: the concern with engines, the availability of conversion processes, and what he called the philosophy of nature. Kuhn saw these factors as central elements of European scientific thought during the period that were "able to guide receptive scientists to a significant new view of nature." There can be little doubt that the conservation of energy is one of the more crucial generalizations in the history of science—or at least of the physical sciences. It was at the very core of physics as it developed during the second half of the nineteenth century. In a slightly modified form, the principle still plays a central role in modern physics. Trying to specify the cultural circumstances that led to the development of the conservation of energy can therefore tell us a great deal about the origins of modern science.

The first question we need to ask ourselves, however, is whether a theoretical generalization like the conservation of energy is really a candidate for discovery? When we think of discoveries, we usually think of discoveries of objects or places. The discovery of America by Western Europeans comes to mind as an obvious example. Another example might be the discovery of a new planet, such as William Herschel's discovery of Uranus.

Stretching the idea, it might make sense to talk about the discovery of a theoretical entity—the discovery of the electron, say. The conservation of energy, by way of contrast, is not a place or an entity; it is a theoretical generalization. It is worth considering, at least, what it might mean to think of the conservation of energy as something that can be discovered. It does seem to commit us, for example, to the view that the conservation of energy is something that really exists in nature, rather than just in our theories about nature. This is not just a philosophical quibble since even some of the principle's "discoverers" had doubts about whether energy or its conservation were things that could be said really to exist in nature. The second question we should ask ourselves concerns the object and the simultaneity of the discovery. For the discovery to be simultaneous, all the discoverers should have discovered the same thing at about the same time. We will see, however, that our historical protagonists described their findings in a number of different ways. In particular the word "energy" was not used to describe the quantity being conserved until rather late in the day.

We shall start our survey with a look at the first two of Kuhn's elements, though we shall suggest that they can easily be regarded as different aspects of the same concern. We will commence with the French engineer and natural philosopher Sadi Carnot and his theory of heat engines, in which he sought to find a relationship between heat and work. We will suggest that this might be regarded as an aspect of a broader interest, during the period, in getting one kind of force from another—what Kuhn calls conversion processes. We will then move on to consider some of the terms used to discuss the relationships between these forces—words such as "conversion" and "correlation" as well as "conservation." In particular, we will look at the ways in which these issues were played out in the contributions of James Prescott Joule and Julius Robert Mayer. Finally, we will follow the ways in which the principle of the conservation of energy was taken up by natural philosophers in Britain and Germany, in particular, during the second half of the nineteenth century and was used as the basis for the development of a whole new way of doing physics. It should become clear that the idea of energy and its conservation had a number of uses to its discoverers. It was a way of formalizing concerns about efficiency—in both economic and physical terms—for example. It provided a way of emphasizing the authority of physics over other sciences and of demonstrating the relevance of physics to industrial progress.

Water Wheels, Steam Engines, and Philosophical Toys

During the opening decades of the nineteenth century, increasing numbers of natural philosophers across Europe were becoming increasingly interested in the relationships between the different forces or powers of nature. Specifically, they were interested in finding out how to cause any of these forces to produce any of the others. In one sense, there was nothing particularly novel about this interest. Since the beginning of the eighteenth century, natural philosophers—particularly those who described themselves as Newtonians—had been keen to investigate the properties of powers such as chemical affinity, electricity, heat, light, magnetism, and what they often called motive force. Natural philosophers such as the Scotsmen William Cullen and Joseph Black, for example, studied the properties of caloric, the substance of heat. Their researches were particularly celebrated in some circles, at least, since they were widely held to have been the inspiration behind the engineer James Watt's steam engine improvements (see chap. 17, "Science and Technology"). This was just when the burgeoning Industrial Revolution was focusing many people's attention on the question of work—and on how to exploit the forces of nature to power machinery. To some people, this seemed to be just what James Watt had done with Black and Cullen's researches. Studying the philosophical principles that underlay the operations of different kinds of machinery, as well as looking at how to turn the different powers of nature to produce motive force (or work), seemed an increasingly profitable line of inquiry (Cardwell 1971).

Some of these speculations centered on the intriguing possibility of creating perpetual motion (fig. 4.1). The German natural philosopher Hermann von Helmholtz (of whom more later in this chapter) highlighted interest in this issue as one of the driving forces that led to the conservation of energy. Many natural philosophers (as well as any number of hopeful inventors and speculators) were interested in the prospect of getting an indefinite amount of work from a finite input. To take a hypothetical example, might it not be possible to construct a water wheel that would produce enough power to pump the water falling from one level to another in order to turn it, back to the upper level? If this could be done, then the wheel should turn forever with no need for any outside source of power. It would be a machine that produced work (and therefore money) for nothing. By the end of the eighteenth century, most natural philosophers were convinced that this was simply impossible. As Helmholtz noted, however, it did focus attention on just where the work came from in such systems.

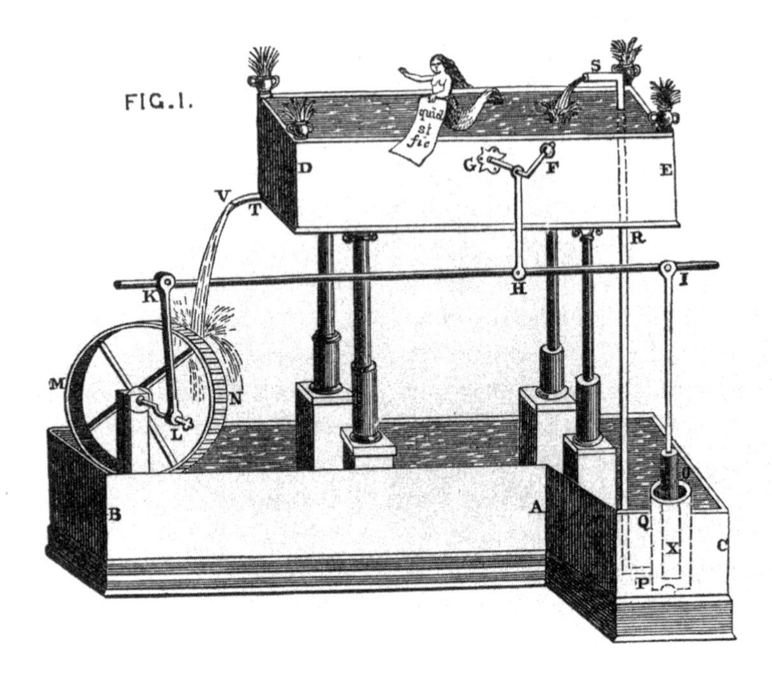

FIGURE 4.1 An example of a hypothetical perpetual motion engine. In this case, water from the upper reservoir pours down over a water wheel that, in turn, powers a pump that returns enough water to the upper reservoir to keep the motion going indefinitely. By the end of the eighteenth century, it was widely believed that engines like this were impossible.

The French engineer and revolutionary, General Lazare Carnot, for example, did some work on water wheels, showing how the amount of work produced was a function of the distance the water fell between levels in making the wheel turn.

Lazare Carnot's son, Sadi, was as interested as his father in questions about the origins of productive motive force. A committed Republican like his father as well, he wanted to find ways of putting his engineering knowledge at the service of humankind. Sadi Carnot focused his attention on the steam engine—the engine that seemed to be playing an increasingly prominent role in powering the rapid industrial expansion of France's great rival Britain. In his *Reflexions sur la puissance motrice du feu* (1824) Carnot carefully analyzed the workings of a hypothetical heat engine. Carnot regarded heat as the "immense reservoir" of nature's economy. It was the force that caused the weather, earthquakes, and volcanic eruptions. His assumption was that by understanding the operations of the actual steam engine he could gain an insight into the principles underlying the properties of the abstract heat engine as well. That, in turn, would help him to work out how to make more efficient engines. His strategy was to follow the movements of caloric—the immaterial fluid of heat—through the engine, trying to pinpoint how and where in the system motive power (or work) was produced. If he could make his hypothetical heat engine simple and general enough, he would be able to use it to "make known beforehand all the effects of heat acting in a determined manner on any body."

Carnot interpreted what happened in a steam engine in terms of the transfer of caloric from one part of the engine to another. As he saw it, that was what the steam did in the engine. The caloric developed in the furnace incorporated itself with the steam. It was then carried into the cylinder and on into the condenser. There the caloric was transferred from the steam to the cold water it found there, which was heated by the intervention of the steam as if it had been placed directly over the furnace. The steam throughout the process was only a means of transporting the caloric. This was the crucial fact for Carnot. What mattered in a steam engine—and in any other kind of heat engine, for that matter—was the movement of caloric from a hot to a cold body rather than its consumption. That was where the work came from: "The production of motive power is then due in steam-engines not to an actual consumption of caloric, but *to its transportation from a warm body to a cold body.*" Crucially, none of the caloric itself was lost in the process. As far as Carnot was concerned, caloric was conserved, just as water was conserved while producing work in the water mills that his father had analyzed. In a water mill, water did work by falling from one level to a lower level. In a heat engine, caloric did work by falling from one temperature to a lower temperature.

In 1820 the Danish natural philosopher Hans Christian Oersted made the dramatic discovery of a long-suspected link between electricity and magnetism. He found that when a magnetized needle was held near a copper wire through which a current of electricity was flowing, the needle twitched. Oersted was an exponent of *naturphilosophie*—a Romantic philosophy of nature particularly prevalent in German-speaking lands at about the beginning of the nineteenth century. Followers of *naturphilosophie,* such as the German poet Johann Wolfgang von Goethe, believed in the fundamental unity of nature. They often argued that the universe as a whole should be regarded as a single organic cosmic entity. Like a living thing, the universe was best approached and appreciated by seeing it as a connected, animated unity. Rather than being taken as separate objects of study, the various phenomena and powers of nature were to be understood as different manifestations of a single underlying and all-embracing cause. Such thinkers as Johann Wilhelm Ritter or F. W. J. Schelling often used terms like "World Soul" or "All-animal" to describe the universe. They emphasized the importance of intuition as a means of discovery and were often vociferously opposed to what they regarded as the dry sterility of analytic Newtonian natural philosophy. Coming from this perspective, Oersted was convinced that a link between electricity and magnetism must exist in nature; it was simply a matter of finding it.

A year following Oersted's discovery, the English experimenter Michael Faraday, then still a laboratory assistant at the Royal Institution, found a way to make a current-carrying wire actually rotate around a magnet. It seemed that electricity and magnetism combined could be used to produce motive force. In France, André-Marie Ampère showed that a current-carrying wire arranged as a helix acted like an ordinary magnet. He argued that magnetism was actually the result of electricity in motion and that magnets were made up of an array of electrical currents circulating around its constituent particles. It took Faraday, by now elevated to the position of Fullerian Professor of Chemistry and director of the laboratory at the Royal Institution, more than another decade to find the reverse effect. In 1832, he showed that when a bar magnet was moved inside a wire coil it produced a current of electricity. Similarly, when electricity was passed through a wire coiled around an iron ring, it produced, as it was switched on and off, a momentary current in another coil wrapped around the same ring. In the meantime, experimenters were exploiting the English instrument-maker William Sturgeon's invention of the electromagnet in 1824 to construct electromagnetic engines. With a variety of ingenious arrangements to switch arrays of electromagnets on and off consecutively, they could produce rotation. Caloric was no longer the only natural power that could be used to produce useful work.

Throughout the first few decades of the nineteenth century, experimenters were busily finding new ways of using one force to produce another. By one interpretation, Alessandro Volta's electric battery, invented in 1800, was an example—at least if one accepted Humphry Davy's explanation that it worked by transforming chemical affinity into electricity rather than its inventor's claim that the electricity was simply produced by the contact of different metals (see chap. 3, "The Chemical Revolution"). In the German state of Prussia, Thomas Johann Seebeck, inspired by Oersted's breakthrough, set out to examine the connections between electricity, magnetism, and heat. His aim was to produce magnetic phenomena by heat. Instead he found a way of producing electricity from heat. He found that if he constructed a circuit partly of copper, partly of bismuth and heated one of the junctions where the two metals joined, a current registered on a magnetized needle suspended nearby. The development of photography during the 1830s also seemed to many observers to be an example of one natural force being used to produce another. The images being produced were the result of light—one kind of force—producing a chemical reaction—the outcome of another kind of force, usually known at the time as chemical affinity. By the 1840s, more and more of these examples were building up.

In lectures at the London Institution, the Welsh natural philosopher William Robert Grove gave an experimental example of the ramifications. He demonstrated an experiment in which a photographic plate was placed in a glass-fronted box filled with water, along with a grid of silver wire connected to the plate to form a circuit along with a galvanometer and a Breuget helix. When light fell on the plate following the removal of a shutter covering the glass front, the galvanometer needles moved and the Breuget helix expanded. The light produced chemical forces on the plate, which produced electricity in the circuit, which produced magnetism in the galvanometer, which produced motion in the galvanometer needle while the electricity also produced heat in the Breuget helix, causing it to expand (more motion). Motion—motive force—was what many experimenters wanted to produce from these kinds of experiments. From the 1820s onward, they invented devices such as Barlow's Wheel, in which a copper wire rotated between the poles of a magnet when a current passed through it, and a variety of electromagnetic engines. On one level these were philosophical toys, designed to demonstrate the powers of nature to lecture audiences. At the same time, however, many natural philosophers recognized that toys such as these had the potential to provide new ways of producing motive force—of putting nature to work (Morus 1998).

The concern with engines and the interest in conversion processes were both aspects of the same preoccupation with getting work out of nature as efficiently as possible. As Helmholtz had noted, that was the concern that motivated enthusiasts for perpetual motion engines. It was what concerned Sadi Carnot in his efforts to analyze the workings of heat engines as well. He wanted to find out what the underlying principles were so that he could find ways of making engines that worked more efficiently. In just the same way, many of the researchers investigating ways of producing motion from other kinds of natural force were concerned to do so as efficiently as possible. At one level there was a theological motive to all this. It made sense that the Creator had designed the natural economy as efficiently as possible. At least as important, however, was the fact that this was a period when the question of work—and how to get as much of it as possible as cheaply as possible—was an issue of increasing concern. Making machines more efficient was an economic and moral imperative. Sadi Carnot was by no means alone in his view that working toward a better understanding nature's economy might prove to be a fruitful way of improving society's economy as well.

Conversion, Conservation, or Correlation?

By the 1830s and 1840s many natural philosophers were starting to come around to the view that these various examples of one force being used to produce another should be regarded as examples of actual transformation. That is, one force (say, electricity) was actually consumed in the process of producing another (say, heat or light). Remember that this was not a self-evident proposition—Sadi Carnot in his published work argued that caloric was not consumed in the process of producing work (though his unpublished manuscripts indicate that he later changed his mind on the issue). Even where experimenters did agree that what was going on was best understood in terms of some kind of transformation from one kind of force to another, there was a great deal of disagreement over just what kind of transformation was taking place. Natural philosophers might talk in general terms about the unity of nature—as they had done since the previous century—but there was little consensus as to how the details of that unity might be understood. Discussions about the issue are a good example of the ways in which early nineteenth-century natural philosophers crossed intellectual boundaries between areas of inquiry that we consider to be widely separated. Their arguments ranged across engineering, metaphysics, and theology as well as natural philosophy (see chap. 15, "Science and Religion").

The example of James Prescott Joule is a good one in this respect. A brewer's son from industrial Manchester, Joule's early natural philosophical enthusiasm was for electromagnetism. He made a name for himself designing and constructing electromagnetic engines during the late 1830s and formed part of the largely London-based circle of electricians around William Sturgeon (fig. 4.2). Joule was particularly concerned, however, to work out just how good his electromagnetic engines were. He applied engineering know-how and principles to the problem. He wanted to know what the duty of his engines was—this was an engineering term used to describe

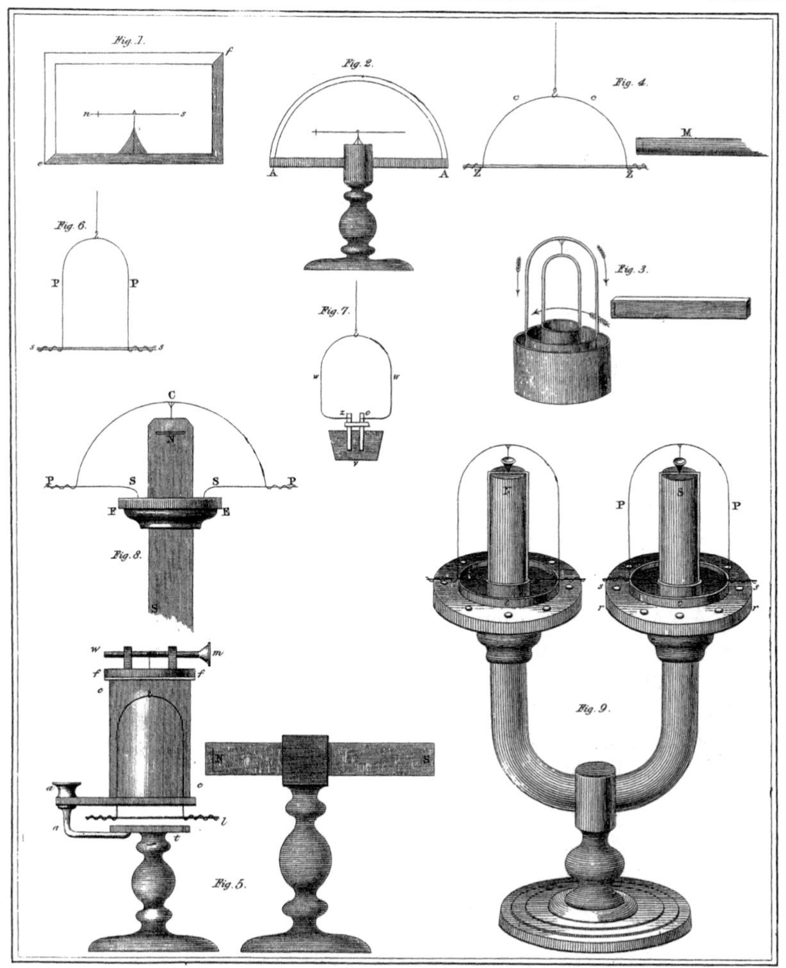

FIGURE 4.2 Instruments illustrating electromagnetism from William Sturgeon, *Scientific Researches.* Instruments such as these were meant for use in popular lectures to demonstrate the relationship of electricity and magnetism.

the efficiency of a steam engine and measured in terms of the weight in pounds that an engine could raise at a rate of one foot per second. What Joule wanted to know, quite specifically, was how much zinc was consumed in the process. Just like a steam-engine engineer, he wanted to know how much fuel was consumed to produce a given amount of work. Joule's experiments on the economic efficiency of electromagnetic engines led him

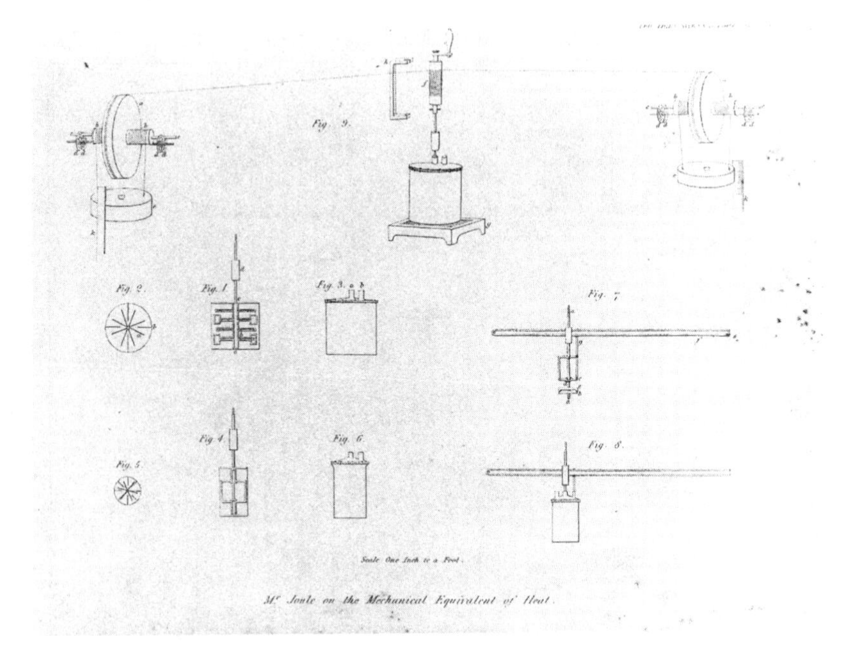

FIGURE 4.3 A diagram of James Joule's famous paddlewheel experiment demonstrating the mechanical equivalent of heat. As the weights fall, they cause the paddles inside the cylinder to rotate, heating the wafer contained in it. Joule argues that the congruent relationship between the distance the weights fell and the increase in temperature of the wafer in the cylinder demonstrated the relationship of work and heat.

to consider more general issues to do with the relationship between heat and work. By the mid-1840s, he was engaged in a series of experiments designed to work out just what that relationship was.

Joule was particularly concerned to try to find ways of quantifying the relationship between heat and work—the mechanical equivalent of heat, as he called it. In 1845 he produced the results of what is now known as his "paddle wheel experiment" (fig. 4.3). In this experiment, weights attached through pulleys to a paddle wheel enclosed in a container of water caused the paddle wheel to rotate as the weights fell. As the paddle wheels rotated, the water in the container heated up. With his background in the brewing industry, Joule had access to just the kind of sophisticated thermometric apparatus and know-how that was needed to perform delicate measurements like these (Sibum 1995). Joule argued that his results showed that the motion of the weights was transformed into heat in the water. This conversion could be accurately measured as well. According to Joule, when the temperature of a pound of water was increased by one degree Fahrenheit, it had acquired a quantity of *vis viva* (as he termed motive force) equal to that acquired by a weight of 890 pounds after falling from the height of one foot. Joule called this number the mechanical equivalent of heat and argued that his experiments showed conclusively that heat was literally turned into motive force in the process of producing work.

As far as Joule was concerned, his experiments carried a theological as well as an engineering message. They provided evidence of the way God had organized creation. Joule was convinced that his experiments were proof not only that one force could be converted into another but of the conservation of force as well. He gave his most comprehensive defense of the conservation of force at a public lecture at St. Anne's Church School in Manchester in 1847. Joule argued for the reality of conservation and conversion processes in nature, that "the phenomena of nature, whether mechanical, chemical or vital, consist almost entirely in a continual conversion of attraction through space, living force and heat into one another." This was an explicitly theological argument. Joule's claim essentially was that God had created force and matter and that since God had created them, neither of them could be created or destroyed. Any apparent loss of living force, as he translated the eighteenth-century Latin mathematical term, *vis viva,* was simply the result of the conversion of one kind of force into another, just as happened in the paddle wheel experiment with the transformation of work into heat. This was a highly controversial claim and not even all those sympathetic to Joule and the general idea of the conservation of force were convinced by it. Michael Faraday, for example, insisted that Joule revise the conclusion of his paper in the Royal Society's *Philosophical Transactions* announcing his claim to reflect Faraday's own doubts on the matter.

Joule was not the first to make a grand metaphysical principle out of the results of experiments on the transformation of force. In a series of lectures at the London Institution, William Robert Grove laid out his views on what he called the correlation of physical forces. Grove argued that all the physical forces were correlated to each other—that is to say, that any one of these forces could be used to produce any of the others, interchangeably. He used the idea to mount a metaphysical assault on the philosophical idea of causality, arguing that experiment showed that no one force could be shown to cause another since they were all mutually correlative. Michael Faraday made similar claims in lectures concerning what he called the conservation of force and occasionally borrowed Grove's vocabulary of correlation. It was not at all obvious that they meant the same thing, however.

Despite his own defense of the conservation of force, Faraday disagreed with Joule's claims on the matter. Faraday argued that all Joule had shown was that the loss of a certain amount of heat always resulted in the same amount of motion. Faraday was happy with the conservation of force but was unconvinced of the conversion of force. This was largely because he shared Joule's theological commitment to the belief that anything created by God (force in this case) could not be destroyed in any natural process. In his view, turning one kind of force into another was tantamount to destroying it.

While debates like these were occupying British natural philosophers, the German doctor Julius Robert Mayer was making his own observations aboard the ship *Java,* sailing for the Dutch East Indies in 1840. In the course of his duties as ship's doctor, Mayer noticed the unusual color of the venous blood of his shipmates. It was unusually red, appearing more like arterial than venous blood, the implication being that the heat of the tropics bore some relationship to the oxygenation of the blood. It was to this observation that he attributed his interest in heat, work, and the body. Pondering the matter back on dry land, Mayer published "Remarks on the Forces of Inanimate Nature" in the *Annalen der chemie und pharmacie* in 1842. He argued for a relationship among what he called "fallforce," motion, and heat. He suggested that heat was necessarily produced during the fall of any body toward the earth's surface since such a fall was the equivalent to a slight compression in the earth's volume and it was known that compression resulted in heating. He argued that the amount of heat produced by such a fall must be proportional to the weight of the falling body and the height from which it fell.

According to Mayer, his observations aboard the *Java* had convinced him "that motion and heat are only different manifestations of one and the same force." From this he had concluded that mechanical work and heat must be capable of being converted into one another. Like Joule he was able to come up with a specific figure as well. He calculated that the fall of a given weight from a height of around 365 meters corresponded to the heating of an equal weight of water from 0° to 1° centigrade. Mayer's work had little impact at the time, though he was later to be hailed as a German pioneer of the conservation of energy. To many of his German contemporaries, Mayer's work looked obscure and out-of-touch. The silence that greeted his work, like the skepticism with which even some friendly critics regarded Joule's experiments, illustrates the difficulties surrounding the issue of force and its transformations. Experimenters disagreed as to just what their experiments showed and what their implications were. The use of different terms, such as "conservation," "conversion," and "correlation," indicated more than just semantic quibbles; they indicated real disagreements concerning the nature of the phenomena. Philosophical concerns about the nature of causality and theological issues to do with God's place in creation were at stake here as well as the more prosaic concern to build more efficient engines.

British Energy

Joule was not alone in his combination of economic, engineering, and theological concerns. Other British natural philosophers also took the view that understanding how to make machines more efficient was a way of understanding nature, too. The pursuit of efficiency, that is, the effort to minimize waste and dissipation, was both an economic and a moral imperative. For young natural philosophers such as William Thomson, born in Presbyterian Belfast and raised in the industrial city of Glasgow, natural philosophy was all about understanding nature as if it were a vast steam engine. Thomson studied natural philosophy at Glasgow University, where his father was professor of mathematics, before departing for Cambridge to study for the mathematics tripos, or final exam. Cambridge, for much of the nineteenth century, provided probably the best mathematical education available, and Thomson was a star student (Harman 1985). Thomson's natural philosophical interests, such as those of his engineer brother James, centered on work, efficiency, and the elimination of waste. He wanted to understand how nature did it so that he could apply the lessons to human endeavor. Thomson was already familiar with Carnot's theory of heat engines. He had read the mathematical version published by Emile Clapeyron while studying steam engines at the experimenter Victor Regnault's Paris laboratory after leaving Cambridge. In 1847, two years after being appointed professor of natural philosophy at Glasgow, he attended a meeting of the British Association for the Advancement of Science and heard Joule present his findings.

Thomson was impressed by Joule's experiments, but as a follower of Carnot's theory they also presented him with a problem. According to Joule, heat was lost in the production of work. According to Carnot, caloric was conserved. This was the conundrum with which Thomson would struggle for the next several years. To produce his own theory, either he was going to have to show that one of them (Carnot or Joule) was wrong or he was going to have to find a way of reconciling two apparently irreconcilable theories. (Thomson was unaware of Carnot's later and unpublished doubts concerning the material nature of heat.) Thomson shared Joule's theological conviction that nothing God created could be destroyed. He was convinced that "nothing can be lost in the operations of nature—no energy can be destroyed." This was exactly where the problem lay, however. If, as Carnot argued, work was simply the result of heat falling from one temperature level to another, what happened to the work that would have been produced if there was no engine there for it to operate on? At the same time, if, as Joule would have it, the production of work meant the absolute loss of heat, where did the heat go in cases where no useful work was being done, as in the case of straightforward heat conduction, for example?

It took Thomson until 1851 to come up with an answer. In a series of papers titled "On the Dynamical Theory of Heat," published between 1851 and 1855, he laid the framework of the new science of heat—thermodynamics. The theory rested on two central propositions. The first was a straightforward assertion of Joule's claim concerning the mutual convertibility of heat and work. This was the first law of thermodynamics—the principle of the conservation of energy. The second proposition rested on his reading of Carnot. In essence, it stated that a perfectly reversible engine—in other words, an engine that produced exactly as much work as the equivalent amount of heat lost or that would take precisely that amount of work to recover the lost heat—was the best possible kind of engine. He had abandoned his earlier commitment to Carnot's insistence that heat was conserved during the process while keeping the insistence that work could only take place when there was a transfer of heat from a higher to a lower temperature. In any process of heat transfer that did not fulfill Carnot's criterion of perfect reversibility—in other words in any real engine—Thomson concluded that there was "an absolute loss of mechanical energy available to man." This was the second law of thermodynamics.

Over the next few years Thomson worked with like-minded allies such as Peter Guthrie Tait and W. J. Macquorn Rankine to make his new dynamic theory of heat into a whole new way of doing natural philosophy, with the new concept of energy, not force, at its very core. Along with P. G. Tait (they jokingly referred to themselves as T & T′), Thomson wrote the monumental *Treatise on Natural Philosophy* to demonstrate the possibilities of the new science of energetics. It was an ambitious project, with the two men self-consciously regarding themselves as stepping into Newton's shoes and writing the new *Principia*. Thomson was the first to start using the term "energy" in a new and precise mathematical sense. Its previous usage had been as a loosely defined synonym for force or power. It now meant simply that mathematical entity which was quantitatively conserved in force transformations. Many of Thomson's critics were unhappy with this new emphasis on energy. The veteran English natural philosopher John Herschel (son of William Herschel, discoverer of Uranus) argued that energy did not really exist, that it was a mere mathematical fiction. He argued for the retention of force as the key concept in natural philosophy since force at least had a tangible and intuitively obvious meaning. In Herschel's view, the introduction of energy deprived natural philosophy of physical meaning.

Thomson and his cohorts were confident that energy and its ramifications went much further than thermodynamics. Energy and its components would serve to unify natural philosophy. Electricity, light, and magnetism could all be understood as energy. The conservation of energy had a role to play in chemistry as well, explaining how chemical reactions took place. It even had a role to play in geology and biology. Thomson was a fervent opponent of new Darwinian ideas about the origins of species, for example (see chap. 5, "The Age of the Earth"). He used the new science of energy to show just how wrong those theories were, demonstrating how thermodynamics proved that neither the earth nor the sun could possibly be old enough to sustain the long and slow geological and evolutionary changes needed by the latest theories. What Thomson was doing in these debates—and what he and Tait were doing in their *Treatise*—was largely about demonstrating the superiority of their kind of natural philosophy. They were showing how energy could be used to solve other disciplines' problems. Energetics was also an example of the usefulness of natural philosophy. It provided a recipe for building better steam engines. It also captured and reflected the industrial culture of Victorian Britain, providing a model in nature for a society that wanted to maximize efficiency and minimize waste (Wise 1989–90).

One enthusiast for the new science of energy was James Clerk Maxwell. He placed energy at the heart of the new theories of electromagnetism that he started developing from the 1850s onward. Having taken William Thomson's advice to read Michael Faraday's *Experimental Researches in Electricity and Magnetism* carefully, Maxwell produced his first paper, "On Faraday's Lines of Force," in 1855. There and in subsequent contributions, he provided a mathematical elaboration of Faraday's explanations of electrical and magnetic phenomena in terms of the distribution of hypothetical lines of force through space. Conscious of critics' complaints about the intangibility of energy, Maxwell laid out a complex mechanical model of molecular vortices and idle wheels to represent his theory. His mathematical theory described a real existing medium—the ether—where energy was stored and transformed from one sort to another (fig. 4.4). Maxwell's electro-

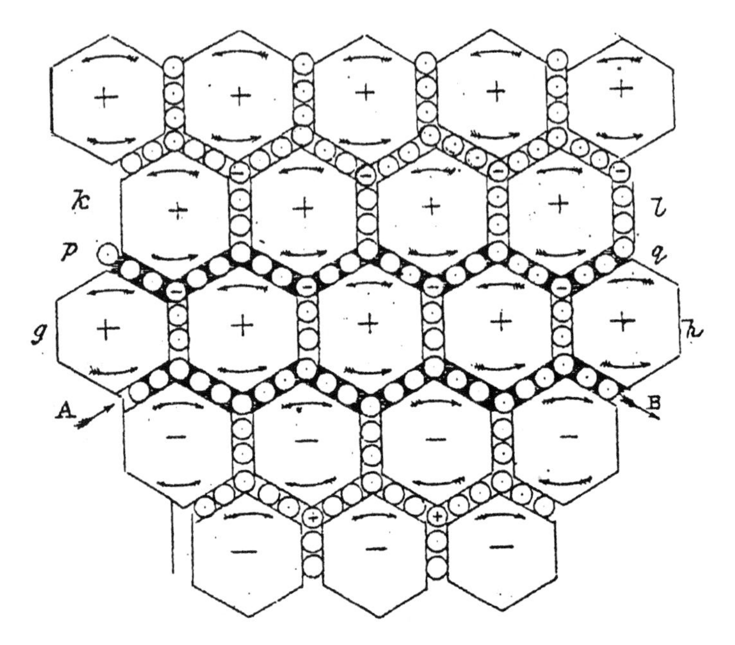

FIGURE 4.4 James Clerk Maxwell's model of a possible mechanical structure for the ether.

magnetic theorizing culminated with the *Treatise on Electricity and Magnetism* of 1873, published just two years after Maxwell had been appointed as first Cavendish Professor of Physics at the University of Cambridge. Just like Thomson and Tait, he was trying to build the foundations of a comprehensive new science based on the concept of energy. He was adamant that electromagnetic energy and the ether were not hypothetical entities. They were as real as anything else in the universe.

The ether rapidly became the embodiment of energy for nineteenth-century British physicists. As far as many of them were concerned, the physics of energy was practically synonymous with the physics of the ether. Physicists, including Oliver Heaviside, Oliver Lodge, and George FitzGerald, took the main business of physics to be working out the physical and mathematical properties of the ether. In 1885, Fitzgerald, for example, developed what he described as a "vortex sponge" model of the ether, with the ether visualized as a three-dimensional network of spongy, compressible vortices filling all space. The aim was to be able to rewrite Maxwell's electromagnetic equations in purely mechanical terms as descriptions of a real mechanical system. Electromagnetic waves, for example, would be understood quite literally as mechanical vibrations in a physical medium. If Thomson's thermodynamics was the physics of the steam engine, Maxwellian electrodynamics was the physics of the Victorian telegraph system. The electric telegraph had been one of the major achievements of Victorian engineering, and Maxwellian physicists were keen to demonstrate their science's capacity to explain its operations. They regarded the discovery by Heinrich Hertz—one of Hermann von Helmholtz's students—of electromagnetic waves in 1888 as a massive vindication of Maxwellian theory and as a victory over practical electrical engineers such as William Preece, head of the post office's telegraph department, who denied the applicability of Maxwellian physics to practical engineering issues.

Men such as Joule, Thomson, and Maxwell were particularly keen to make the science of energy practical and tangible. Not everyone agreed with this perception of what physics should be about. The French physicist Pierre Duhem was scathing about the way the physics of energy seemed to be the physics of the factory, too. He did not understand the British obsession (as he saw it) with making sure that the concept of energy was firmly anchored in reality. He regarded physics as a far more abstract business and had no problem with the prospect of theoretical entities that had no physical counterpart. British physicists, aware maybe of the criticisms aimed at them by opponents such as John Herschel, wanted to make sure that energy was recognized as a real entity, however. The physicist Oliver Lodge went so far as to say that the existence of the ether was as firmly established as was the existence of matter. This was a feature of their concern with the practicality of their science too. Most British physicists would not have been as insulted as Duhem would have wanted them to be by his comment that their physics was tainted by the factory floor. They were proud of the fact that their physics was above all practical.

The German Science

In the German lands during the second quarter of the nineteenth century, there were also moves by a new generation of natural philosophers to reform the practice and the key concepts of their science. In particular, many of this new generation were keen to disassociate themselves from what they perceived as the metaphysical excesses of the previous generation's *naturphilosophie.* They castigated their predecessors' science for being too speculative, obsessed with the unity of nature and treating the universe almost as if it were a living thing. Rising practitioners such as Emile du Bois Raymond, Carl Ludwig, and Hermann von Helmholtz embraced materialism and rationalism instead. Helmholtz studied medicine as a student at the University of Berlin during the early 1840s. Over the next few years he served as a staff surgeon in the Prussian army while carrying out experiments on the role of heat in muscle physiology and making a name for himself in physiological circles. In 1849, with the help of his former teacher, the physiologist Johannes Müller, Helmholtz got a job as professor of physiology at the University of Königsberg. Where their predecessors had wanted to show that the universe could be treated like a living organism, the new generation of physiologists of whom Helmholtz was part wanted to show that living organisms could be treated like machines (fig. 4.5).

In 1847, two years before he took up his professorship, Helmholtz published a little pamphlet called *Über die Erhaltung der Kraft* (On the conservation of force). Helmholtz based his theory of conservation on the denial of perpetual motion. If the amount of work done by a system in changing from one state to another were not the same as the amount of work that would be needed to change it back, then perpetual motion would be possible. He then proceeded to show how his theory applied in mechanical systems—those involving motion under the influence of gravity, the motion of elastic bodies, wave motion, and so on. In dealing with mechanical systems in which it had been supposed previously that an absolute loss of force took place, such as those involving friction or the collision of inelastic bodies, Helmholtz raised the possibility of the mechanical equivalence of heat, citing some of Joule's early experiments as evidence. He argued that heat could not be a species of matter, as the caloric theory suggested, since experimental evidence suggested that there were ways (like mechanical friction or magneto-electricity) of producing indefinite amounts of heat in a system. If heat were a kind of matter, then it would seem, according to Helmholtz, that it could be produced out of nothing.

Helmholtz applied the same kind of mechanical principles to the phenomena of electricity and magnetism. He went through a thoroughgoing analysis of motion under the influence of electrical and magnetic forces. He picked up on Joule's experiments on the relationship between electricity and heat and provided detailed consideration of the action of different kinds of batteries, such as Daniell and Grove cells. Helmholtz concluded his essay with an examination of the conservation of force in organic bodies. He was, after all, a physiologist—and one who was committed to showing that physiology could be studied on materialist principles. Helmholtz's earlier physiological work had been aimed at showing how the heat of animal bodies and their muscular action could be traced to the oxidation of

FIGURE 4.5 The German physicist and pioneer of the conservation of energy, Hermann von Helmholtz (The Wellcome Trust, London). By the time of his death in 1894, Helmholtz was widely regarded as the leading figure in German science.

food—their fuel. His research was following in the footsteps of the German chemist Justus von Liebig, who had pioneered research into the connections between the chemistry of nutrition and vitality. He argued that experiments by physiologists comparing the amount of heat produced by the combustion and transformation of the substances taken in as nutrition equaled the amount of heat given off by living things. In other words, there was no missing vital force to be accounted for. Organic bodies obeyed the conservation of force like every other natural system.

Helmholtz published his essay in pamphlet form since it had been rejected for publication in the prestigious *Annalen der Physik*. The editor, the physicist Johann Christian Poggendorff, turned it down on the grounds that it was too speculative and did not contain enough new experimental material. Helmholtz was, moreover, a physiologist not a physicist by both training and profession. His position at Königsberg brought him into contact with mathematically trained physicists such as Carl Neumann, however. Gradually, physicists started paying attention to Helmholtz's speculations concerning the conservation of force, and Helmholtz acquired expertise in experimental physics and mathematics. Throughout the 1850s, his researches increasingly bridged the gap between physiology and physics, many of them like his experiments with Neumann on the propagation of electricity through nerves being aimed at working out the physical properties of physiological systems. By the 1860s he was increasingly recognized as a physicist, and he ended his career as director of the prestigious Berlin Physikalisch-Technische Reichsanstalt. He produced a new generation of German physicists, including Heinrich Hertz, who would apply and extend Helmholtz's own theoretical researches on the conservation of energy into new areas. One of the first physicists to take Helmholtz's work seriously was, however, Rudolf Clausius, a young schoolteacher recently graduated, like Helmholtz himself, from the University of Berlin.

Clausius had written his doctoral dissertation under the supervision of the physicist Gustav Magnus on the light-dispersing and luminous effects of the atmosphere, looking in particular at the ways in which tiny particles in the atmosphere reflected light. He moved on to the study of the motion of gases and elastic bodies. It was this research that focused his attention on the problems of heat and work, through his reading of the French experimenter Regnault's work and of Clapeyron's interpretation of Carnot's theory. In 1850 he published "On the Moving Force of Heat, and the Laws regarding the Nature of Heat Which Are Deducible Therefrom" in Poggendorff's prestigious *Annalen der Physik*. His argument was based on his reading of a paper on Carnot's theory by William Thomson in 1849. He argued that it was possible to reconcile Carnot's claim that work was the result of heat flowing from one temperature level to another, lower, temperature level, with Joule's assertion that work was the product of conversion from heat. All that was needed was to drop Carnot's assumption that heat was conserved during the production of work. Clausius suggestion was that the production of work by heat required both the flow of heat from one temperature level to another and the conversion of a certain proportion of the heat into work. Both Carnot and Joule were therefore correct, so long as Carnot's claims concerning the conservation of caloric were relegated to the status of a superfluous subsidiary statement. This was much the conclusion at which Thomson would arrive in his 1851 paper "On the Dynamical Theory of Heat."

Clausius continued to work on his theories of heat throughout the 1850s and beyond. In 1853 he dealt with Helmholtz's essay, praising it for its "many beautiful ideas" but criticizing it for its mathematical inexactitude. Clausius's main concern was to try to find connections between the dynamic theory of heat and the work on gases in motion that had originally drawn his attention to the issue. Clausius was interested in the kinetic theory of gases—the idea that the large-scale properties of gases could be understood as the results of the small-scale movements of the particles, or molecules, of which the gases were made up. In his view, heat was simply the outcome of the motion of these particles. Hot gases were made up of fast-moving particles while colder gases were made up of slower particles. Since the molecules in hot bodies were moving faster, they tended to be further apart from each other, and Clausius argued that heat could therefore be expressed in terms of this distance. In 1865, Clausius introduced a new concept—entropy—into the dynamic theory of heat, so that he could rewrite the second law of thermodynamics as the assertion that the entropy of the universe tends to a maximum. The Austrian physicist Ludwig Boltzmann later argued that this meant that the second law of thermodynamics was statistical in nature and that entropy should be understood as a statistical term defining the relative order or disorder of the system. This was a big step, implying as it did that the law of cause and effect only had a statistical, rather than absolute validity at molecular levels.

Thermodynamics and energetics as they developed in German hands were very different affairs from the British version, particularly in the case of Clausius's work. The science that Clausius produced was self-consciously abstract and rationalist. It was avowedly and deliberately the antithesis of the previous generation's wildly metaphysical *naturphilosophie.* Like Helmholtz, in papers produced during the 1850s and 1860s he expanded his work on heat to consider electrical phenomena as well. The basis for his comparison of electricity with heat was, however, explicitly mathematical rather than experimental. In many ways, the kind of research that Clausius and his students pursued was a direct precursor of twentieth-century theoretical physics. It was a tradition that regarded mathematical theorizing about nature as an autonomous activity in its own right. By the 1860s, it was rapidly becoming clear that however much it might appear to the casual observer as having much in common with it, this German science was the direct antithesis of the kind of practical natural philosophy that William Thomson and other similar-minded British physicists practiced. As Clausius's researches developed during the 1860s, James Clerk Maxwell complained that they bore less and less reference to material, physical reality. As far as he was concerned, even the most abstract mathematical concept had to have a measurable component if it was to be a part of a physical theory. Theoreticians such as Clausius had no such scruples. Unlike the British, German physicists had little interest in working out the mechanical structure of the ether. What mattered to them was the mathematics.

Conclusions

In many respects, Thomas Kuhn was clearly right. There was a simultaneous discovery of the conservation of energy during the second quarter of the nineteenth century. The personages highlighted here as well as several others came up with versions of what now look like the conservation of energy. Kuhn names twelve of them (while somehow missing Thomson and Clausius), and it would not be too difficult to come up with others. That what was being discovered by these various protagonists was in any sense the same thing—or indeed that anything was being discovered at all—is, however, the product of retrospection. It is only with hindsight that the various experimental claims and theoretical generalizations discussed here seem to add up to the principle we now recognize as the conservation of energy. When they were originally made, they might just as easily seem to pertain to whole different sets of concerns and issues. What we now regard as a straightforward piece of empirical science was regarded by Joule or Thomson—or Michael Faraday, for that matter—as a fundamentally theological issue. It was not simply over matters of detail that many of the simultaneous discoverers disagreed with each other about just what had been found. They disagreed about the fundamental meaning of what had been discovered and how it fit into the general scheme of natural philosophy.

None of this basic disagreement prevented vociferous priority disputes later in the century when it had been decided that a fundamental discovery had indeed been made. A number of figures laid claim to the discovery of the conservation of energy during the second half of the nineteenth century. William Robert Grove, for example, proclaimed his 1846 publication *On the Correlation of Physical Forces* as the key text, a claim that P. G. Tait dismissed as "humbug." Many British natural philosophers did, however, continue to use the term "correlation of forces" interchangeably with "conservation of energy" until at least the 1880s. In Britain, most commentators pointed to James Prescott's Joule's experiments on the mechanical equivalent of heat as the crucial discovery. In Germany, likewise, historians of the new doctrine of energy pointed to Robert Mayer as its originator. There were dissenters—the Anglo-Irish natural philosopher John Tyndall, a vociferous opponent of Thomson's and Tait's form of physics, agreed with the

Germans that Mayer rather than Joule was the real discoverer. The American physicist Josiah Willard Gibbs gave the laurels to Clausius, while P. G. Tait claimed that Clausius's excessive mathematical abstraction debarred him from consideration. The British and the Germans were particularly vociferous in their claims and counterclaims. Laying claim to having originated nineteenth-century physics' key theory was a matter of national pride.

The principle of the conservation of energy did, regardless, play a key role in the nineteenth century, institutionally as much as intellectually. On the one hand, it provided a new and powerful theoretical tool for understanding nature. On the other, it provided an equally powerful resource for the institutional reorganization of natural philosophy. If we are looking for origin points, it might not be unreasonable to argue that the conservation of energy marks the end of natural philosophy and the beginning of physics as we know it. The principle of the conservation of energy provided a focus for the emergence of physics as a discipline. It gave physicists a common set of experimental and theoretical practices and theories—though as we have seen it took some time for this common perspective to appear. Historians have argued that it was during the nineteenth century that science became a profession in the modern sense. In that case, the conservation of energy certainly provided common ground for the forging of a professional identity for physicists. It provided a way of demonstrating the intellectual and practical power of the new discipline. With its connections to steam engines and telegraphs, it signaled the important role that physics could play in industrial society.

References and Further Reading

Cahan, David, ed. 1994. *Hermann von Helmholtz and the Foundations of Nineteenth-Century Science.* Berkeley: University of California Press.

Caneva, Kenneth. 1993. *Robert Mayer and the Conservation of Energy.* Princeton, NJ: Princeton University Press.

Cardwell, Donald. 1971. *From Watt to Clausius: The Rise of Thermodynamics in the Early Industrial Age.* Ithaca, NY: Cornell University Press.

———. 1989. *James Joule: A Biography.* Manchester: Manchester University Press.

Carnot, Sadi. 1986. *Reflections on the Motive Power of Fire.* Translated and edited by Robert Fox. Manchester: Manchester University Press.

Elkana, Yehuda. 1974. *The Discovery of the Conservation of Energy.* London: Hutchinson.

Harman, Peter. 1982. *Energy, Force and Matter: The Conceptual Development of Nineteenth-Century Physics.* Cambridge: Cambridge University Press. .

———, ed. 1985. *Wranglers and Physicists.* Manchester: Manchester University Press.

———. 1998. *The Natural Philosophy of James Clerk Maxwell.* Cambridge: Cambridge University Press.
Hunt, Bruce. 1991. *The Maxwellians.* Ithaca, NY: Cornell University Press.
Jungnickel, Christa, and Russell McCormmach. 1986. *The Intellectual Mastery of Nature: Theoretical Physics from Ohm to Einstein.* Vol. 1. Chicago: University of Chicago Press.
Kuhn, Thomas. 1977. "Energy Conservation as an Example of Simultaneous Discovery." In *The Essential Tension: Selected Studies in Scientific Tradition and Change.* Chicago: University of Chicago Press.
Morus, Iwan Rhys. 1998. *Frankenstein's Children: Electricity, Exhibition and Experiment in Early-Nineteenth-Century London.* Princeton, NJ: Princeton University Press.
Rabinbach, Anson. 1990. *The Human Motor.* Berkeley: University of California Press.
Sibum, Otto. 1995. "Reworking the Mechanical Value of Heat: Instruments of Precision and Gestures of Accuracy in Early Victorian England." *Studies in the History of Philosophy of Science* 26:73–106.
Smith, Crosbie. 1998. *The Science of Energy.* London: Athlone.
Smith, Crosbie, and M. Norton Wise. 1989. *Energy and Empire: A Biographical Study of Lord Kelvin.* Cambridge: Cambridge University Press.
Williams, L. Pearce. 1965. *Michael Faraday.* London: Chapman & Hall.
Wise, M. Norton (with the collaboration of Crosbie Smith). 1989–90. "Work and Waste: Political Economy and Natural Philosophy in Nineteenth-Century Britain." *History of Science* 27:263–301, 27:391–449, 28:221–61.

CHAPTER 5

THE AGE OF THE EARTH

The enormous expansion in the timescale of earth history is one of the more formidable conceptual revolutions produced by modern science. The biblical timescale, based on a literal reading of the creation story in the book of Genesis, places the origin of the earth (and indeed of the whole universe) only a few thousand years ago. In this story, there is no prehistory because human beings are there from the start, and we know something about their activities from the sacred record. Contrast this with the picture established by the modern earth sciences, in which the earth is several billions of years old, with the human species appearing only at the very end of a vast sequence of events. Without this extended timescale, the theory of evolution is unthinkable, and it is no accident that modern "young earth" creationists seek to undermine the plausibility of the worldview established by the earth sciences. The biblical timescale was widely accepted in the late seventeenth century when the first efforts were being made by naturalists to understand the geological and the fossil records. Over a period of a century or more, continued work in this area made it increasingly difficult to sustain a theory of the earth that did not contain an extended sequence of physical events stretching over a vast period of time. Just how vast that period was would remain controversial until the early twentieth century. To the young earth creationists it is still controversial today.

The history of the earth sciences has tended to focus on issues that highlight the supposed "warfare" between science and religion. This has had a distorting effect on our interpretation of the theoretical debates, an effect that has been slowly dispelled by more recent historical studies. The older model of how these sciences developed, still visible in C. G. Gillispie's *Gen-*

esis and Geology (1951), adopted a "heroes and villains" approach in which a few key scientists were identified as the founders of the modern timescale. Those who opposed these pioneers were dismissed as bad scientists who allowed their work to be distorted by their religious beliefs. The two most important heroes were James Hutton and Charles Lyell, who promoted the geological methodology of "uniformitarianism." This method ruled out any appeal to unknown causes and saw the earth's history as an almost eternal cycle of slow, gradual changes. Significantly, Charles Darwin was one of Lyell's greatest disciples. Opposed to uniformitarianism was a geological theory called "catastrophism," which sought to limit the necessity for a vastly extended timescale by invoking violent events in which whole continents could be created or destroyed almost instantaneously. This not only limited the need to challenge the Genesis timescale but also allowed Noah's flood to be seen as a real geological event. Lyell and Hutton were portrayed as the founders of the modern earth sciences, while the catastrophists were ridiculed as religious bigots who manipulated their science to defend narrowly defined religious beliefs.

Modern historians have almost completely overturned this simple black-and-white model of how geology developed. Far from being poor geologists, the catastrophists made major contributions to our understanding of the sequence of geological periods making up the earth's history. They had no interest in reducing the age of the earth to a few thousand years, and most of them had no intention of portraying the last catastrophe as the flood recorded in Genesis. At the opposite end of the scale, it has been shown that Hutton and Lyell had their own religious and cultural values, which significantly influenced their scientific thinking. Although their models of earth history look superficially modern, they contain elements that no modern geologist could accept. Outside the English-speaking world they were largely ignored. The geologists of the late nineteenth century continued to work with a timescale that was much shorter than we accept today, although it was still immense by human standards. Lyell's impact was more on the popular imagination—his books were widely read—than on science. Only in the early twentieth century did new evidence from physics force the geologists to begin working with a timescale extending to billions of years.

Studying the controversies over the age of the earth thus provides us with a good illustration of how the history of science has developed. New insights have been arrived at by challenging the myths established by the scientists themselves (and sometimes by their opponents). The older historiography was based on a tendency to manufacture heroes and villains according to a superficial estimate of how closely their theories approximated to what scientists accept today. And when apparently "bad" science was identified, external forces such as religious beliefs were called in to explain why those involved were deflected from the true path of scientific objectivity. The influence of the heroes was greatly exaggerated, giving the impression that they were able to precipitate a sudden revolution establishing the modern theoretical paradigm. We now see that the whole process was far more protracted and that the emergence of the modern view of earth history required the synthesis of different theoretical and methodological perspectives once thought to be mutually hostile to one another.

Stephen Jay Gould—himself a paleontologist—eloquently captured the need to rethink the conceptual differences between uniformitarians and catastrophists. His *Time's Arrow, Time's Cycle* (1987) shows how Lyell's apparently modern viewpoint rested on a "steady state" view of the past in which the earth could have no beginning and no end. By this standard, the modern view of geological time is more closely related to that of the catastrophists because they saw the earth as a planet that had a beginning and underwent a sequence of developments leading toward the earth we know today. Simply having more time in his theory did not ensure that Lyell got all the rest of his geology right. The catastrophists who resisted his arguments may have had good reasons for doing so, although this does not rule out the possibility that some of their reasons may have come from outside the bounds of science (for other modern surveys of the history of geology, see Greene [1982], Hallam [1983], Laudan [1987], Oldroyd [1996], Porter [1977], and Schneer [1969]).

Seventeenth-Century Theories of the Earth

One consequence of the so-called Scientific Revolution (see chap. 2) was that by the middle decades of the seventeenth century the earth itself became an object of study, and its origins a topic of theoretical speculation. Some of the resulting ideas sound bizarre by modern standards, but they helped to identify issues and problems that would shape the subsequent history of geology. One characteristic of these early theories that seems particularly odd today is the fact that they were almost all shaped within a conceptual framework defined by the biblical timescale. The seventeenth century was the period in which Protestant theologians and scholars established the "young earth" chronology based on a literal reading of Genesis.

(Paradoxically, the Church Fathers who established the foundations of Christian thought in the early centuries did not take the creation story literally.) In the mid-seventeenth century it was James Ussher, archbishop of Armagh, who published the now widely ridiculed calculation that the earth was created in 4004 B.C. His technique established the date of Adam's creation by working back through the Hebrew patriarchs. By taking the seven days of creation literally, it was then only a matter of adding on those seven days to arrive at the date of the creation of the earth and the universe itself. Ussher's scholarship was widely respected at the time and the naturalists who studied the structure of the earth at first saw little reason to challenge it. So their "theories of the earth" were framed in such a way that any changes they postulated could be fitted into this short timescale (see chap. 15, "Science and Religion").

Some of these early theories arose out of efforts to situate the origin of the earth within the new cosmologies proposed by Descartes and Newton (for details, see Greene [1959], Rappaport [1997], and Rossi [1984]). Thomas Burnet's *Sacred Theory of the Earth* (1691) followed Descartes in depicting the earth as a dead star and explained Noah's flood as the consequence of a massive collapse of the originally smooth surface (fig. 5.1). William Whiston's *New Theory of the Earth* (1696) appealed to Newton's theory to explain the flood as a due to water deposited from a near-collision with a comet. Both followed the biblical timescale, although Burnet—whose theory was criticized for departing from the literal text of Genesis—warned against tying the veracity of the sacred record too closely to a single theory. Burnet was aware that there were forces of erosion that could wear away mountain ranges but argued that the continued existence of mountains was evidence that they had been formed quite recently as fragments of the original crust.

What was new about these theories was their willingness to explain events of deep spiritual significance, such as Noah's flood, as a consequence of purely physical events. More disturbing in the long run was the evidence accumulated by naturalists who began to study the structure of the rocks and the fossils they contained. After some debate, it became widely accepted that fossils were the remains of once-living creatures petrified within the rocks (Rudwick 1976). The anatomist Nicholas Steno showed how fossil shark's teeth were almost indistinguishable from those of a living shark he had dissected. Robert Hooke showed that fossil wood was similar to its modern equivalent even under the microscope. Both Steno and Hooke noted the appearance of fossils within layers or strata of rock that gave every appearance of being deposited under water, even though they were now exposed on dry land.

FIGURE 5.1 The frontispiece to Thomas Burnet's *Sacred Theory of the Earth* (1691). Christ stands at the top, astride the beginning and the end of the sequence of events making up the history of the earth. Beginning as a dead star (*top right*) the earth acquires a smooth crust, which then breaks up in Noah's flood—the ark is just visible—to give the irregular surface of today's continents. Eventually, the planet will reignite and become a star again.

One possible explanation for this, expounded by the fossil collector John Woodward in his *Essay toward a Natural History of the Earth* (1695), was that all the sedimentary rocks were laid down from sediment created when Noah's flood covered the whole surface (this is the theory still advocated by young earth creationists). But Steno and Hooke were already aware of problems with this view. The twisting and faulting of the strata gave the strong impression that they had been massively transformed after having been laid down; indeed there seemed to have been a whole sequence of events by which the present structure of earth's surface had been formed. Hooke postulated earthquakes that had raised new areas of land surface from the depths of the ocean. But, unwilling to challenge the short timescale proposed by the theologians, he assumed that these events had been catastrophic. Here we see the origins of the legend that a "catastrophist" position was designed to shorten the timescale by invoking violence rather than gradual processes such as those observed today. Yet Hooke was as interested in the legend of the sinking of Atlantis as he was in the biblical flood. He also noted that some fossils seemed to represent creatures no longer alive today, raising the disturbing prospect that species created by God might have gone extinct in the course of time (fig. 5.2).

Buffon and the Dark Abyss of Time

The worrying implications of these observations were articulated more actively during the eighteenth-century Age of Enlightenment. Philosophers, especially in France, now felt that human reason could hope to understand the nature of the physical universe and humanity's place within it. They were impatient with the Church, which they held to be an agent of social conservatism, and were willing to exploit any avenues offered by science to discredit its teaching. The potential challenge to the Genesis creation story offered by the earth sciences did not go unnoticed. Already in the early years of the new century Benoît de Maillet wrote his *Telliamed,* an account of the earth's history that took it for granted that vast amounts of time had been needed to shape the rock formations we observe. There was no mention of a universal flood—instead de Maillet opted for the increasingly popular retreating-ocean theory, later called "Neptunism" after the Roman god of the sea. He supposed that the whole planet had once been covered with a vast ocean, which had gradually diminished in depth, exposing dry land and the sedimentary fossil-bearing rocks we see today. Far from being an attempt to preserve the credibility of Noah's flood, *Telliamed* pushed the

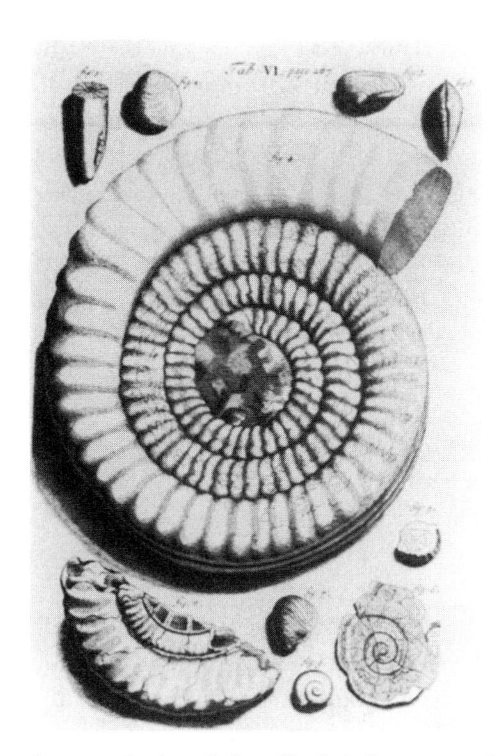

FIGURE 5.2 Fossil ammonite from Robert Hooke's "Lectures and Discourses of Earthquakes," in *The Posthumous Works of Robert Hooke* (London, 1705), plate 6. The plate also shows other common fossil seashells, but Hooke notes in his text that no shell exactly resembling the ammonite is found in the modern seas, raising the possibility that these creatures are now extinct.

great ocean back into the distant past and made no mention of any more recent inundation. Although not published in his own lifetime (it circulated in manuscript) de Maillet still though it prudent to pretend that his nonbiblical theory was suggested to him by an Egyptian wise man whose name just happened to be his own, read backward.

The most famous assault on the biblical timescale came from the leading Enlightenment naturalist, Georges Louis Leclerc, comte de Buffon (see Roger 1997). Buffon's *Natural History,* the first three volumes of which appeared in 1749, eventually expanded to become the most comprehensive account of the living world then available. As a follower of Newton, Buffon wanted to explain the origins of the present world in a purely materialist way. His first volumes included a comprehensive theory of the earth from its beginnings to the present. According to Buffon, the best way of explaining the orbits of the planets was to assume that they were all derived from globules of molten material struck off from the sun by a glancing blow by a comet. Each planet, the earth included, then gradually cooled down, and Buffon made observations of how rapidly large bodies cooled after removal from a furnace to estimate how long it might have taken the earth to cool to its present temperature. The answer he reported was 70,000 years, a figure that seems trivial today, but that expanded the old timescale by an order of magnitude. Privately he thought it might be much longer than this, and even he expressed fear at gazing into the "dark abyss of time" (Rossi 1984).

Buffon was censured by the Church authorities and forced to print a retraction of his assault on Genesis. But as superintendent of the Royal Gardens (the modern Jardin des Plantes) in Paris he was relatively secure from persecution, and in 1778 he published a revised version of his theory as a supplementary volume to the *Natural History* with the individual title *The Epochs of Nature*. He still began with his theory of planetary origins but now traced a definite sequence in the events leading from the earth's initial molten state to the present. The only concession to tradition was that there were seven epochs, which could be identified loosely with the seven "days" of creation in Genesis. Buffon's cosmological theory gave his history an obvious "direction" defined by the cooling of the earth. Originally too hot to support life, our planet eventually cooled enough to allow the appearance of species adapted to high temperatures. These died off as the cooling proceeded, to be replaced by the ancestors of the present species. These had been forced to migrate toward the equator as the earth cooled—Buffon pointed to the fossils of "elephants" (we now call them mammoths) as evidence that tropical creatures had once flourished in Siberia.

There was, however, another "direction" built into the theory. Like de Maillet, Buffon could not follow Hooke in his supposition that earthquakes could elevate the land surface. He assumed that once the earth had solidified it was completely rigid. The only way of explaining how sedimentary rocks are now exposed on dry land was to invoke the retreating ocean theory (although for Buffon the ancient ocean had at first been boiling hot). Once dry land appeared, however, it was attacked by wind, rain, frost, and the other agents of erosion, which wore down the surface. The debris was washed down rivers and into the sea, where the sediment was laid down to form younger rocks on top of those deposited while the whole earth was covered in water. In this respect, Buffon anticipated the most important techniques exploited by the geologists of the late eighteenth century. But he made little progress in identifying the sequence of rock formations, and his theory remained embedded in an older tradition in which theories of the earth took their origins from cosmological speculation.

Stratigraphy and the Fossil Record

The empirical study of rocks, minerals, and fossils had not been just a matter of curiosity. In an age where Francis Bacon's philosophy had been used to promote the claim that science would allow us to control nature by understanding its operations, the study of the earth's surface had obvious potential benefits to the mining industry. If we could know which rocks held the best prospect of yielding useful minerals, the economic benefits would be enormous. By the late eighteenth century, this pragmatic approach to the study of the earth had become well established in Germany, where many of the small independent states drew their income from mining. Mining academies were set up to train people in the skills needed to locate and extract minerals, and here the practical implications of a detailed knowledge of the earth's crust first became apparent. Out of this practical study of minerals came a methodology for identifying the sequence in which the successive rocks had been deposited in the course of the earth's history. This was the science of stratigraphy, based on the principle of superposition, that is, the assumption that newer rocks were always laid down on top of existing rocks. The assumption was necessarily historical because the identification of a rock's position in the sequence of deposits implied identifying the period in the earth's history when it was laid down. From the early efforts to define the sequence of formations (and hence the sequence of geological periods) came the modern outline of the earth's history.

In its earliest version, this program was associated with the name of Abraham Gottlob Werner, who taught at the mining school in Freiburg. Although he published little, Werner attracted students from all over the world and thus achieved immense influence. He concentrated on identifying the mineral character of rocks and then assumed that each type of rock was laid down at a particular period in the earth's history. He felt justified in making this assumption because he accepted the Neptunist theory—as the great ancient ocean dried up, the chemicals in it were precipitated out in a particular sequence. Eventually, erosion of the land surface would add a regular sequence of sedimentary rocks.

Although this theory was widely accepted in the late eighteenth century, it was soon refuted by evidence that the same types of rocks can be laid down at different periods of history. Later scientists ridiculed Werner and expressed astonishment that anyone could be taken in by so obviously false a theory. Because some of Werner's followers tried to link the theory with a reemergence of the waters that could be identified with the biblical flood, it was argued that Neptunism was bad science maintained by those with an interest in defending religion against materialism. It is certainly true that some Neptunists, including Richard Kirwan and Jean-André Deluc, tried to link the theory with the flood. These were conservative thinkers who, in the aftermath of the French Revolution, wanted to make sure that the New Science did not endorse an assault on the Church as a bastion of the social order. But such attitudes were largely confined to Britain. Werner himself expressed no interest in the Genesis story, nor did his Continental followers. They followed the theory because it offered hope of providing an ordering principle by which the complex sequence of rock formations could be understood. If they oversimplified in their anxiety to make order out of apparent chaos, they nevertheless conceived the basic program by which geology would advance, that is, the program of identifying the rock formations by the order in which they were laid down. And because the sequence was a long one, there was no question of it being compressed within the biblical timescale.

By the early nineteenth century, it was becoming clear that the Neptunist theory could not be sustained. The famed traveler Alexander von Humboldt saw for himself the immense power of volcanoes and earth movements when he studied the Andes Mountains in South America. Humboldt and many others abandoned Neptunism, but they continued to regard themselves as followers of Werner because they saw their key task as the identification of the successive rock formations. It was Humboldt who named the Jurassic formation, after characteristic rocks found in the Jura Mountains on the French-Swiss border. Earth movements replaced the retreating ocean as an explanation of how the sedimentary rocks were elevated to form dry land.

It was now recognized that since similar rocks could be formed at different periods in the earth's history, the best way of identifying the sequence was through the fossils embedded in the strata. The fossils of each period were characteristic, whatever the type of rock they were embedded in. Stratigraphy was firmly linked to the establishment of a series of geological periods, each of which was assumed to have its own population of animals and plants quite unlike those that are alive today (fig. 5.3). The fossil-based stratigraphy was pioneered in England by the canal-builder William Smith

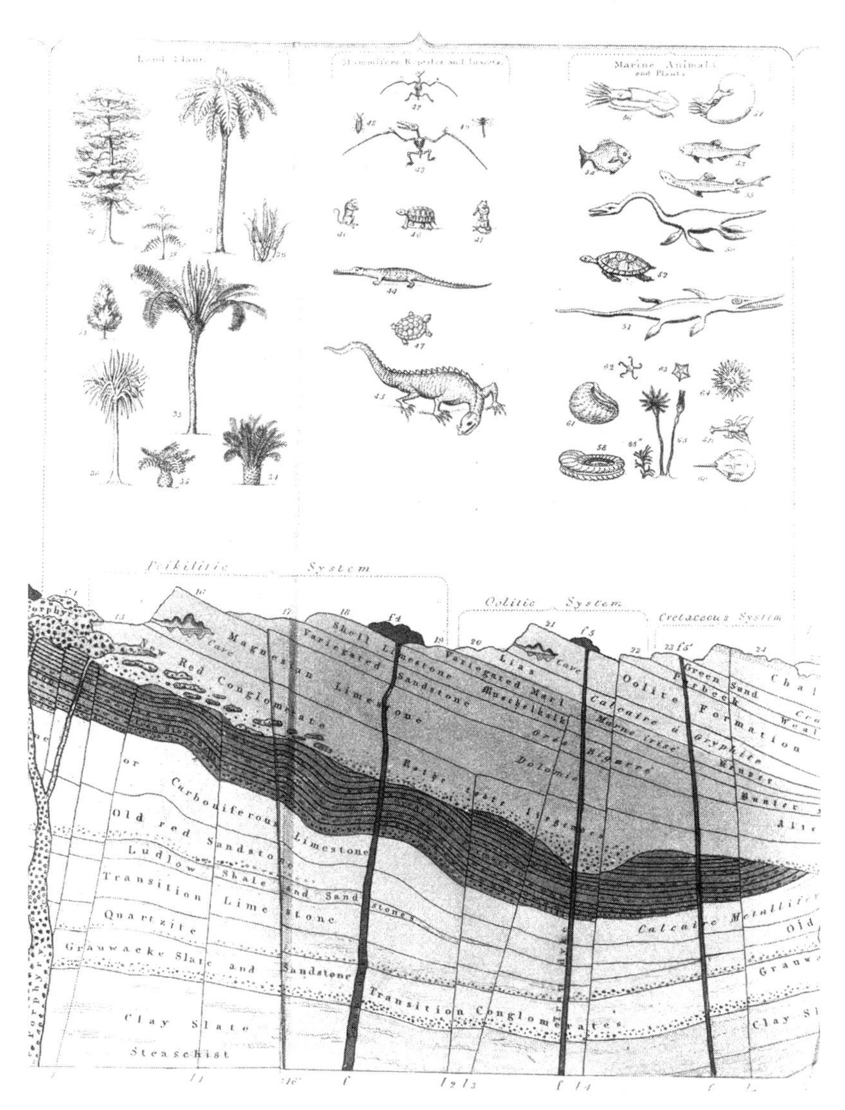

FIGURE 5.3 Part of a hypothetical cross section of the earth's crust from William Buckland's *Geology and Mineralogy Considered with Reference to Natural Theology* (London, 1837), vol. 2, plate 1. The cross section shows beds of sedimentary rocks distorted by later earth movement and with veins of igneous (volcanic) rock intruding from below. The figures at the top show creatures typically found as fossils in the Secondary rocks (the Mesozoic era) including a dinosaur looking remarkably like a dragon. Compare this with fig. 5.5 below.

and in France by the paleontologist Georges Cuvier and the geologist Alexandre Brongniart. Historians of geology still debate the relative significance of their contributions: Smith's geological map of England and Wales of 1815 was a pioneering work, but he was to some extent marginalized by the elite scientists of the time. Cuvier was at the heart of the French scientific establishment, a leading figure in the creation of comparative anatomy and the reconstruction of vertebrate fossils. He studied the structure of different species of animals in order to work out the underlying principles on which the different types of organization were based, and he used his skills to put together the often-fragmentary bones being dug out from the rocks all over Europe. It was Cuvier who established the reality of extinction beyond all reasonable doubt—no one could believe that the mammoths and mastodons he described were still alive in some remote part of the world. From this point on, scientists could take it for granted that each new formation would have distinctive fossils of its own, many of the earlier species having died out and been replaced. But it was Brongniart's work with the fossil invertebrates that proved the more useful guide in establishing the sequence of rocks, as in their collaborative survey of the formations making up the Paris basin, published in 1811.

Over the next couple of decades, geologists extended the sequence of formations down to the oldest fossil-bearing rocks (fig. 5.4). It was in Britain that some of the oldest and hence most distorted formations were sorted out. Working in Wales, Adam Sedgwick and Roderick Impey Murchison named the Cambrian and Silurian systems, respectively (significantly, Darwin got his first geological training on a field trip with Sedgwick). In 1841 John Phillips named the three great eras in the history of life: Paleozoic, Mesozoic, and Cenozoic (the eras of ancient, middle, and new life). The Mesozoic was already becoming known as the "age of reptiles" thanks to the discovery of dinosaurs and other extinct reptile species (fig. 5.5), although it was again the invertebrate fossils that formed the basis of the technical classification. Defining the boundaries between the systems was far from straightforward and required a good deal of negotiation between experts. Sedgwick and Murchison fell out over the Cambrian-Silurian boundary, while the overlying Devonian also caused a great deal of controversy (see Rudwick [1985] and Secord [1986] on these debates). Yet by the 1830s, no one could ignore the fact that the earth's crust was composed of a vast series of deposits, each of which represented a whole epoch of geological time. As yet no one would hazard an estimate of just what length of time was involved, but clearly the amount was immense by the standards of human history.

	Modern names	Old names (c. 1850)	
Cenozoic era (Age of mammals)	Recent	Recent	Tertiary series
	Pleistocene	deposits	
	Pliocene	Pliocene	
	Miocene	Miocene	
	Oligocene		
	Eocene	Eocene	
	Paleocene		
Mesozoic era (Age of reptiles)	Cretaceous	Cretaceous	Secondary series
	Jurassic	Wealden	
		Oolitic	
		Lias	
	Triassic	New Red Sandstone	
Paleozoic era (Age of fishes and invertebrates)	Permian		Transition series
	Carboniferous (Pennsylvanian Mississippian)	Carboniferous	
	Devonian	Old Red Sandstone	
	Silurian	Silurian	
	Ordovician		
	Cambrian	Cambrian	
	Precambrian	Primary rocks	

FIGURE 5.4 The sequence of geological formations established by the mid-nineteenth century (*right*) and their modern equivalents. The sequence of formations corresponds to the succession of geological periods in the earth's history. The complete sequence is never observed in any one location but is built up by using fossils and other clues to identify rocks of the same age in different areas.

Catastrophism and Uniformitarianism

Cuvier noticed that the boundaries between successive formations seemed abrupt, so that the transition from one fossil population to the next appeared to have been more or less instantaneous. In his *Discourse on the Revolutions of the Surface of the Globe,* first published in 1812 as the introduction to his survey of fossil vertebrates, he attributed the sudden extinction of species to catastrophic earth movements and tidal waves. There did seem to be a lot of evidence for a dramatic transformation of the landscape in the re-

FIGURE 5.5 Life-sized reconstruction of the carnivorous dinosaur *Megalosaurus,* originally described by William Buckland. Richard Owen, who created the name "dinosaur," helped to design this and other models in the 1850s. They can still be seen at Crystal Palace in Sydenham, south London. The dinosaur is depicted as a giant lizard walking on four legs, although later discoveries of more complete fossils showed that *Megalosaurus* actually walked on its hind limbs.

cent geological past. Vast mounds of boulder-clay, and gravel, along with large "erratic" boulders, littered the landscape of northern Europe. There was no observable cause that could have transported this material across the face of the earth, so it seemed natural to postulate a great flood. Cuvier made no effort to identify this last catastrophe with the biblical deluge, but his British followers had no such qualms. William Buckland, the reader in geology at the fiercely conservative University of Oxford, sought to vindicate his science from the charge that it aided irreligion by showing how it could provide evidence that Noah's flood was a real event. His *Reliquiae diluvianae* [*Relics of the Flood*—only the title was in Latin] of 1823 described a cave at Kirkdale in Yorkshire that had been filled with mud, in which was buried the bones of hyenas and their prey (fig. 5.6). How else except by a universal flood could a cave in the hills have been filled in this way? And the event seemed to have been accompanied by major transformation of the climate, since hyenas are no longer to be found in Europe. For Buckland this was evidence of a geological catastrophe that would fit in with the Genesis record.

Older histories of geology describe this theory of "catastrophism" as a disaster for the development of the science. Wildly improbable events, possibly of a miraculous nature, were postulated to make the theory fit into a preconceived model defined by Genesis. By invoking violent events as agents of transformation, the need to extend the age of the earth much beyond the traditional estimates was avoided. On this model, catastrophism is a classic example of the kind of bad science that is done when external forces such as religion interfere with scientific objectivity. The rival uniformitarian model of Hutton and Lyell (discussed below) showed the real way forward through the study of observable causes and the postulation of vast amounts of time in which they could have transformed the earth.

The uniformitarians' model of the history of geology has now been profoundly modified if not rejected outright. It is a vision of the science's

FIGURE 5.6 Cross section through a cave similar to the one described by William Buckland at Kirkdale in Yorkshire, northern England, from Buckland, *Reliquiae diluvianae* (London, 1824), plate 27. The cave is partially filled with hardened mud containing the remains of animals of kinds no longer found in Europe. Buckland argued that a global flood was the only explanation of how caves like this, well above sea level, could have been filled with mud. The material is now thought to have come from lakes formed when valleys were dammed by glaciers during the ice age.

history first sketched in by Lyell himself—and he was hardly an objective scholar on this topic. Lyell insisted that both Neptunism and catastrophism were implausible theories supported solely for nonscientific (i.e. religious) reasons. Modern studies reveal how distorted this condemnation is. We have seen how catastrophist geologists such as Cuvier, Humboldt, Sedgwick, and Murchison played key roles in establishing the stratigraphical sequence still accepted today. Most Neptunists and catastrophists had no interest in linking their theories to the flood story—only a few conservative writers in the English-speaking world followed this line. Cuvier went out of his way to insist that the last catastrophe was not universal, as Genesis implies, and even Buckland eventually gave way on this point. For all of them, the most recent catastrophe was only the last in a vast sequence of violent transformations, all separated by periods of relatively normal conditions. All the earlier periods lay completely outside the biblical story of creation. There was good evidence that something anomalous had happened in the recent geological past, and the uniformitarians struggled to explain the mud deposits studied by Buckland and related phenomena. Only in the 1840s was it suggested that this material might have been transported by glaciers in an "Ice Age" when much of northern Europe had been buried in ice, and this theory took several decades to gain wide acceptance (Hallam 1983).

There was another factor that made catastrophism plausible and that incidentally made geologists reluctant to accept a cold spell in the past. Lyell did his best to imply that the catastrophists invoked supernatural causes (miracles) to explain their hypothetical upheavals. But they had no intention of appealing to anything but natural causes—they just thought there was evidence that earthquakes had once occurred on a scale far beyond anything we have observed in the few thousand years of recorded human history. In fact, the catastrophists relied on the assumption that the earth's history is much vaster than human history to argue that what little we have observed is not necessarily typical of the whole. Their theory also had a sound basis in physics. Everyone now accepted that the center of the earth was very hot. This explains the origin of the molten rock expelled by volcanoes, and the concept of a reservoir of molten or at least very hot rock, under enormous pressures, deep in the earth also seemed to explain the instability of the solid crust revealed by earthquakes. If the center of the earth is hot, however, both common sense and the physicists' studies of the behavior of hot bodies suggest that it must cool down. Heat will be conducted up to the surface (or brought up by molten lava) and radiated into space.

The early nineteenth century thus saw a reinvigoration of Buffon's cooling-earth theory.

The implications for the cooling-earth theory for catastrophism were explored by geologists such as Léonce Elie de Beaumont. If the central heat of the earth diminishes, then volcanic activity would also be expected to diminish in the course of geological time. More significant, earthquake activity would diminish as the crust got thicker and the rate of cooling slowed down. An analogy suggested by Constant Prévost compared the earth with the wrinkling of an apple: the skin wrinkles because its surface area remains constant while the volume of the apple is reduced by evaporation. A cooling earth would also diminish in volume, so mountain building will be caused by a similar crumpling of the skin. But as Elie de Beaumont pointed out, the earth's crust is rigid, so the crumpling will be expected to take place in sudden catastrophic events when the pressures building up beneath finally cause the crust to give way. Since the planet was hotter in the past, it was natural to assume that past episodes of mountain building involved earth movements on a scale far beyond anything observed in the modern world. The cooling-earth theory thus provided catastrophism with a plausible physical mechanism to complement the evidence geologists had for discontinuities in the past.

The uniformitarian alternative to this model has been hailed as the foundation stone of modern geology because it adopts a methodological precept based on the claim that a true science operates only with those causes it can actually observe. In fact, the catastrophists were quite happy with this method of "actualism" because their upheavals were supposed to be the same as modern earthquakes only bigger. But to uniformitarians, only observable causes acting at observable intensities can be employed by a truly scientific geology. Anything else opens up the way to wild speculation and even the postulation of supernatural causes. This was the methodology pioneered by James Hutton and articulated most fully by Charles Lyell in the 1830s. It seems very modern because our current geological theories include little room for catastrophes (although asteroid impacts are now widely accepted to have interrupted the steady changes produced by internal processes linked to continental drift). The uniformitarian approach also seems modern in its appeal to vast periods of time. Because all past changes, including the elevation of mountain ranges and the excavation of valleys, are to be explained by modern-scale earth movements and erosion, vast amounts of time are needed for such slow-acting agents to produce the effects we observe. It would be quite wrong to accuse the catastrophists of opting for a young earth along the lines proposed by Archbishop Ussher, but there is no doubt that the demands made by uniformitarians for an extension to the timescale went far beyond anything that had been imagined before.

The uniformitarian method was not without its problems, however. In their anxiety to rule out speculation, the uniformitarians were forced to opt for what Gould (1987) calls a "cyclic" model of earth history. There can be no arrow of time defined by cooling or the retreating ocean—past geological periods have seen only an eternal cycle of events similar to those we observe today. It is outside the realm of science to postulate a period when things were radically different—let alone a process by which the planet itself was given its modern form. These are restrictions that no geologist could accept today, so the claim that uniformitarianism forms the sole basis of our modern science is flawed. Modern geology draws on both the uniformitarian and the "directionalist" model of the catastrophists. Once we realize this, we see that neither side in the debate should be presented as "pure" scientists working only on objective principles. It is as important to know why Hutton and Lyell were motivated to propose a steady state theory of the earth as it is to know why some catastrophists were tempted by biblical ideas about the flood.

The first effort to put this program into operation was made by the Scottish geologist James Hutton (Dean 1992). In a paper published in 1788, and again in his two-volume *Theory of the Earth* of 1795, Hutton took on the Wernerianism promoted in his native Edinburgh by Robert Jameson. Hutton dismissed the retreating ocean theory by pointing out (as Hooke had done a century earlier) that earth movements could explain how the sediment laid down on the seabed could be elevated to dry land. He could draw on studies of volcanoes that had begun to suggest that they derived their lava from reservoirs of molten rock deep in the earth. The idea that the earth's central heat was responsible for most geological activity came to be known as "Vulcanism" after the Roman god of fire. Hutton linked this theory with his belief that the earth's crust was unstable—for him it was the central heat that was responsible not only for volcanoes but also for earth movements and mountain building. He also argued that many of the so-called primary rocks, including granite, were of igneous origin: they had crystallized out from a molten state, not from solution in water. When challenged to explain why these rocks look so different from the lavas expelled by modern volcanoes, he showed how molten rock could intrude between the strata deep in the earth and there cool very slowly. This gave time for the crystals seen in rocks such as granite to form. For Hutton, granite could be produced at various points in the earth's history—it was not necessarily the most ancient of rocks as the Wernerians had claimed.

What made Hutton's theory different from any other form of Vulcanism was his insistence that the processes responsible for forming the rocks have all occurred at the same rate as we observe today. Although it was hot inside, the earth was not cooling down, so earth movements were not diminishing in intensity. Hutton also went to great lengths to show how the ordinary agents of erosion—wind, rain, flowing streams, and so on—can have sculpted out the valleys within mountain ranges. There was no need to postulate violent tidal waves, provided one allowed for the vast amounts of time needed for a flowing stream to carve its own valley through the mountain rocks. The debris from this erosion was washed out to the seabed, where it was laid down as sediment, baked into rock, then eventually elevated to produce more dry land. There was a perfect cycle here, in which the elevation of new land exactly balanced the destruction of the old surface by erosion. Hutton was accused of being irreligious by conservative Wernerians because his theory had no room for a flood and demanded vast amounts of time. More seriously, from the conservative point of view, it had no room for a creation: Hutton's earth was eternal, a perpetual motion machine that never ran down. He wrote that "we find no vestige of a beginning—no prospect of an end" (Hutton 1795, 1:200) Yet in fact, Hutton's motivation for setting up such a theory was his own religious beliefs, which were deist rather than Christian. His god was the perfect workman who had designed a machine that could work forever without His superintendence. The purpose of the whole system was to maintain the earth as a habitat for living things because without the perpetual rebuilding of the land surface all the soil on which life depends would eventually be washed out to sea.

Hutton's theory generated controversy in Edinburgh but attracted little attention elsewhere. It was more widely disseminated by John Playfair's *Illustrations of the Huttonian Theory* of 1802. In Britain, at least, his work played a role in the conversion of geologists from Neptunism to Vulcanism—but it was the catastrophist version of Vulcanism based on the cooling-earth theory that benefited. Continental geologists had their own reasons for moving toward catastrophism. The uniformitarian model was eventually revived in Charles Lyell's *Principles of Geology* (1830–33) as the basis for an explicit attack on catastrophism (Wilson [1972], but see also Rudwick's introduction to the modern reprint of the *Principles*). It was the introductory historical chapters of the *Principles* that created the negative image of both Neptunism and catastrophism accepted by later scientists. Lyell's assault was explicitly methodological, accusing the catastrophists of betraying science by opting for wild speculation rather than careful observation. His book played an important role by providing evidence of just how much change is actually occurring through the action of modern volcanoes, earthquakes, and erosion (fig. 5.7). Lyell had studied Mount Etna in Sicily and had shown how this immense volcano had been built up from a vast series of eruptions, only the last few of which had been witnessed by humans. It was ancient by human standards, yet it stood on the youngest sedimentary rocks. Lyell dismissed all the so-called evidence for catastrophes in the past as illusory: it was always possible to imagine a long sequence of ordinary changes that could have produced the effect, given enough time. The apparently sudden transitions from one stratum to another were the result of vast periods being unrepresented in the sedimentary record. Lyell made his own contribution to stratigraphy by naming the Eocene, Miocene, and Pliocene formations—but he showed that the fossil populations did not completely change from one to the other. There were always some species that survived, undermining the plausibility of catastrophic extinctions.

Although he accepted the conventional sequence of geological formations, Lyell had revived Hutton's cyclic or steady state model of history. He assumed that even the earliest strata we see had been formed under conditions essentially similar to those of today. The known geological record is only the last part of an endless sequence, all the earliest phases of which have been destroyed or distorted beyond recognition. Science cannot hope to find evidence of a "primitive" phase of earth history dating from the planet's purely hypothetical formation. To maintain his steady state theory, Lyell attacked the evidence used to support the cooling of the earth, arguing that there had been only a fluctuation in the climate as continents were created and destroyed. He also insisted that the apparent progressive development of life was an illusion—eventually we would find mammalian fossils in even the oldest rocks. Here we see the ways in which Lyell's position went far beyond what geologists can accept today. In effect, his methodology became a straightjacket confining him to an ahistorical view of the earth. His position can to some extent be linked to his religious and political beliefs. Lyell was a liberal in politics, and he resented the way in which conservatives like Buckland were using catastrophism to defend Christianity and by implication the Church as pillars of aristocratic privilege. His own religious beliefs—so strongly held that he could never accept Darwin's view of human origins—were more like Hutton's, a form of deism in which a wise and benevolent Creator has designed a universe that can operate for ever without renewal.

FIGURE 5.7 The Roman temple of Serapis at Puzzuoli, outside Naples, the frontispiece to Charles Lyell, *Principles of Geology* (London, 1830–33), vol. 1. The dark bands on the columns have been formed by the action of marine creatures, showing that earth movements have submerged the temple beneath the sea and then elevated it again but without actually destroying the columns. Lyell argued that if noncatastrophic earth movements could have this much effect in the two thousand years since Roman times, over a longer time span they could elevate mountain ranges or even whole continents.

Lyell was a popular writer and was influential in convincing the general public that the earth was immensely old. His impact on geology is more debatable. His greatest disciple was Charles Darwin, who saw evidence on his voyage aboard HMS *Beagle* that the Andes Mountains were still being elevated by earthquakes. Darwin applied the uniformitarian method where Lyell would not: to the organic world and the process by which species change in the course of time (see chap. 6, "The Darwinian Revolution"). But even he would not follow Lyell in his rejection of the progressive development of life. Most geologists acknowledged the power of modern causes and scaled down the catastrophes they postulated in the distant past. But they continued to believe that there were episodes of mountain building in which earth movements were much more intensive than we see today. These form the natural "punctuation marks" allowing us to define the geological periods (for Lyell these were merely gaps in the record that we use for convenience). More seriously, most geologists continued to support the cooling-earth theory, seeing this as an essential foundation to explain the crumpling of the crust and the violence of at least some past events. They also tended to limit the age of the earth to around a hundred million years—a vast period by any human standard but far less than Lyell and Darwin wanted and far less than we accept today.

Physics and the Age of the Earth

This last point leads us to a final controversy but one whose significance has often been overestimated. Lyell's steady state theory had a fatal inconsistency: it assumed that the center of the earth is hot, yet it denied that the planet cools down in the course of almost endless geological time. This point was noted in the controversies of the 1830s, but it became crucial as the physicists began to refine their ideas on energy and create the science of thermodynamics (see chap. 4, "The Conservation of Energy"). In the 1860s, the physicist William Thomson, later Lord Kelvin, began to attack Lyell and, by implication, Darwin (Burchfield 1975). In Kelvin's worldview, God had created only so much energy, and as it became slowly less available, the universe was inevitably running down. The cooling of hot bodies was the most obvious expression of this irreversible process, and to Kelvin it was unthinkable that the earth could be treated as an exception. A hot earth must cool down, so Lyell was wrong and the catastrophists right—geological processes must have gone on more rapidly in the past when the earth's interior was hotter. Kelvin then did some calculations to suggest how much time it would have taken an initially molten earth to cool down to the state in which we see it today. The answer came out to at the most a few hundred million years, far less than Lyell and Darwin were demanding.

It has often been assumed that this assault by the more fundamental science of physics came as a blow to the geologists of the time. But this assumption is based on the mistaken belief that all of the geologists had followed Lyell's uniformitarianism. Kelvin's attack was certainly important for Lyell and for Darwin and the evolutionists. But in fact most geologists were perfectly happy with Kelvin's timescale, indeed they had estimates of their own, based on the rate of sedimentation and on the accumulation of salt in the oceans, which limited the age of the earth to a hundred million years. It was only when Kelvin reduced his estimate to twenty-five million years that the geologists began to complain that the physicists were getting too big for their boots and must have got something wrong. There was simply no way that the convoluted history of the earth revealed by the rocks could be fit into so short a time.

The physicists had got something wrong, of course, and this was already apparent by the end of the century. Radioactivity was discovered in 1896, and its implications soon began to overturn Kelvin's whole worldview (see chap. 11, "Twentieth-Century Physics"). By 1903, Pierre Curie had noted that radioactive elements give off heat, and three years later Lord Rayleigh pointed out that since such elements are distributed throughout the earth in small but significant quantities, a substantial amount of heat would be generated in the interior. This would be more than enough to offset the cooling predicted by Kelvin. Moreover, the rate of radioactive decay of some natural elements is so slow that this source of heat could last for billions of years. In a sense, Lyell was vindicated, since the evidence for radioactive heating now more or less required geologists to extend their timescale enormously and made catastrophes superfluous. Indeed, the new physics precipitated a crisis in the earth sciences by undermining the idea that mountain building was due to the crumpling of the crust on a gradually shrinking earth. This would eventually lead to the postulation of the theory of continental drift and modern plate tectonics (see chap. 10, "Continental Drift").

Radioactivity also provided something that the geologists had always lacked, a way of measuring geological time in absolute terms (as opposed to the relative sequencing of formations). Since the decay products of each radioactive element are known, it is possible to compare the proportion of the original element and its decay product in a mineral and—knowing the half-life (a measure of the rate of decay)—to calculate how old the mineral is. The first technique used the decay of radium to lead, although others such as the potassium-argon method eventually became better known. Within a few years, pioneers of radioactive dating such as Arthur Holmes were estimating the age of the earth to be several billions of years (Lewis 2000). The consensus eventually established the age as around 4.5 billion years, a figure that has remained secure despite numerous refinements through the twentieth century and into the twenty-first.

Conclusions

Geologists have become used to dealing with periods of time that beggar the imagination. Modern young-earth creationists reject the latest figures and dismiss radioactive dating along with the whole apparatus of the modern earth sciences. For them, as for the naturalists of the late seventeenth century, the earth is only a few thousand years old and all the fossil-bearing rocks were laid down beneath the waters of Noah's flood. Nothing could more strikingly indicate the extent of the conceptual revolution involved in scientists' efforts to provide the earth with a history. The full extent of this revolution only became apparent with the emergence of radioactive dating shortly after 1900, although Lyell had made an important effort to extend the timescale to this order of magnitude in the 1830s. In another sense, however, we can see that the main leap of the imagination had already been made before Lyell published. The Neptunist and catastrophist geologists who created modern stratigraphy in the decades around 1800 had already accepted a sequence of geological periods that they know extended into an antiquity far exceeding that of human history. They would not have advertised the age of a hundred million years accepted by their later followers, but they were probably aware that something of this order of magnitude was required. To this extent, the modern concept of geological time had already taken shape, even though it would take the efforts of Lyell and the atomic physicists to complete the final extension of the timescale to the figure we accept today.

References and Further Reading

Burchfield, Joe D. 1974. *Lord Kelvin and the Age of the Earth.* New York: Science History Publications.
Dean, Dennis R. 1992. *James Hutton and the History of Geology.* Ithaca, NY: Cornell University Press.
Gillispie, Charles Coulson. 1951. *Genesis and Geology: A Study of the Relations of Scientific Thought, Natural Theology and Social Opinion in Great Britain, 1790–1850.* Reprint, New York: Harper, 1959.

Gould, Stephen Jay. 1987. *Time's Arrow, Time's Cycle: Myths and Metaphor in the Discovery of Geological Time.* Cambridge, MA: Harvard University Press.
Greene, John C. 1959. *The Death of Adam: Evolution and Its Impact on Western Thought.* Ames: Iowa State University Press.
Greene, Mott T. 1982. *Geology in the Nineteenth Century: Changing Views of a Changing World.* Ithaca, NY: Cornell University Press.
Hallam, Anthony. 1983. *Great Geological Controversies.* Oxford: Oxford University Press.
Hutton, James. 1795. *Theory of the Earth, with Proofs and Illustrations.* 2 vols. Reprint, Codicote, Herts: Weldon & Wesley 1960.
Laudan, Rachel. 1987. *From Mineralogy to Geology: The Foundation of a Science, 1650–1830.* Chicago: University of Chicago Press.
Lewis, Cherry. 2000. *The Dating Game: One Man's Search for the Age of the Earth.* Cambridge: Cambridge University Press.
Lyell, Charles. 1830–33. *Principles of Geology: Being an Attempt to Explain the Former Changes of the Earth's Surface by Reference to Causes now in Operation.* 3 vols. Introduction by Martin J. S. Rudwick. Reprint, Chicago: University of Chicago Press, 1990–91.
Oldroyd, David. 1996. *Thinking about the Earth: A History of Geological Ideas.* London: Athlone.
Porter, Roy. 1977. *The Making of Geology: The Earth Sciences in Britain, 1660–1815.* Cambridge: Cambridge University Press.
Rappaport, Rhoda. 1997. *When Geologists Were Historians, 1665–1750.* Ithaca, NY: Cornell University Press.
Roger, Jacques. 1997. *Buffon: A Life in Natural History.* Translated by S. L. Bonnefoi. Ithaca, NY: Cornell University Press.
Rossi, Paolo. 1984. *The Dark Abyss of Time: The History of the Earth and the History of Nations from Hooke to Vico.* Chicago: University of Chicago Press.
Rudwick, Martin J. S. 1976. *The Meaning of Fossils: Episodes in the History of Paleontology.* New York: Science History Publications.
———. 1985. *The Great Devonian Controversy: The Shaping of Scientific Knowledge among Gentlemanly Specialists.* Chicago: University of Chicago Press.
Schneer, Cecil J., ed. 1969. *Toward a History of Geology.* Cambridge, MA: MIT Press.
Secord, James A. 1986. *Controversy in Victorian Geology: The Cambrian-Silurian Debate.* Princeton, NJ: Princeton University Press.
Wilson, Leonard G. 1972. *Charles Lyell: The Years to 1841: The Revolution in Geology.* New Haven, CT: Yale University Press.

CHAPTER 6

THE DARWINIAN REVOLUTION

THE POPULARITY OF THE TERM "DARWINIAN REVOLUTION" (Himmelfarb 1959; Ruse 1979) suggests that we are dealing with a scientific theory with major consequences. If Darwin's naturalistic theory of evolution were accepted, then a host of beliefs and values that had been integral to Christian culture would have to be rejected or renegotiated. Living things, including the human species, could no longer be regarded as divine creations. At best God might be supposed to play some indirect role in the process of evolution, but even that was difficult to imagine if it worked through as harsh a mechanism as natural selection. Equally seriously, the status of the human soul was threatened. If we are just improved animals, then it is hard to believe that we have an immortal soul if the lower animals do not. And to abandon the concept of a spiritual dimension to human existence would undermine traditional concepts of morality and threaten the stability of the social order.

What lines of evidence could have been so persuasive that they required scientists such as Darwin to take so bold a step? On the model of history preferred by scientists such as Gavin De Beer (1963) it is possible to see how Darwin was led to his theory by an accumulation of new information from areas as diverse as the fossil record and the study of animal breeding. If there were problematic consequences of the theory, then people simply had to cope with them if they wanted to live in the real world. But even today there is no shortage of critics who maintain that the Darwinian theory is not good science, so Darwin and his followers must be driven by something more than the desire to study nature. To modern creationists, Darwinism is the agent of a materialist philosophy that wants to destroy traditional values and beliefs and plunge the world into anarchy. They argue that the materialists manipulate dubious scientific evidence to support a theory whose real purpose is much more ambitious and much more dangerous.

There is another line of argument, however, that has also been used in an effort to undermine Darwinism's scientific credibility. Socialist critics from Marx and Engels onward have noted the analogy between Darwin's "struggle for existence" and the competitive free-market economy in which individuals struggle to gain a living. Can it be a coincidence, argue the critics, that such a theory was proposed in the heyday of Victorian capitalism? Darwin simply projected the ideology of the class to which he belonged onto nature so that he and his followers could maintain that a competitive society was "only natural." This is a very different argument questioning the theory's scientific credentials. Cautious observers may, however, reflect on the fact that the creationists who decry Darwinian materialism are among the most vociferous supporters of the free-enterprise system—so can they, too, be unwitting social Darwinists?

These rival perceptions of modern Darwinism are reflected in the vast range of historical literature on the theory's origins. De Beer's interpretation of Darwin as the courageous scientist is followed by that of other scientist-historians such as Michael Ghiselin (1969) and Ernst Mayr (1982). The values of those who dislike the implications of Darwinism can be seen in the far less flattering portraits created by Jacques Barzun (1958) and Gertrude Himmelfarb (1959). The sociological interpretation of the origins of Darwinism is explored in the writings of the Marxist historian Robert Young (1985) and in a biography of Darwin by Adrian Desmond and James Moore (1991). Other historians have tried to balance the conflicting pressures. Few would now deny that Darwin was influenced—perhaps creatively—by the ideology of his time, but there is a widespread suspicion that we cannot make sense of his contribution unless we see those creative insights through the medium of his scientific work (for surveys, see Bowler [1983b, 1990]; Eiseley [1958]; and Greene [1959]). The historians' task is made more complicated by the vast archival record of Darwin's activity that is being edited for publication (e.g. Darwin 1984–, 1987).

The temptation for both supporters and critics to focus on the work of Darwin himself may have distorted our image of the Darwinian revolution. It is all too easy to assume that there must have been a sudden transition from a more-or-less stable creationism to a rabidly materialistic Darwinism that has continued undaunted (if not unchallenged) to today. This perception feeds off a peculiar combination in Darwin's achievements: he converted the world to evolutionism, and he was also the discoverer of what most modern biologists take to be the correct explanation of how evolution works, natural selection. There is an obvious temptation to believe that he must have successful because his contemporaries realized that he had got the mechanism right. On this model, only a limited "tidying up" of the theory was needed to generate modern Darwinism. Yet an increasing number of studies suggest that natural selection was not accepted by Darwin's fellow scientists. Rival mechanisms of evolution flourished through into the early twentieth century. We need to see the emergence of modern Darwinism as a much more protracted process requiring major transformations long after the basic idea of evolution was accepted (Bowler 1988).

These points feed into the work of historians who are developing a more complex model of how the first generation of Darwinians succeeded in dominating the scientific community. Darwin was not the first to initiate widespread discussion of evolutionism. Long before he published the *Origin of Species* in 1859, radical writers were promoting the theory as a foundation for a political philosophy that demanded social progress. By undermining the traditional beliefs that sustained the Church, evolution opened up the prospect that nature itself was founded on a law of progress—which then made human progress seem inevitable. Such ideas made little impression on the scientific elite, but they paved the way for the reception of Darwin's theory and may have shaped the popular assumption that it, too, was the basis for a philosophy of universal progress. If this is so, many of the philosophical, theological, and ideological consequences normally attributed to Darwinism may be a reflection of this wider cultural movement.

At the same time, we need to look more carefully at what made scientists take Darwin more seriously than they did earlier writers. They certainly saw his book as a new initiative that would transform many areas of science, especially in morphology (the comparative study of animal structures) and paleontology. And even though most of them did not accept natural selection as the main mechanism of evolution, they thought it was a plausible and scientifically testable theory that went far beyond the earlier speculations. It has been suggested that younger professional scientists, such as T. H. Huxley (who became known as "Darwin's bulldog"), were attracted to the theory because it helped their campaign to convince the public that science rather than the Church was the best source of expertise in a modern economy. All this suggests that the impact of Darwinism must be evaluated both in terms of its scientific advantages (which were real enough even to those who had doubts about the detailed theory of selection) and its appeal to the values and prejudices of potential supporters both inside and outside science.

Design in the Natural World

The worldview still accepted by modern creationists does not date back to the foundations of Christianity. As noted in chapter 5, "The Age of the Earth," a literal reading of the Genesis creation story first became widely accepted in the seventeenth century. If the earth was only a few thousand years old, any gradual process of development became unthinkable. The only explanation for the origin of plants, animals, and human beings was that their first ancestors were created directly by God. The naturalists of the period were only too happy to exploit this insight to provide a justification for science's exploration of the natural world. After all, there were critics who warned against the materialism of the new science being promoted by Galileo, Descartes, and Newton. If the whole world was to be treated as a giant machine, then the only way to preserve a role for the Creator was to insist that the machine needed a wise and intelligent Designer. Even if they did not believe in the Garden of Eden, seventeenth-century naturalists could appeal to a "natural theology" in which the study of living things would reveal God's handiwork. The "argument from design" sought to convince the skeptics that the best explanation for the existence of such complex structures as living things was a God who had, in the analogy used later by William Paley, designed them just as a watchmaker designs a watch (see chap. 15, "Science and Religion.").

A leading advocate of this view was the English naturalist John Ray, whose *Wisdom of God Manifested in the Works of Creation* appeared in 1691 (Greene 1959). Ray appealed to the structure of the human body, especially the eye and the hand, to argue that here were complex mechanisms exquisitely designed so as to provide us with the instruments we need to conduct our lives. But Ray did not believe that the whole world was created for our benefit alone. Each animal species has its own structures designed to allow the individuals to gain a livelihood and enjoy their lives in a particular environment. The argument from design thus focused attention onto the adaptation of structure to function. God is not only wise, he is also benevolent because he gives each species exactly what it needs to live in the place where he created it. The argument presupposes a static creation, in which species and their environments remain just as they were when first created. It has often been said that Darwin would turn the argument from design on its head by showing that adaptation is a process by which species are adjusted to changing environments.

Ray's vision of a designed world was not without its applications in the scientific world of the time. It encouraged the detailed study of species and their relationship to the environment. But it was also the basis for the first efforts to provide a biological taxonomy, a system for classifying the animals and plants so that we can try to make sense of the bewildering diversity of species. Each individual species has its own special adaptations, but there are relationships between species that surely imply that there must be some rational pattern in God's creation. The lion and the tiger are both "big cats"—we can see the relationship between them, and a more distant resemblance to the domestic cat. If these and other degrees of similarity can be ordered and related together, we might be able to see the whole plan of creation displayed in our natural history museums or textbooks. There will also be an immense benefit for scientists who need to refer unambiguously to any one of the vast number of living species, a problem made the more acute as European naturalists confronted the vast array of new species discovered in remote parts of the world.

Ray made important contributions toward establishing such a system, but it was the Swedish naturalist Carl von Linné, better-known by the Latinized form of his name, Linnaeus, who laid down the foundations of the modern system of biological taxonomy (Farber 2000). His *System of Nature* (1735) eventually expanded into a multivolume work that attempted to classify every plant and animals species into a rational system. Linnaeus also founded the system for naming species still in use today, the binomial nomenclature. The most closely related species are linked into a genus (plural "genera") and each is given two Latin names, always italicized: the first is the name of the genus, the second of the individual species. Thus the lion was *Panthera leo,* the tiger *Panthera tigris.* The big-cat genus *Panthera* was then included within the family Felidae (the cat family), which in turn belongs to the order Carnivora (the flesh eaters) of the class Mammalia (the mammals). Although much has changed in how we assess the relationships and in the details of some of the groupings, this is still how scientists classify species. Darwin's theory of evolution explains the grouping of species as a result of common ancestry: in the branching "tree of life," the more recently two species share a common ancestor, the more closely they are related. It is worth remembering, however, that when Linnaeus set up the system he believed it represented the divine plan of creation—the relationships existed only in the mind of God. He thought that most species were created exactly as we see them today.

The pattern of relationships that Ray and Linnaeus sought to represent consists of groups nesting within larger groups, which is why it is consistent with Darwin's model of branching evolution. The system undermined a much older vision of natural order known as the "chain of being," founded on the commonsense notion that some animals are higher or more advanced than others. Most of us think that humans are superior to the other animals, and we tend to see mammals as being superior to fish, and fish to invertebrates. From the ancient Greeks onward, this natural hierarchy had been visualized as a linear chain in which the species were the links, stretching down from humans to the lowest form of life. A spiritual hierarchy also stretched up through the angels to God, so that humans occupied the crucial boundary between the animal and spiritual realms. The chain of being was still exploited by eighteenth-century poets such as Alexander Pope (see Lovejoy 1936), but Linnaeus and the naturalists had now shown that as a practical system of classification it did not work. However, the broader notion of an animal hierarchy was too deeply rooted for it to be abandoned, and the theory of evolution would be shaped by a widespread assumption that the history of life must represent the ascent of life toward higher forms (Ruse 1996). The tree of life retained a main trunk, equivalent to the chain of being, but with a host of minor side branches (see fig. 6.5 below).

Forerunners of Darwin?

The naturalists who believed that the universe was a divine creation did not find this a very precise guide, given the detailed nature of their work, and such ambiguities would get worse as the life sciences became more sophisticated. But by the mid-eighteenth century there was a growing movement to reject the whole idea of design and look for more materialistic explanations of how things came to be in their present state. Some of the resulting theories do include an element of transformism, or what we would today call evolution, and the naturalists who proposed them have sometimes been hailed as the "forerunners of Darwin" (Glass, Temkin, and Straus 1959). Later historians have become suspicious of this search for the precursors of the modern theory, because it fails to take into account the very different context within which these early ideas were articulated. It is easy to find isolated passages that give the impression that eighteenth-century thinkers were coming close to Darwinism, but a more careful reading suggests that they were usually thinking of something quite different from the modern theory. There are many different ways of imagining how the universe might change through time, and Darwinism is only one of them. The so-called forerunners were actually exploring very different models of how new forms of life might appear. We should be aware of the growing willingness to challenge the idea of a static creation, but to twist these early ideas to fit our modern theories can only distort them beyond recognition.

The motivation behind many of these speculations lay in the philosophy of the Enlightenment, which celebrated the power of human reason to understand the world and dismissed all traditional religions as superstition. The Church was seen as a barrier to social reform, so undermining the credibility of the Genesis creation story had an ideological as well as an intellectual purpose. Some of the Enlightenment philosophers became outright atheists and materialists, and they sought an explanation of the origin of life that did not depend on the supernatural (Roger 1998). For Denis Diderot, the world was a ceaseless round of material transformations that formed and reformed material structures without any predesigned plan or purpose. He challenged the assumption that species are constant and emphasized the unplanned nature of natural change by speculating that monstrosities might sometimes be born with new characters that by chance enabled the creature to survive and establish a new species. But materialists like Diderot developed no detailed theory of transformism because they also thought that inorganic nature could produce even complex living things directly by a process know as "spontaneous generation."

This alternative also occurs in the thought of the most influential naturalist of the Enlightenment, Georges Louis Leclerc, comte de Buffon (Roger 1997). It was Buffon who promoted the new timescale of earth history on which these speculations about the origin of life rested (on developments in geology and paleontology, see chap. 5, "The Age of the Earth"). He proposed a theory that postulated that the earth is not only very old but was also hotter and hence more energetic in the distant past. His multivolume *Natural History,* which began publishing in 1749, also provided an overview of all the known animal species and included several (not altogether consistent) speculations about their origin. Buffon ridiculed Linnaeus's search for the divine plan of creation, although he too accepted the reality of species. But he became increasingly convinced that the species have a good deal of flexibility to adapt to the new conditions they encounter in an ever-changing world. In a 1766 chapter titled "On the Degeneration of Animals," he argued that the species making up a modern genus have all descended from a single ancestor—so the lion and the tiger are not true species, only varieties of a single big cat species. But the ancestral forms have not evolved from anything else, and it is clear from his other writings that Buffon thought they were originally produced by spontaneous generation. In his supplementary volume *The Epochs of Nature* (1778) he suggested two episodes of spontaneous generation in the course of the earth's history, one to produce creatures adapted to the early, very hot conditions and a second to produce the ancestors of the modern forms. This was certainly a bold alternative to Genesis, but it involved only very limited amounts of transmutation.

At the end of the century there were two thinkers whose ideas included a more substantial element of what we might call evolution. One of them, the English physician and poet Erasmus Darwin, has attracted much attention because it was his grandson, Charles Darwin, who proposed the modern theory of evolution. Erasmus endorsed the idea of a gradual development of life through time in his poems (which were quite popular at the time) and in a chapter of his *Zoonomia* of 1794–96. But far more influential was the parallel theory developed by the French naturalist J. B. Lamarck (Burkhardt 1977; Jordanova 1984). Lamarck studied the invertebrate animals at the Museum of Natural History established in Paris by the revolutionary government and made important contributions to invertebrate taxonomy. Around 1800 he abandoned his original commitment to the fixity of species and began to develop the theory he published in his *Zoological Philosophy* (1809). He accepted spontaneous generation, appealing to electricity as a force that could vivify nonliving matter, but assumed that only the simplest forms of life could be produced in this way. The higher animals evolved in the course of time by a progressive trend that made each generation slightly more complex than its parents. Lamarck thought that this progression would in theory generate a linear scale of animal organization—in effect, a chain of being with humans as the last and highest products. Note, however, that this "ladder" model of evolution included no branching—there were many parallel lines ascending the scale starting from different acts of spontaneous generation. Lamarck denied the possibility of extinction and the reality of species. He thought the scale was absolutely continuous, with no gaps marking off distinct species (the gaps we see are due to lack of information—the missing links are all out there somewhere).

This is a model of evolution quite unlike anything we accept today. But Lamarck was an experienced naturalist and he knew that we cannot in fact fit the various forms of life into a linear pattern. He supposed that there was a second evolutionary process at work, which distorted the chain and produced an irregular arrangement. It is this second process for which he is remembered because it was taken seriously by biologists through until the emergence of modern genetics. Lamarck knew that species were adapted to their environments, but he could not attribute this to design by God. He supposed instead that species are adapted to changes in their surroundings by a process called the "inheritance of acquired characteristics" or "use inheritance." An acquired character is one developed by the organism after birth as a result of it exercising its body in an unusual way. The weightlifter's bulging muscles are an acquired character because they would be much smaller if it were not for all the exercise. Lamarck (and many others) supposed that such acquired characters might have a very slight tendency to be inherited, so that the weightlifter's children would be born with muscles slightly larger as a result of their parents' efforts. This process would produce adaptive evolution if the new habit directing the exercise was adopted to cope with a change in the environment. In the classic example, the giraffe's long neck is a consequence of generations of its ancestors reaching up to feed off the leaves of trees.

Lamarck's theory was the last product of the age of Enlightenment speculation, and historians of science used to think that it had been dismissed as nonsense by a new generation of conservative naturalists working in the Napoleonic era. It was certainly dismissed by some of the elite, but as we shall see in the next section there were still radicals willing to use the idea of evolution to challenge traditional beliefs. For these radicals, there were elements of Lamarck's theory that fit well with their continued calls for social reform.

Interpreting the Fossil Record

The scientific elite of the early nineteenth century was anxious to distance itself from Enlightenment materialism. In Britain, this meant a revival of natural theology. On the Continent there were fewer explicit appeals to religion, but new approaches to the life sciences tended to reinforce belief in the fixity of species and in some cases presented the world of life as an orderly pattern that expressed some rational principle at the heart of nature. But there was a new factor that had to be taken into account by all these theoretical approaches: the history of life as revealed by the fossil record (see, for an outline of the impact of the fossil record, chap. 5). However conservative in outlook, naturalists had to see the modern species as the last stage of a historical process. They had to transform the older traditions to incorporate this element of change without supporting transmutation as the agency by which new species appeared. At one time, it seemed easy for historians to dismiss these efforts as mere stopgaps desperately trying to hold back the emergence of Darwinian evolutionism. But modern studies suggest that in some cases these early theories had important results that helped to create the worldview to which Darwin also contributed. Recent work also confirms the point noted above: the radicals did not go away, and to some extent the antievolutionary philosophies of the scientific establishment were designed to combat the threat from this source.

The work of Georges Cuvier and his followers on vertebrate fossils established that the present order of nature was merely the last in a long series. To reconstruct the fossilized remains of extinct animals, Cuvier drew on his skills in comparative anatomy (see chap. 7, "The New Biology"). He showed that the earth had passed through a number of geological epochs, each with its own distinct population of animals and plants. How was this insight to be accommodated without giving ground to Lamarck and the evolutionists? Cuvier was convinced that geological catastrophes wiped out the populations of whole continents, leaving room for an entirely new population to occupy the area after things had settled down. He went out of his way to ridicule Lamarck's theory, arguing that the structure of each species is so carefully balanced that any significant disturbance would make the organism nonviable. Yet he did not appeal to design, and he evaded the need to postulate successive creations to explain the appearance of new species, suggesting instead that they migrated in from areas not affected by the catastrophe. To his British followers, however, the idea of successive creations was irresistible. The Genesis story would have to be modified to include a series of miraculous creations in the course of the earth's history (Gillispie 1951). They applauded William Paley's *Natural Theology* (1802), which restated the argument from design using the analogy of the watch and the watchmaker, and saw themselves as modifying this traditional view in the light of the new knowledge of the fossil record. William Buckland contributed to a series known as the *Bridgewater Treatises,* commissioned to promote natural theology, using his volume to show how the species comprising each successive population were all adapted to the prevailing conditions. By supposing that the earth was gradually cooling down so that the environment moved step by step toward the one we enjoy today, he could explain why it was necessary for God's creations to be wiped out periodically in order to leave room for newer populations approximating more closely to the creatures we see today.

In Germany there was a more innovative challenge to materialism associated with the Romantic movement in the arts and idealism in philosophy. Idealists believed that the material world is an illusion created by the sense impressions in our minds, and since the world is orderly, the laws of

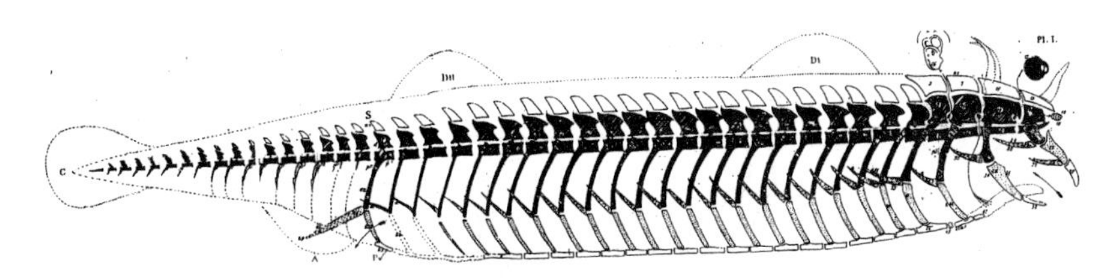

FIGURE 6.1 The vertebrate archetype, from Richard Owen, *On the Archetype and Homologies of the Vertebrate Skeleton* (1848). This is an idealized representation of the simplest imaginable backboned animal, with all the specializations of real species stripped away. It does not correspond to a real animal, although evolutionists would later try to identify the simplest and most primitive vertebrate form from which the whole phylum had developed by divergent evolution.

nature must represent some ordering principle in whatever ultimate reality is the source of those impressions. Whether one calls this ordering principle God, or some more abstract term such as the "Absolute," the implication is that the apparent complexity of nature conceals a deeper underlying pattern. Inspired by such beliefs, a group of *Naturphilosophen* (nature philosophers) sought to explain the orderly groupings among species revealed by taxonomy as just such a pattern. This viewpoint was imported into Britain by Richard Owen, who made creative use of it in his concept of the archetype defining the basic form of each major taxonomic group (Rupke 1993). Owen's vertebrate archetype, proposed in 1848, defined the essence of what it was to be a backboned animal. It was an idealized model of the simplest conceivable vertebrate—all real vertebrate species were more or less complex adaptive modifications of the archetypical form (fig. 6.1). This idealist approach allowed Owen to define the important concept of homology: the fact that the same combination of bones can be modified for different purposes in species adapted to different environments (fig. 6.2). The archetype did not, however, undermine the idea of progress—primitive fish were the simplest modifications, human beings the most complex. To Owen this offered a better form of the argument from design because it implied that underneath the bewildering variety of different species described in the *Bridgewater Treatises* was an ordering principle that could only arise from the mind of the Creator. Owen saw the successive expressions of the archetype as a progressive pattern unfolding through time, something that at times brought him perilously close to transformism, although he always insisted that each species was a distinct unit in the divine plan. Darwin's theory of branching evolution drew on a similar model of development, albeit for Darwin the archetype was replaced by the common

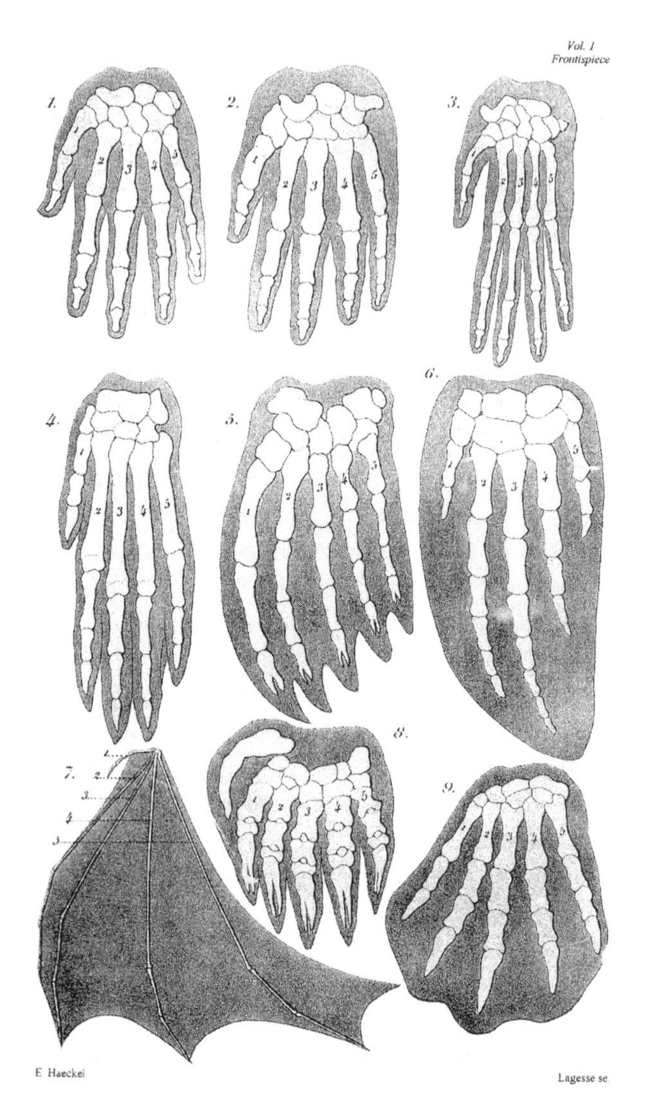

FIGURE 6.2 Homologies of the mammalian "hand" as depicted in Ernst Haeckel's *History of Creation* (New York, 1876), vol, 2, plate 4. The same bones as found in the human hand (*1, top left*) are adapted to different purposes in the forelimbs of the gorilla (*2*), the orangutan (*3*), and the dog (*4*); for swimming in the seal (*5*) and the porpoise (*6*); for flying in the bat (*7*); for digging in the mole (*8*); and again for swimming in a primitive mammal, the duck-billed platypus (*9*). The modification of the same basic structure for different purposes in different animals was described by Richard Owen as an illustration of the rational foundations of the plan of creation, but for Haeckel it was evidence that all the mammals had all descended from a common ancestor.

ancestor from which the various members of the group diverged in the course of evolution.

Other idealists, including the Swiss naturalist Louis Agassiz—who became one of the founding fathers of American biology—focused on the development of the human embryo as an illustration of how the pattern of creation unfolded (Lurie 1960). The embryo was seen to develop from a simple uniform substance in the fertilized egg, gradually acquiring the more complex structures it needed in order to become an adult. It was widely believed at the time that the new structures were added in a way that paralleled the taxonomic hierarchy: the human embryo passed through stages resembling a fish, a reptile, and a simple mammal, before adding on the final characters that defined it as human. But this was also the sequence embodied in the ascent of life revealed by the fossil record, and to Agassiz this parallelism must be God's way of telling us that we humans are the goal of his creation. Here an element of the old chain of being crept back into naturalists' thinking, although Agassiz was well aware that there would have to be many branches off the main line. Like Owen, he also went out of his way to deny an evolutionary interpretation of his model. Every species was a distinct element in the divine plan, supernaturally created at the appropriate point in time.

These models of the history of life have been central to most histories of the period leading up to the publication of the *Origin of Species*. Later studies have shown, however, that they are not the whole story. There were more radical alternatives under discussion, sometimes within the scientific community itself but also among interested laypersons. In France, Cuvier was challenged by Étienne Geoffroy Saint-Hilaire, who proposed a materialist interpretation of the archetype concept (Appel 1987). He envisioned a form of transmutation based on saltations, or sudden leaps, by which one species could be transformed into another instantaneously through the appearance of "monstrosities" that could survive and breed. In Britain, Geoffroy Saint-Hilaire's ideas, along with those of Lamarck, were favored by radicals seeking to discredit the traditional perspective as part of their plan to reform the medical profession (Desmond 1989). The Lamarckian anatomist Robert Grant was discredited by Owen after he moved to London in the 1830s. Although blocked from serious influence in the scientific community, these transformists kept the idea alive and to some extent forced the elite to liberalize its views in order to defend them within a context that increasingly took the idea of progressive development for granted.

Perhaps the most important move in this campaign came from the Edinburgh publisher Robert Chambers, who published his anonymous

Vestiges of the Natural History of Creation in 1844 (Secord 2000). Chambers wanted to sell the idea of progressive evolution to the middle classes because it would offer them an ideology in which their demands for reform would seem part of nature's own development. Social progress would be merely a continuation of the history of life on earth. But to do this he had to sidestep the image of Lamarckism as a dangerously radical idea. His tactic was to argue that the progressive development of life was central to God's plan but was engineered not through a succession of miracles but through laws built into nature by its Creator. The normal law of reproduction (like produces like) was occasionally interrupted by the operation of a higher law that jumped the embryo one stage further up the hierarchy of organization. Here the law of parallelism between embryological development and the history of life on earth was transformed into a law of evolution by progressive saltations. Nor did Chambers shrink from extending the law to the human species: we were merely the highest animals, our superior mental powers the result of an expansion of the brain through successive saltations. He appealed to the science of phrenology in which different parts of the brain were supposed to be responsible for different mental functions—if new parts of the brain were added by evolution, then new mental functions would appear.

The conservative establishment condemned *Vestiges* as dangerous materialism that would undermine moral values and the fabric of society. Outside the scientific community the book was widely read, and it seems that many were prepared to take the basic philosophy of "progress by law" seriously (see chap. 16, "Popular Science"). The book thus prepared the world for Darwin's far more radical ideas and shaped the way that the *Origin of Species* would be read. There was no built-in progressive trend in Darwin's theory, although he did not doubt that natural selection would produce progress in the long run. But people automatically assumed that evolution did mean progress, and this was the legacy of *Vestiges.* Even some members of the scientific elite began to concede that God's purpose might be worked out through predesigned laws rather than a series of miracles. In his analysis of *Vestiges'* impact, James Secord (2000) suggests that the book should be regarded as the real starting point of the public debate about evolution that was resolved by the controversy sparked by Darwin's *Origin.*

The impact of *Vestiges* on scientists was less conclusive, and this left the whole issue still in the air. It is interesting to note the reaction of younger, more radical scientists such as Thomas Henry Huxley, soon to become Darwin's leading advocate (Desmond 1994; Di Gregorio 1984). Huxley condemned *Vestiges* in a review that even he later conceded was unjustly vitriolic. This was partly because Chambers's science was sloppy. He had slurred over real difficulties in the fossil record, which did not support the linear model of progress. But more seriously, Chambers's theory was not radical enough for Huxley. He was a professional scientist anxious to demolish the image of the clergyman-naturalist and he was looking for a theory that would eliminate all trace of the argument from design. Chambers's book left the reader to believe that the only explanation of progress was God's purpose. If Huxley was to accept evolution, it would have to be based on a mechanism driven solely by observable effects, not by mysterious trends designed by God. Fortunately for him, Darwin would soon publish a theory that fulfilled exactly this requirement.

The Development of Darwin's Theory

Darwin had conceived his theory in the late 1830s, but he had not published and only gradually allowed a few close contacts to know what he was doing. Thus the publication of the *Origin of Species* in 1859 came like a bolt from the blue as far as most scientists were concerned. Here was a major new initiative on the cause of evolution, backed up by a wealth of evidence and insight accumulated by Darwin over twenty years. As noted in the introduction to this chapter, historians disagree radically over how to interpret the process by which Darwin put his ideas together. For some he worked as a pure scientist, and if he gained insights from social debates this does not undermine the credibility of his theory (De Beer 1963). Others stress the parallel between natural selection and the competitive ideology of Victorian capitalism and see Darwin as someone who projected the social values of his own class onto nature itself (Desmond and Moore 1991; Young 1985). Many historians seek to balance these two positions, acknowledging the inspiration provided by social theories but recognizing that we can only explain the unique character of Darwin's thinking if we take note of how he applied his insights to a particular set of scientific questions (Bowler 1990; Browne 1995; Kohn 1985).

Darwin was born into a prosperous middle-class family in 1809. He was sent to Edinburgh for medical training, where he met and worked with the Lamarckian anatomist Robert Grant (although he subsequently claimed to have been unimpressed by Grant's evolutionism). He abandoned medicine and went to Cambridge to study for an arts degree, as a prelude to becoming an Anglican clergyman—an ideal career for an amateur naturalist. All his scientific training at Cambridge was thus outside the curriculum, but he impressed the professors of botany and of geology, John Stevens Henslow and Adam Sedgwick, respectively. Henslow then helped to gain him the opportunity that would transform his life: he was accepted as the gentleman-naturalist to travel with the survey vessel H.M.S. *Beagle,* about to sail for South America. The voyage of the *Beagle* lasted five years (1831–36), and while the ship was charting the coastal waters, Darwin had ample opportunities to travel to the interior. Here he made discoveries in geology and natural history that would make his reputation as a scientist and give him the insights that made him an evolutionist.

Sedgwick had trained Darwin as a catastrophist, interpreting the discontinuities in the geological record as evidence of vast upheavals in the past. But Darwin had been given the first volume of Charles Lyell's *Principles of Geology,* and his own observations soon made him a uniformitarian (see chap. 5, "The Age of the Earth"). He saw how the Andes Mountains were still being raised by earthquakes, as well as evidence that the whole range had been elevated gradually over a vast period of time, not in a single catastrophe. From that point on, Darwin felt it necessary to explain the distribution and adaptations of animals and plants in Lyellian terms: the present situation must be the outcome of slow changes driven by natural causes. At Cambridge he had read Paley's *Natural Theology* and been impressed with the claim that adaptation was an indication of God's design. But Paley's argument didn't work in a world of gradual change. As Lyell himself recognized, if geology is constantly modifying the environment by elevating and destroying mountains, species must either migrate to find conditions they can survive in or gradually go extinct. Lyell remained convinced that species are fixed, leaving it for Darwin to raise the possibility that they might be transformed by a process that adapts them to changes in their environment.

In South America, Darwin saw evidence that species competed with one another to occupy territory, a struggle whose outcome might be influenced by changes in the environment. But the most crucial observations came when the *Beagle* called at the Galapagos Islands, a group of volcanic islands lying five hundred miles off the coast in the Pacific. Although he nearly missed the evidence, Darwin was able just in time to appreciate that the animals were different on different islands. The giant tortoises on each island had significantly different shells, while the birds, especially the mocking birds and finches, displayed an immense variety. The finches could be found in a range of forms with entirely different beak structures adapted to different ways of finding food (fig. 6.3). Darwin only noticed the significance of this fact just before he left the islands, but he pondered on its implications while on the way home, and when he was told by the orni-

FIGURE 6.3 Heads of four of the Galapagos ground finches, from Darwin's *Journal of Researches into the Geology and Natural History of the Countries Visited during the Voyage of H.M.S. Beagle* (reprint, London 1891), chap. 17. The variation in the beak structures shows adaptation to different ways of obtaining food, such as cracking seeds or picking up insects. Darwin was told that these forms should be classified as distinct species, but he was convinced that they must have evolved from a common ancestor that had adapted to different ways of life on the various Galapagos Islands.

thologist John Gould that the various finches had to be counted as distinct species, he was faced with a dilemma. He could not accept that God had independently created a range of distinct species to occupy each of these tiny islands. It was more reasonable to believe that small populations derived from South America had been able to establish themselves on each island and had there changed to adapt themselves to their new environment. Transmutation, what we call evolution, could create not just new varieties but also new species, and if it could create species, why not—given time—new genera, families, and even classes?

Dissatisfied with the explanations offered by Lamarck and earlier writers (although he did not deny a limited role for the inheritance of acquired characteristics), Darwin set out to discover a plausible mechanism. His ideas were constrained by the Lyellian principle that the mechanism must be based on a combination of observable processes. Evolution is essentially an adaptive process, and it cannot be predetermined because the branching effect seen in the Galapagos implies that when a population is subdivided by geographical barriers, each group is able to adapt in its own way. There is no automatic ladder of progress—although Darwin did not deny that in the long run some branches of the tree of life had advanced to higher levels of organization than others. Many branches have evidently ended in extinction, while others have multiplied by subdivision.

In search of clues, Darwin turned to one area where animals could actually be observed to change: the production of artificial varieties by human breeders. The path of discovery revealed by his notebooks (reprinted as Darwin 1987) is complex, but in the end the breeders taught him certain important principles. All populations exhibit individual differences: no one organism is identical to another (just as no human being is identical to another). And there seems to be no obvious pattern or purpose to this variation (just as there seems no obvious purpose in, for instance, the variation of hair color in humans). How do the breeders use this random variation to create a new variety of dogs or pigeons? The answer, Darwin eventually realized, was selection—they pick out the very few individuals who happen to vary in the direction they want and breed only from them. The rest are rejected and probably killed.

Could there be a natural equivalent of this artificial selection, a process that would pick out only those better adapted variants to breed for the next generation? Darwin realized that there could be a natural form of selection when he read the clergyman Thomas Malthus's *Essay on the Principle of Population.* This work on political economy was intended to challenge Enlightenment optimism by showing that human progress was impossible. All efforts at social reform were doomed because poverty was not a consequence of social inequality—it was natural because the reproductive capacity of any population always exceeds the food supply. The consequence was that in every generation many must starve, and when writing of the wild tribes of central Asia (not, significantly, of his own society), Malthus argued that there must be a "struggle for existence" to determine who would live and who would die. Darwin picked up this idea and realized that the variability of the population would give some individuals an edge in the struggle. Those best adapted to any change in the environment would be most likely to survive and breed, those less well adapted would starve, and the result would be that the next generation would be bred largely from better adapted parents. Repeated over innumerable generations, this process of natural selection would modify organs and habits and, in the end, produce new species. It is the influence of Malthus that is often singled out to argue that natural selection reflects the values of free-enterprise capitalism. There can be little doubt that Darwin did think of the species in individualist terms, as a population not as a type. But he applied this insight in a unique way shaped by his scientific observations—Malthus had not seen his principle as source of change, and it was only after Darwin published his findings that people began to think seriously of struggle as the driving force of progress.

In an essay Darwin wrote in 1844 to outline his theory (intended for publication only in case of his death), he described the effect thus, using the example of a population of dogs forced to chase faster-running prey (hares instead of rabbits):

> Let the organization of a canine animal become slightly plastic, which animal preyed chiefly on rabbits, but sometimes on hares; let these same changes cause the number of rabbits very slowly to decrease and the number of hares to increase; the effect of this would be that the fox or dog would be driven to try to catch more hares, and his numbers would tend to decrease; his organization, however, being slightly plastic, those individuals with the lightest forms, longest limbs and best eyesight (though perhaps with less cunning or scent) would be slightly favored, let the difference be ever so small, and would tend to live longer and to survive during that time of the year when food was shortest; they would also rear more young, which young would tend to inherit these slight peculiarities. The less fleet ones would be rigidly destroyed. I can see no more reason to doubt but that these causes in a thousand generations would produce a marked effect, and adapt the form of the fox to catching hares instead of rabbits, than that greyhounds can be improved by selection and careful breeding. (Darwin and Wallace 1958, 120)

Over the next twenty years this was the theory that Darwin would explore in all its ramifications. He continued to work with animal breeders. He corresponded with a vast range of naturalists, sounding them out on detailed questions without revealing his true purpose. He undertook a massive study of barnacles, then a little-known group, which helped him to understand how branching evolution could be mapped onto the taxonomic hierarchy. This study also showed him that, on many branches of the tree of life, adaptive evolution has led to parasitism and degeneration. Perhaps inevitably, given its source in Malthus's principle, this was not a theory of inevitable progress—better adapted to a particular environment does not mean "fitter" in any absolute sense. Yet in the end Darwin did believe that higher animals, and ultimately the human species itself, had been produced. Struggle did tend to set in motion improvement, at least some of the

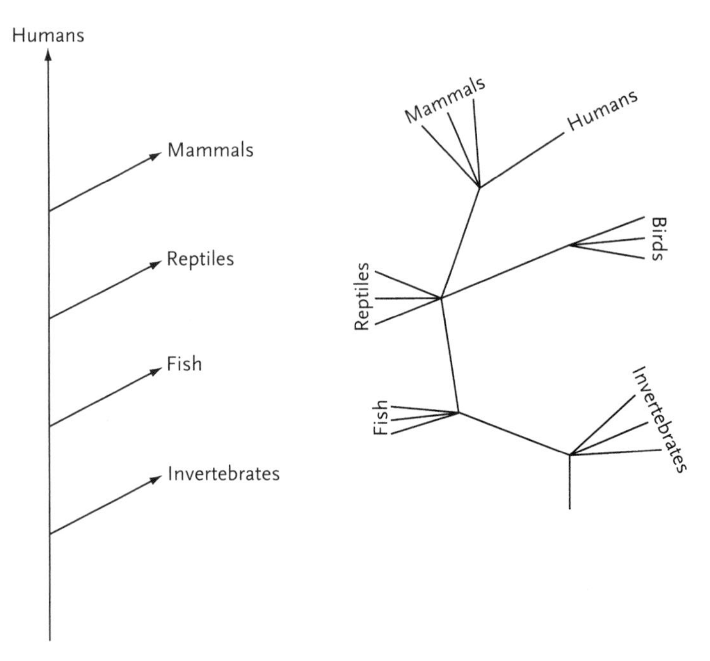

FIGURE 6.4 Diagram illustrating the difference between a linear model of evolution (*left*) and a branching model (*right*). The linear model treats evolution as a progressive advance along a linear hierarchy toward the human species. The "lower" forms of life thus appear as the rungs of a ladder that life has climbed to reach its goal in humanity. This model is easily compatible with the recapitulation theory in which the human embryo passes through stages corresponding to the lower animals. In the branching model, the emphasis is on adaptation and divergence, not progress. Each class splits into a range of different adaptations, and later classes are derived from a single branch of a previous class. Progress has to be defined in terms of distance from the simplest common ancestor, but there are many different lines of advance and no living form can be treated as a stage in the development of another. This diagram focuses on the vertebrates, but note that in fact the invertebrates form a range of phyla fully equivalent to the vertebrates in diversity.

time, and this viewpoint would eventually be incorporated into "social Darwinism." Yet Darwin was very careful not to link his theory to the linear model of progress. There was no main line of evolution, and most adaptive trends have nothing to do with the ascent of life. Darwin also admitted that the imperfection of the fossil record would make it difficult to reconstruct the detailed course of evolution, although the general outline of the record fit a theory of branching, adaptive evolution in which each branch was specializing for a different way of life (fig. 6.4).

By the mid-1850s Darwin had let a few colleagues, including Lyell and the botanists Joseph Hooker and Asa Gray, know the details of his theory and had begun writing. He was interrupted in 1858 by the arrival of a paper written in the Far East by another naturalist, Alfred Russel Wallace, outlining a theory similar to his own. Historians have disagreed enormously over the significance of Wallace's discovery. Some accept Darwin's initial reaction at face value and treat Wallace as the codiscoverer of the theory, implying that the subsequent events were designed to rob Wallace of his credit. Others have taken a closer look at Wallace's 1858 paper and point out that there are significant differences that Darwin seems to have overlooked. Wallace had no interest in artificial selection, and it is quite possible that his paper was really intended to describe a form of natural selection acting between varieties or subspecies, not between the individuals of the same population (for an overview, see Kottler 1985). This may not be a case of independent discovery at all, but of two naturalists with similar, but not identical, backgrounds exploring different aspects of the same problem. Whatever the differences and similarities, Darwin saw enough of a parallel with his own work to fear the loss of his twenty-year priority. Lyell and Hooker arranged for the publication of two extracts of Darwin's writings along with Wallace's paper (reprinted in Darwin and Wallace 1958). No one paid much attention, but Darwin now rushed to complete the account of his theory, which was published at the end of 1859 as *On the Origin of Species.*

The Reception of Darwin's Theory

The *Origin* sparked a renewed debate over evolution. Darwin was an eminent scientist, and natural selection was an important new initiative backed up by a wealth of new evidence. The debate was rendered all the more emotional because the theory seemed to undermine any hope of seeing evolution as the unfolding of a divine plan. In these circumstances, both scientists and laypersons were forced to assess the theory at various levels: their evaluation of the evidence would almost certainly be influenced by their wider beliefs. Debates raged over the plausibility of both evolution in general and natural selection in particular. Darwin had important new lines of argument, but there were also technical arguments against his theory. Some of these focused on the area of heredity, where his thinking did not anticipate modern genetics and left him vulnerable to arguments that would not be plausible today. In these circumstances, there was little hope of a clear-cut debate that would end decisively with the rejection or acceptance of the new theory. No one was going to be converted by scientific arguments alone, and to some extent the outcome would depend on the politics of the scientific community and the possibility of a wider change in public opinion. In the end, after a few years of uncertainty, the general idea of evolution came to be widely accepted, but natural selection remained controversial.

To younger, radical scientists such as T. H. Huxley, Darwin's theory offered immense opportunities (Desmond [1997]; on the scientific debate, see Hull [1973]). As professional scientists they were anxious to discredit natural theology, which in their eyes left science subservient to religion (see chap. 14, "The Organization of Science"). Darwin's theory certainly did this and, thus, fit well into the philosophy that Huxley called "scientific naturalism"—although to his opponents it was little better than materialism. The whole world, including the human mind, was to be explained in terms of the operations of natural law. Here Huxley could make common cause with the philosopher Herbert Spencer, who presented evolution as the underlying principle of both nature and society. Spencer welcomed the individualism of Darwin's theory, since it fit his view that the general progress of nature was the product of innumerable acts by individuals, each seeking its own wellbeing. This points the way to the social applications of Darwin's theory (see chap. 18, "Biology and Ideology"), although it is important to realize that natural selection was not the only model of evolution available. Spencer favored Lamarck's theory of the inheritance of acquired characters because it better fit his ideology of self-improvement. Huxley would not accept natural selection as the sole mechanism of evolution, preferring to believe that variation was directed in a few consistent directions, instead of being random as Darwin supposed.

Even within the scientific community there were many who rejected naturalistic philosophy, often because they retained deep religious beliefs. Outside science, religious and moral problems influenced many people's response to the theory (see chap. 15, "Science and Religion"). A survey of the popular press reaction by Alvar Ellegård (1958) shows how the more conservative periodicals lagged behind in acceptance of evolution, their authors worrying that the theory undermined both divine providence and the spiritual status of the human soul. Huxley's confrontation with the smooth-talking Bishop "Soapy Sam" Wilberforce at the 1860 meeting of the British Association has become a symbol of the confrontation between evolutionism and conservative religion, although we now know that Huxley was by no means as successful as the popular image of this event implies (see fig. 15.3, p. 356). In the long run, however, conservatives reluctantly accepted the basic idea of evolution. But they needed to see the process as the expression of God's purpose and, thus, remained hostile to the trial-and-error model of natural selection. The reemergence of a sustained creationist opposition occurred only in the 1920s, however.

There were certainly scientific arguments to be deployed. Darwin made much of the difficulties naturalists often encountered in deciding whether a particular form was a distinct species or merely a variety of another species. He showed how geographical distribution could be explained far more easily in terms of branching evolution rather than as arbitrary acts of creation. The botanists Joseph Hooker and Asa Gray supported Darwin here, while A. R. Wallace undertook a major study of animal distribution, publishing an important synthesis in 1876. Yet increasingly, the emphasis began to fall on an area that Darwin had tried to avoid: the detailed reconstruction of the history of life on earth using fossil and anatomical evidence. Darwin thought the fossil record was so incomplete that it would be impossible to reconstruct the ancestry of any known species in detail. But this left him vulnerable to critics who insisted that unless the "missing links" could be found, evolution remained implausible. By the 1870s, important new fossils had been discovered that seemed to fit the evolutionists' predictions. In Germany, the remains of *Archaeopteryx* provided clear evidence of a form intermediate between reptiles and birds. From America came a series of fossil horses showing a line of specialization leading toward the modern horse that Huxley proclaimed as "demonstrative evidence of evolution" (on these developments, see Bowler [1996]).

Even where fossils were not available, enthusiastic evolutionists such as Ernst Haeckel in Germany used anatomical and embryological evidence to reconstruct the links between the major branches of the tree of life. Haeckel was a leading proponent of the recapitulation theory, which built on the old law of parallelism by assuming that the development of the embryo offered a speeded-up model of the organism's whole evolutionary ancestry. He and his followers (Huxley included himself in this group) proposed hypothetical genealogies to explain the origin of all the vertebrate classes, and even of the vertebrates themselves. Michael Ruse (1996) dismisses this whole movement as inferior science driven by an overenthusiastic support for the idea of progressive evolution. It is certainly true that these evolutionists ignored some of the most important lessons that could have been learned from Darwin. By using the embryo as a model for evolution, they highlighted the progressive development of life in a way that portrayed the human species as its intended goal. Haeckel's version of the tree of life had a main trunk leading through to humans, with everything else dismissed as side branches—a linear model more reminiscent of the old chain of being (fig. 6.5). He had little interest in exploring the kind of adaptive pressures

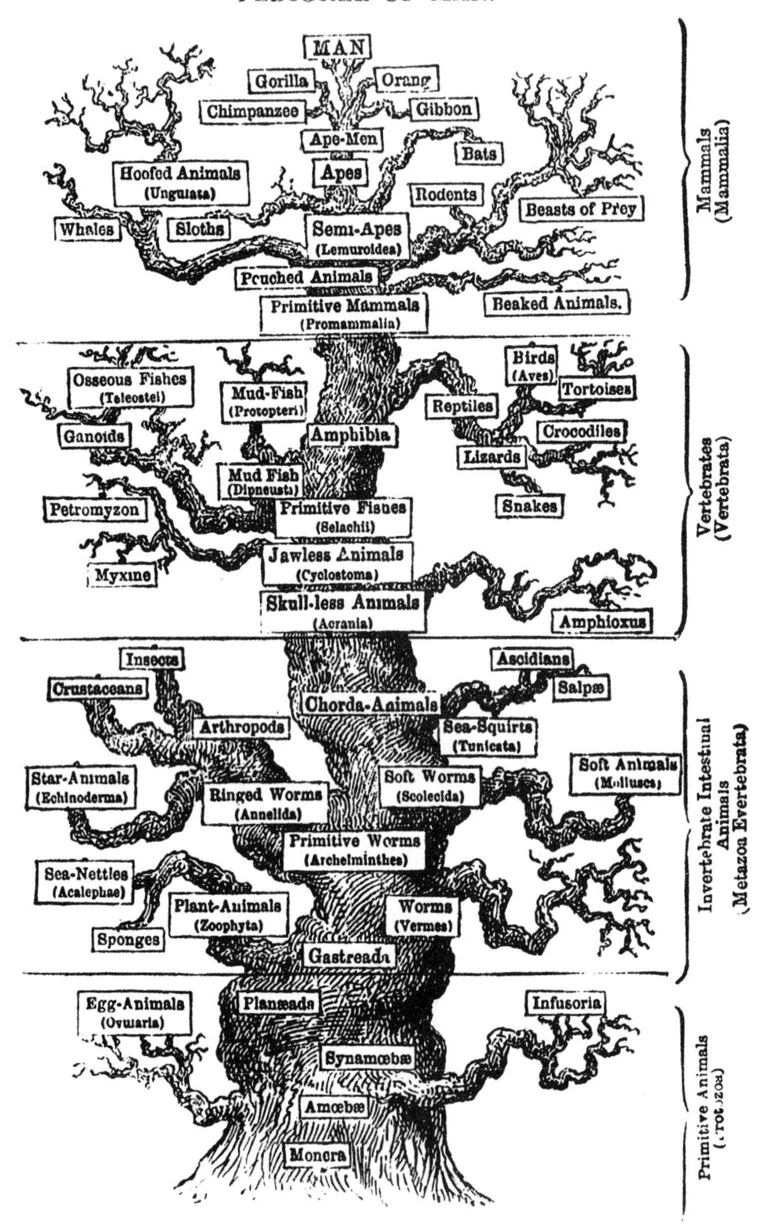

FIGURE 6.5 The tree of life from Ernst Haeckel's *History of Creation* (New York, 1876), vol. 2, facing p. 188. Note how Haeckel combines the linear and branching models of evolution (fig. 6.3, above) by deliberately giving his tree a main trunk with humanity at the top. He thus retains something of Darwin's emphasis on divergence and adaptation but superimposes this onto a linear ascent by treating all those creatures that do not lie on the "main line" as side branches leading off to stagnation.

that might have brought about the changes he was postulating. It is also true that this project to create an evolutionary morphology (the science of animal form) got bogged down as rival hypotheses emerged, with little hope of fossil evidence to determine which was right (see chap. 7, "The New Biology"). But to dismiss this whole generation of evolutionary biology as a waste of time misses the point that it was perceived as the most exciting application of the theory at the time. It certainly confirms that evolution was welcomed because it seemed to endorse the idea of progress, but the debates that were engendered raised substantive issues that are only now being resolved as the techniques of molecular biology (to say nothing of a wealth of later fossil discoveries) are brought to bear on them.

Haeckel called himself a Darwinian, but he combined the selection theory with a generous dose of Lamarckian use-inheritance and a commitment to the idea of progress that owed much to the *Naturphilosophie* of an earlier generation. The selection theory had, in fact, encountered substantial criticism from a host of scientists who found it difficult to believe that a process based on random variation could ever have a purposeful outcome (Gayon 1998; Vorzimmer 1970). Richard Owen accepted evolution but insisted that its course was predetermined by a divine plan (Rupke 1993). The anatomist St. George Jackson Mivart's *Genesis of Species* (1871) outlined a host of objections, some of which are still in use by modern creationists. How, he asked, could natural selection force a transition through the intermediate phase where a structure had lost its old function but was not yet efficient at the new one, for example, when a limb no longer worked as a leg but was not yet a proper wing? Some naturalists shared Mivart's belief that many structures have no adaptive function at all, indicating the existence of predetermined trends not controlled by natural selection. There was also the problem of geological time (see chap. 5, "The Age of the Earth")—by the late 1860s William Thomson was limiting this to a point where many believed that natural selection would be too slow to have produced the ascent of life up to humans.

Equally serious was an objection raised by the engineer Fleeming Jenkin based on Darwin's model of heredity and variation. Like most of his contemporaries, Darwin had no notion of the discrete genetic units that would be postulated by Gregor Mendel—he thought that the offspring would simply blend together any differences between the parents (although this is self-evidently not true for sex). If a beneficial new character appeared in a single favored individual, Jenkin argued that the offspring would only have half the benefit, the next generation only a quarter, and so on. Within a few generations, the beneficial new character would be diluted to insignificance and could not be acted on by selection. Darwin had no real answer to this, and it was Wallace who pointed out that favorable characters do not appear in single individuals. If we think of the population of ancestral giraffes when it first began to feed off trees, it would have shown a range of variation in neck length, with significant numbers at both ends of the range. There would have been no shortage of individuals with longer than average necks to benefit from the action of selection.

By the 1880s, Wallace was one of a relatively small number of biologists still defending the Darwinian selection theory. The theory of evolution itself was secure, but Darwinism was increasingly under fire as critics sought alternatives to the selection theory. This was the period that Julian Huxley later referred to as the "eclipse of Darwinism" (Bowler 1983a). Building on Mivart's work, many argued that evolution was driven by nonadaptive trends somehow built into the nature of life itself. Those who accepted a role for adaptation saw the Lamarckian theory as an alternative rather than a supplement to Darwinism. In America there was a strong neo-Lamarckian movement led by such paleontologists as Edward Drinker Cope. They were sure that the almost linear trends they found in the fossil record could only be the result of some directing agent, in this case the new habit that drove the species toward a more specialized structure. Viewed from the perspective of the late nineteenth century, Darwin's theory was a relic of the past that had played only a fleeting role in forcing scientists to reconsider the case for evolution in the 1860s.

Human Origins

Darwin had avoided discussing the human race in the *Origin of Species,* knowing that this was a particularly sensitive topic. But controversies over the degree of relationship between humans and apes were already underway, and the whole issue had become a battlefield long before Darwin eventually entered the fray with his *Descent of Man* in 1871. Religious thinkers were dismayed that the theory linked us with the animals and thus, by implication, undermined the credibility of the immortal soul. Humans alone had traditionally been endowed with higher mental and moral faculties—so by suggesting that we were only improved animals, evolutionism threatened our unique status and might even undermine the fabric of the social order. In the scientific naturalism favored by Darwin and Huxley, however, it was important to show that there were no supernatural agents in the world, so even the human mind was a product of the activity of the brain, which in turn had been shaped by evolution.

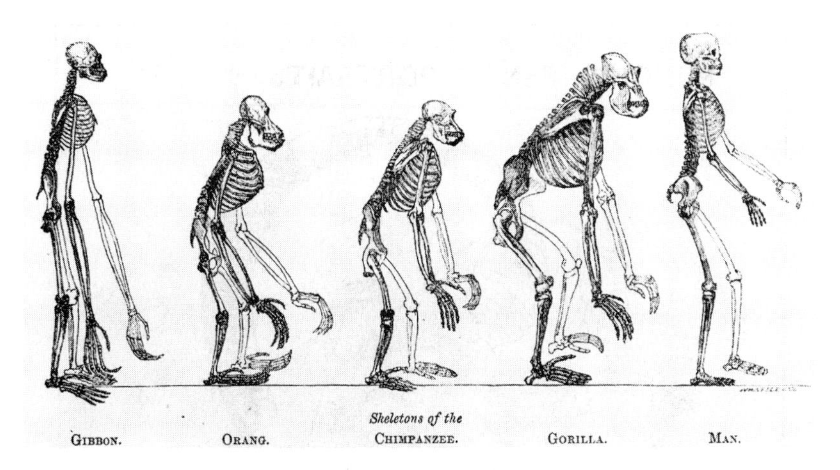

FIGURE 6.6 Comparison of the skeletons of a human (*right*) with those of a gorilla, a chimpanzee, an orangutan, and (twice life-size by comparison) a gibbon; the frontispiece to T. H. Huxley's *Man's Place in Nature* (London, 1863). Huxley argued that the degree of similarity meant that humans had to be classed as Primates and hence, by implication, must share a common ancestry with the apes.

The case for an evolutionary ancestry for humanity was boosted by a revolution in archaeology that took place in the early 1860s. Lyell's *Antiquity of Man* (1863) summed up evidence that Stone Age humans had existed on the earth for tens of thousands of years before civilization emerged. Yet Lyell himself could not accept an evolutionary link between those primitive humans and apes. There was as yet no plausible fossil evidence for a missing link between humans and apes, so those who wanted to argue for an evolutionary connection had to stress the anatomical similarities between humans and the living great apes. Huxley was already engaged in a debate with Richard Owen on the degree of similarity between the human and ape brains. He summed up his arguments for a close link in his *Man's Place in Nature* in 1863 (fig. 6.6). But it was the mental, not the physical, comparison that was crucial, and already philosophers such as Herbert Spencer were beginning to create an evolutionary psychology by which they hoped to explain how the higher mental faculties had been added in the course of evolution (Richards 1987).

Darwin offered his *Descent of Man* as a contribution to this enterprise. He wanted to show that the apparent gulf between animal and human mentalities was not as great as traditionally assumed (fig. 6.7). Like many of his contemporaries, he was increasingly inclined to treat those modern races that the Victorians regarded as "savages" as surviving relics of earlier stages in the ascent from the ancestral ape. They were the equivalent of the

CHARLES ROBERT DARWIN, LL.D., F.R.S.

IN HIS *DESCENT OF MAN* HE BROUGHT HIS OWN SPECIES DOWN AS LOW AS POSSIBLE—*I.E.*, TO "A HAIRY QUADRUPED FURNISHED WITH A TAIL AND POINTED EARS, AND PROBABLY *ARBOREAL* IN ITS HABITS"—WHICH IS A REASON FOR THE VERY GENERAL INTEREST IN A "FAMILY TREE." HE HAS LATELY BEEN TURNING HIS ATTENTION TO THE "POLITIC WORM."

FIGURE 6.7 A caricature of Darwin from the magazine *Punch* in 1881. The caption refers to Darwin's theory that humans are descended from a "hairy quadruped," but the picture links him to an even lower animal, the earthworm—the subject of Darwin's last book. He was fascinated by the ability of worms to regenerate the soil and even transform the landscape over a long period of time, retaining an interest in detailed natural history even while dealing with the broadest of theoretical issues.

Europeans' Stone Age ancestors, surviving into the present and in effect showing us what the "missing link" would have been like (see chap. 18, "Biology and Ideology"). Darwin also tried to exaggerate the mental powers of animals: there were as yet no scientific studies of animal behavior, so he could use anecdotal evidence from travelers and zookeepers that often presented an anthropomorphic interpretation of animals' actions. For Darwin the human conscience was merely an expression of the social instincts that our ancestors had been endowed with by evolution. Far from generating instincts for pure selfishness, natural selection (coupled with a Lamarckian inheritance of leaned habits) could promote social instincts in species that normally lived in groups. Our moral values were just rationalizations of instincts imprinted on our ape ancestors.

Darwin saw that it was important to explain why humans gained a higher level of mental powers than their ape relatives. He suggested that perhaps our ancestors stood upright when they moved out of the forests onto the plains of central Africa. This freed their hands for tool making and thus promoted extra intelligence. Most nineteenth-century evolutionary psychologists simply assumed that evolution would steadily add on new stages of mental activity. Their work thus expanded the developmental model of evolution promoted in biology by Haeckel. Darwin's chief disciple in this area, George John Romanes, wrote a series of books on the mental powers of animals and humans, trying to reconstruct the exact sequence in which new mental powers were added. He used the recapitulation theory to portray the mental development of the human child as a model for the whole evolution of animal life. Although fossil discoveries toward the end of the century would challenge this linear model of evolution (see Bowler 1986), its influence on late nineteenth-century thought was profound. And in the end it was turned on its head by Sigmund Freud, who recognized that the animal instincts buried in the unconscious may often be too much for the overlying rational mind to control (Sulloway 1979).

The Resurgence of Darwinism

In the decades around 1900 most biologists remained evolutionists, but they believed that Darwinism was dead. New developments in the life sciences were, however, challenging the foundations on which late nineteenth-century evolutionism had been built. To enhance their status as professional scientists, many biologists turned to experimental work and began to look down on the comparative anatomists and paleontologists who had tried to reconstruct the ascent of life on earth. One product of this move was a program of research on heredity and variation that would lead to the foundation of modern genetics (see chap. 8, "Genetics"). The geneticists repudiated the Lamarckian effect and the developmental trends that had upheld the recapitulation theory. They gradually eroded support for neo-Lamarckism, and with hindsight we can see that this paved the way for a reemergence of the Darwinian selection theory. Yet the first geneticists had no more time for Darwinism than they had for Lamarckism. They thought that large genetic mutations created new species without any need for selection. The final phase of the Darwinian revolution emerged from a complex process of reconciliation by which the geneticists were brought round to the view that selection was indeed necessary to explain the accumulation of favorable genes in a population. It turned out that Darwin had been right after all, even though a generation of biologists had turned their backs on his theory.

The first moves were made by biologists who became convinced that heredity rigidly determines the character of the organism. Environmental effects are powerless to alter the characteristics inherited by the child from its parents. In Germany, August Weismann postulated the "germplasm" that was responsible for transmitting the characters from one generation to the next. He argued that it was isolated from the rest of the body, making the Lamarckian effect impossible. Weismann insisted that natural selection was the only way that the transmission of characters could be affected by the environment. In Britain, the statistician Karl Pearson adopted similar views and tried to detect the effect of selection on the variation of wild populations (fig. 6.8). His views were controversial, and Pearson's support for the selection theory generated antagonisms that would alienate him from the founders of genetics. As far as he was concerned, evolution was a slow, gradual process just as Darwin had assumed—but that was exactly the point being challenged by the biologists who would create Mendelian genetics.

The alternative being explored by several of the biologists involved in the "rediscovery" of Gregor Mendel's long-neglected laws of heredity was the theory of evolution by sudden leaps or saltations (Bowler 1989). William Bateson, who went on to coin the term "genetics" and provide the first English translation of Mendel's paper, openly rejected Darwinism during the 1890s. He insisted that studies of variation within species showed that the distinct varieties within them were created abruptly by saltation, not by gradual adaptive change. The Dutch botanist Hugo De Vries, one of the biologists who first drew attention to Mendel's paper, proposed his "mutation theory" based on the apparently sudden appearance of new types

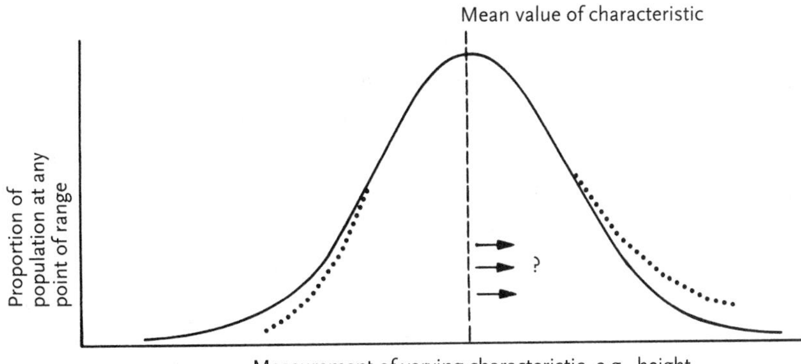

FIGURE 6.8 Diagram to illustrate the distribution of a continuously varying character in a population and the effect of selection on the distribution. The *solid line* is the bell-shaped "normal" curve that would be obtained, for instance, for the variation in height within a human population. The proportion of the population occupying any point in the range (*vertical axis*) is plotted against the measurement of the character (*horizontal axis*). The largest proportion is clumped around the mean value with smaller proportions tailing off to either extreme—most people are of approximately average height, and there are smaller numbers of very tall and very short persons. Biometricians such as Karl Pearson and W. F. R Weldon measured the variation for different characters in wild populations of crabs and snails and obtained curves such as this. But as Darwinists, they then had to show that if the population were subject to selection, there would be a permanent shift in the distribution. If taller individuals were favored in a certain environment, and shorter ones at a corresponding disadvantage, this would generate more tall individuals and less short in the next generation, as indicated by the *dotted lines*. But would the effect of this be to shift the mean value for the population as a whole in the favored direction as shown by the arrows? The measurement seemed to show that such an effect did occur, but it was too small to convince many anti-Darwinian biologists.

within the evening primrose, *Oenothera lamarckiana.* Thomas Hunt Morgan, who eventually established the true nature of mutations, began as a supporter of De Vries's theory and a strong opponent of Darwinism. What drew all of these biologists to the model of heredity that they found in Mendel's laws was their preference for the idea that new characters are created as discrete units. It seemed natural for them to accept a theory in which all hereditary characters are treated as fixed discrete units transmitted from one generation to the next. The fact that Mendel had already worked out the laws governing the transmission of these units—soon to become known as genes—was hailed as a remarkable anticipation of the latest thinking when De Vries and others came across his paper in 1900, more than thirty years after its original publication.

Not surprisingly, the early Mendelians saw their theory as a new alternative to Darwinism, while Pearson rejected the geneticists' model of heredity as incompatible with the continuous range of variation he studied in many wild populations. It took twenty years for a bridge to be built between the two positions by biologists who realized that each side had been looking at only one aspect of the problem. In the meantime, Morgan's studies of true genetic mutations showed that De Vries's large-scale saltations did not reflect the way in which new genetic characters are normally produced (in fact the evening primrose is a hybrid, and the "new" forms De Vries was observing were not true mutations). Genes normally transmit their character without change from one generation to the next, but Morgan and his team showed that every now and again something alters the gene so that it codes for a different character. Large mutations are deleterious and often fatal, but there are many smaller ones that are transmitted to future generations as their carriers breed with other members of the population. By 1920, Morgan had realized that mutations keep up a supply of genetic variation within the species and even began to concede that an effect similar to natural selection would determine what mutations will spread into the population. If a mutated gene corresponds to a character that is beneficial in a new environment, the organisms that carry it will breed more readily and the next generation will contain more organisms with that gene. Conversely, a gene conferring a harmful character will gradually be eliminated. Mutations thus provide the ultimate source of the random variation that Darwin had postulated.

It was also realized that because many characters can be influenced by more than one gene, the genetic model of variation is not incompatible with the continuous range of variation observed by Darwinists such as Pearson. A new science of population genetics emerged to study how genes maintain the variability of populations, and how the range of variation can be altered by natural selection (Provine 1971). In Britain, Ronald Aylmer Fisher published his *Genetical Theory of Natural Selection* in 1930, arguing that all evolution takes place through the slow action of selection on large populations. J. B. S. Haldane also contributed to the theory but realized that the process could work much faster than Fisher supposed when genes conferred major adaptive advantages. In America, Sewall Wright used a different model derived from artificial selection to show that natural selection works best when the species is divided into small subpopulations that only occasionally interbreed. When Wright's mathematical formulas were translated into terms that the field naturalists could understand in Theodosius

Dobzhansky's *Genetics and the Origin of Species* of 1937, the way was open for the final emergence of Darwinism as the dominant model of evolution.

Field naturalists such as Ernst Mayr now began to contribute to the new Darwinism—indeed Mayr has since maintained that he and his co-workers were already finding their way toward a more selectionist model before they became aware of the genetical theory (see Mayr and Provine 1980). In 1942 the British naturalist Julian Huxley, grandson of Thomas Henry, published his *Evolution: The Modern Synthesis,* and the theory has been known ever since as the modern or evolutionary synthesis. Those involved, and a subsequent generation of historians, argued and still argue over exactly what was synthesized to make the theory. Was it a theoretical synthesis bringing together selection and genetics, or a reconciliation between previously hostile areas of biological research made possible by the elimination of rival non-Darwinian ideas? Why was the synthesis more visible in the Anglo-American scientific communities than elsewhere—does this reflect the fact that even genetics developed in a less deterministic way in France and Germany than in Britain and America? These arguments will no doubt continue, fueled in part by the fact that the synthesis has been remarkably successful in holding evolutionism together ever since.

Conclusions

The once-popular notion of a Darwinian revolution following the publication of the *Origin of Species* no longer holds water. Historians have shown that challenges to the idea of divine creation began long before Darwin published and that even the concept of a designed universe could be made more sophisticated so that it could accommodate the idea of development through time. The basic idea of evolution was widely debated following the publication of *Vestiges,* and Darwin's theory was understood in part as a contribution to Chambers's vision of progress. Darwin's more materialistic theory offered some new opportunities to scientists, especially those willing to go along with Huxley's scientific naturalism, but in the end the most radical implications of the selection theory had to wait until the twentieth century before they could be realized. The original Darwinian revolution turned out to be only a transition to an evolutionary interpretation of an already-existing worldview based on faith in the idea of progress as the product of divine providence or of nature's laws. What modern biologists see as most original in Darwin's work served only to shock his readers into acceptance of the general idea of evolution—in the end they could not take natural selection seriously. It took a second revolution associated with the emergence of Mendelian genetics to destroy the developmental view of evolution that had subverted Darwin's proposals and complete the transition to modern Darwinism.

In some senses, of course, the revolution is still not over. The supporters of the modern synthesis did not conceal the difficulties their theory created for traditional beliefs, and in response there was a reemergence of Fundamentalist opposition first articulated in the 1920s. A large number of traditional believers, especially in America, simply reject the theory outright and still look to divine creation. If the Darwinian revolution in science is complete, the revolution in popular attitudes has a long way to go.

References and Further Reading

Appel, Toby A. 1987. *The Cuvier-Geoffroy Debate: French Biology in the Decades before Darwin.* Oxford: Oxford University Press.

Barzun, Jacques, 1958. *Darwin, Marx, Wagner: Critique of a Heritage.* 2d ed. Garden City, NY: Doubleday.

Bowler, Peter J. 1983a. *The Eclipse of Darwinism: Anti-Darwinian Evolution Theories in the Decades around 1900.* Baltimore: Johns Hopkins University Press.

———. 1983b. *Evolution: The History of an Idea.* 3d ed. Berkeley: University of California Press, 2003.

———. 1986. *Theories of Human Evolution: A Century of Debate, 1844–1944.* Baltimore: Johns Hopkins University Press; Oxford: Basil Blackwell.

———. 1988. *The Non-Darwinian Revolution: Reinterpreting a Historical Myth.* Baltimore: Johns Hopkins University Press.

———. 1989. *The Mendelian Revolution: The Emergence of Hereditarian Concepts in Modern Science and Society.* London: Athlone; Baltimore: Johns Hopkins University Press.

———. 1990. *Charles Darwin: The Man and His Influence.* Oxford: Basil Blackwell. Reprint, Cambridge: Cambridge University Press, 1996.

———. 1996. *Life's Splendid Drama: Evolutionary Biology and the Reconstruction of Life's Ancestry, 1860–1940.* Chicago: University of Chicago Press.

Browne, Janet 1995. *Charles Darwin: Voyaging.* London: Jonathan Cape.

Burkhardt, Richard W., Jr. 1977. *The Spirit of System: Lamarck and Evolutionary Biology.* Cambridge, MA: Harvard University Press.

Darwin, Charles 1859. *On the Origin of Species by Means of Natural Selection; or, The Preservation of Favoured Races in the Struggle for Life.* Facsimile of the 1st ed., with introduction by Ernst Mayr. Reprint, Cambridge, MA: Harvard University Press, 1964.

———. 1984–. *The Correspondence of Charles Darwin.* Edited by Frederick Burkhardt and Sydney Smith. 12 vols. Cambridge: Cambridge University Press.

——— 1987. *Charles Darwin's Notebooks, 1836–1844.* Edited by Paul H. Barrett et al. Cambridge: Cambridge University Press.

Darwin, Charles, and Alfred Russel Wallace. 1958. *Evolution by Natural Selection.* Cambridge: Cambridge University Press.

De Beer, Gavin, 1963. *Charles Darwin: Evolution by Natural Selection.* London: Nelson.
Desmond, Adrian 1989. *The Politics of Evolution: Morphology, Medicine and Reform in Radical London.* Chicago: University of Chicago Press.
——— 1994. *Huxley: The Devil's Disciple.* London: Michael Joseph.
———. 1997. *Huxley: Evolution's High Priest.* London: Michael Joseph.
Desmond, Adrian, and James R. Moore. 1991. *Darwin.* London: Michael Joseph.
Di Gregorio, Mario A. 1984. *T. H. Huxley's Place in Natural Science.* New Haven, CT: Yale University Press.
Eiseley, Loren. 1958. *Darwin's Century: Evolution and the Men Who Discovered It.* New York: Doubleday.
Ellegård, Alvar. 1958. *Darwin and the General Reader: The Reception of Darwin's Theory of Evolution in the British Periodical Press, 1859–1871.* Goteburg: Acta Universitatis Gothenburgensis. Reprint, Chicago: University of Chicago Press, 1990.
Farber, Paul Lawrence. 2000. *Finding Order in Nature: The Naturalist Tradition from Linnaeus to E. O. Wilson.* Baltimore: Johns Hopkins University Press.
Gayon, Jean. 1998. *Darwinism's Struggle for Survival: Heredity and the Hypothesis of Natural Selection.* Cambridge: Cambridge University Press.
Ghiselin, Michael T. 1969. *The Triumph of the Darwinian Method.* Berkeley: University of California Press.
Gillispie, Charles C. 1951. *Genesis and Geology: A Study in the Relations of Scientific Thought, Natural Theology, and Social Opinion in Great Britain, 1790–1850.* Reprint, New York: Harper & Row, 1959.
Glass, Bentley, Owsei Temkin, and William Straus, Jr., eds. 1959. *Forerunners of Darwin: 1745–1859.* Baltimore: Johns Hopkins University Press.
Greene, John C. 1959. *The Death of Adam: Evolution and Its Impact on Western Thought.* Ames: Iowa State University Press.
Himmelfarb, Gertrude 1959. *Darwin and the Darwinian Revolution.* New York: Norton.
Hull, David L., ed. 1973. *Darwin and His Critics: The Reception of Darwin's Theory of Evolution by the Scientific Community.* Cambridge, MA: Harvard University Press.
Jordanova, Ludmilla 1984. *Lamarck.* Oxford: Oxford University Press.
Kohn, David, ed. 1985. *The Darwinian Heritage.* Princeton, NJ: Princeton University Press.
Kottler, Malcolm Jay 1985. "Charles Darwin and Alfred Russell Wallace: Two Decades of Debate over Natural Selection." In *The Darwinian Heritage,* edited by David Kohn. Princeton, NJ: Princeton University Press, 367–432.
Lovejoy, Arthur O. 1936. *The Great Chain of Being: A Study in the History of an Idea.* Reprint, New York: Harper 1960.
Lurie, Edward 1960. *Louis Agassiz: A Life in Science.* Chicago: University of Chicago Press.
Mayr, Ernst 1982. *The Growth of Biological Thought: Diversity, Evolution and Inheritance.* Cambridge, MA: Harvard University Press.
Mayr, Ernst, and William B. Provine, eds. 1980. *The Evolutionary Synthesis: Perspectives on the Unification of Biology.* Cambridge, MA: Harvard University Press.
Provine, William B. 1971. *The Origins of Theoretical Population Genetics.* Chicago: University of Chicago Press.
Richards, Robert J. 1987. *Darwin and the Emergence of Evolutionary Theories of Mind and Behavior.* Chicago: University of Chicago Press.

Roger, Jacques 1997. *Buffon: A Life in Natural History.* Translated by Lucille Bonnefoi. Ithaca, NY: Cornell University Press.

——— 1998. *The Life Sciences in Eighteenth-Century French Thought.* Translated by Robert Ellich. Stanford, CA: Stanford University Press.

Rupke, Nicolaas A. 1993. *Richard Owen: Victorian Naturalist.* New Haven, CT: Yale University Press.

Ruse, Michael. 1979. *The Darwinian Revolution: Science Red in Tooth and Claw.* 2d ed. Chicago: University of Chicago Press, 1999.

——— *Monad to Man: The Concept of Progress in Evolutionary Biology.* Cambridge, MA: Harvard University Press.

Secord, James A. 2000. *Victorian Sensation: The Extraordinary Publication, Reception and Secret Authorship of "Vestiges of the Natural History of Creation."* Chicago: University of Chicago Press.

Sulloway, Frank. *Freud: Biologist of the Mind.* London: Burnett Books.

Vorzimmer, Peter J. 1970. *Charles Darwin: The Years of Controversy: The "Origin of Species" and Its Critics, 1859–82.* Philadelphia: Temple University Press.

Young, Robert M. 1985. *Darwin's Metaphor: Nature's Place in Victorian Culture.* Cambridge: Cambridge University Press.

CHAPTER 7

W BIOLOGY

GICAL SCIENCES BEGAN to take on their modern form entury, and indeed widespread use of the term "biology" only came into use during that period (Coleman 1971). Previously, the life sciences had been studied through natural history and through the physicians' use of anatomy and physiology (although the two areas were linked, e.g., by the interest in plants for drugs). In the nineteenth century, however, a determined effort was made to turn the study of living things into a science matching the prestige of the physical sciences. It would no longer be enough just to collect and classify the diversity of species found at home and around the world. Biologists wanted to understand the detailed internal structure of the different forms of life, and they were increasingly concerned with how those structures were built up, both in the individual embryo and in the evolution of life on earth. Natural history was replaced by comparative anatomy and embryology, sometimes unified as the science of "morphology" (the study of form or structure). This science took place within the dissecting room or laboratory, using ever more complex microscopes and analytical techniques. In the drive to create a professional academic community devoted to the life sciences, the old tradition of field study found itself marginalized.

Detailed studies of the structure of living tissues initiated a major transformation of biologists' ideas about the nature of life by pointing the way toward the theory of the cell. The idea that all living structures are composed of cells, specialized for particular functions, would open new pathways to the study of how those functions operated at the chemical level. It would also transform the study of reproduction by showing how egg and sperm were united to form the basis of the developing embryo. To an increasing extent, though, the model to be followed by all these sciences was derived from experimental physiology. Physicians had always been trained in anatomy (the study of the body's structure) and had used theories about how the parts of the body functioned—a study that began to be known as "physiology" in the course of the eighteenth century. But in the nineteenth century, physiology was transformed by the application of experimental methods, providing an entirely new theoretical framework for understanding how the body worked. This study was still expected to be of use in medicine, since the more one knew about normal functions the better one could understand how things could go wrong. But where earlier physiologists had worked within the framework of medical education, now the subject became a scientific discipline in its own right, based in university science departments as well as in medical faculties (for an old-fashioned but factually detailed study of many of the biologists discussed below, see Nordenskiöld [1946]).

This transformation is normally associated with the application of experimental methods in the life sciences, including vivisection—operations performed for scientific purposes on the bodies of living animals. There had been some use of experiment in ancient medicine, and William Harvey had based his theory of the circulation of the blood partly on the use of demonstrations with live animals. But in the nineteenth century, vivisection became the normal process for attempting to understand how the body functioned. The anatomist might use dead bodies to explore structure, but function could only be investigated by interfering in a controlled way with the processes at work in the living organism. There were moral problems here that exerted considerable effect on how the science developed, but physiologists insisted that causing limited suffering to animals was essential to achieve the greater good of understanding and possibly curing human illnesses.

The laboratory now became the central location for the conducting of scientific physiology, and morphology was linked as closely as possible to this new model. Most of the early developments in this direction took place in France and Germany. When Thomas Henry Huxley and his disciples began to establish the modern discipline of "biology" in Britain during the 1870s (borrowing a term introduced at the beginning of the century), they sought to distance it from old-fashioned natural history by linking physiology and morphology as the twin foundations of a laboratory-based science (Caron 1988). Increasingly, though, it was physiology that determined what the new science should look like: mere description of dead animals was not enough to understand how living organisms actually worked. By the end of the century, many areas of the life sciences were affected by a "revolt against morphology" driven by the desire to follow physiology into the realm of experiment (Allen 1975).

The application of experimental methods led to new theories of the nature of life and of living processes that we take for granted today. Harvey's discovery of the circulation had transformed physicians' understanding of anatomy and had undermined the credibility of the medieval tradition of physiology. It did not, however, lead to an immediate displacement of medical treatments such as blood letting, which were based on the logic of the old system. In part this was because there was no new system of physiology to make sense of what the body did in the course of respiration and the absorption of food. Some important steps were made toward identifying the functions performed by different living tissues, but there was little knowledge of how those functions were effected. Efforts to create a new science of physiology were hampered by the lack of an adequate chemistry, and it is no accident that modern physiology came into being during the century following Lavoisier's "chemical revolution" and the first steps in the creation of an organic chemistry (the chemistry of complex carbon compounds, including those that make up living bodies). Lavoisier himself made a start by postulating that the body "burned" chemicals derived from food in the oxygen absorbed into the blood from the air—a proposal that would form the basis for a whole series of research programs in the nineteenth century, including many that are seen as the foundation stones of modern biology.

In addition to the impact of experimentalism, most traditional histories of physiology focus on a major theoretical debate over the nature of life. Until the seventeenth century, physicians had followed those ancient philosophers who argued that the physical body was vivified by a nonmaterial soul or vital force. The mechanical philosophy encouraged the reemergence of materialism: the claim that the living body (and by implication the human body) is nothing more than a complex material structure driven by physical forces (see chap. 2, "The Scientific Revolution"). Further development of this materialist approach was hampered by the lack of a suitable chemistry, which might effectively serve as a bridge between the behavior of atoms and molecules and the complex functions of a living body. The development of physiology in the nineteenth century saw a steady advance of materialism, although some eminent scientists stood out against the trend to reduce life to nothing more than physical processes. The elimination of "vitalism" is often presented as a key conceptual advance in the rise of the modern life sciences, but more recent histories take a less black-

and-white view. Those biologists who resisted materialism often did so for what seemed to them very legitimate reasons, and some of them did important work precisely because they were still inspired by the belief that life was something more than material activity. In the early twentieth century, eminent physiologists such as J. S. Haldane rejected a simple reductionist materialism, although they seldom sought to revive the old idea of a vital force interfering with the physical world in an almost supernatural manner. Some biologists recognized the need to see organic processes as functions of complex systems that could not be explained away by reducing them to the molecular level. This is the philosophy of organicism or holism, the belief that the whole can be more than the sum of its parts and can exhibit higher-order functions even though the operation of each part is governed solely by physical law.

This chapter will provide a selective overview of some key developments in the establishment of the modern life sciences. It will briefly highlight the rise of morphology, linking this to our studies of other sciences including evolutionism. It will then focus on the expansion of knowledge of organic tissues and cell theory. We then move to physiology and the efforts to uncover the operations of the more fundamental functions of the "animal machine," including respiration and nutrition. The roles of both the experimental method and the new materialism in defining the underlying ethos of the New Science will form themes that traverse the whole story.

The Study of Structure

The eighteenth century had seen a vast expansion in naturalists' knowledge of exotic species and a massive focus on the problem of how to classify the diversity of living things, exemplified by the work of Linnaeus (see chap. 6, "The Darwinian Revolution"). By the early nineteenth century, the project to put classification on more "scientific" grounds led Georges Cuvier and others to insist that the true nature of a species, and hence its true position in the plan of nature, could only be determined from its internal structure (Coleman 1964). Comparative anatomy became the key to a new and more technically sophisticated form of natural history. The location where the research was done was increasingly not the field, where collectors still hunted for new species, but laboratories within the great museums or university departments where the specimens sent back to the metropolis were dissected in ever more minute detail (fig. 7.1). Cuvier and his great rival Geoffroy Saint-Hilaire both worked at the Natural History Museum in Paris, while Richard Owen became the leading British morphologist from a

FIGURE 7.1 The gallery of comparative anatomy at the School of Medicine in Paris, created in 1845. This gallery was primarily a center of research where the details of the different skeletal structures could be compared, but similar collections in natural history museums were also used for public display of exotic specimens brought from different parts of the world.

base at the museum of the Royal College of Surgeons (Appel 1987; Rupke 1993). In the later part of the century, though, morphology became increasingly based in the zoology departments of universities, sometimes with overlaps to medicine (on the institutionalization of morphology in Germany, see Nyhart [1995]). Similar developments took place in botany, where the old tradition of classification was replaced by detailed studies of the structure and functions of plants.

Cuvier and his contemporaries revolutionized the science of classification by taking it out of the hands of those who actually studied nature in the wild and bringing it into the carefully controlled world of the laboratory or dissecting room. The old tradition of field study, still visible in Darwin's studies on the *Beagle* voyage, was now becoming marginalized, with a consequent loss of interest in the problems of how organisms actually live in the wild. This would only be regained through the rise of ecology at the end of the century. Darwin himself went on to spend years dissecting a vast collection of barnacles, making his name as a biologist through his publication of the first major study of this group. But even here Darwin was already out of date: he used only a simple hand lens in his study at home. By the middle decades of the century, similar work on other groups was being carried out using an ever more sophisticated array of microscopes, dissecting tools, and staining chemicals, usually within specially constructed laboratories in museums and universities.

Classification was still the main purpose of understanding the internal structure of organisms, but it now formed part of the new science of morphology, the study of form. Cuvier had insisted that to understand the structure of an animal one needed to know about the function the various organs performed, but all too often, the actual function that the structures fulfilled in the life of the organism was ignored. Later critics accused the morphologists of being more interested in dead organisms than living ones. There was an extended debate over the relative significance of form and function, with many morphologists following Geoffroy Saint-Hilaire in insisting that there were "laws of form" that determined the various possible structures independently of their actual function (Russell 1916). It was within this tradition that ideas of nonadaptive evolution flourished as alternatives to natural selection during the "eclipse of Darwinism" at the end of the century (see chap. 6, "The Darwinian Revolution"). Morphologists such as Ernst Haeckel welcomed the theory of evolution because it allowed them to insist that the relationships they were uncovering between different forms of life were real, that is, the product of genealogical descent by natural processes rather than patterns in the mind of the Creator. But they were reluctant to follow Darwin's own close studies of how animals functioned in the wild, including how they were affected by changing climates or the invasion of rival species. Instead, they were more inclined to see evolution as the unfolding of orderly patterns driven by internal biological forces (Bowler 1996).

To understand how life had evolved, the morphologists turned to the study of comparative embryology (fig. 7.2). In Haeckel's terminology, it was assumed that ontogeny (the development of the individual organism) recapitulates phylogeny (the evolutionary history of the species). Embryology had in fact made great strides in the early nineteenth century. The old preformation theory, in which the embryo simply expands from a preformed miniature in the fertilized egg, was replaced by a sophisticated model of epigenesis, in which the very simple form of the egg undergoes a complex series of transformations by which the various structures of the organism are gradually built up. In 1828 Carl Ernst von Baer, who had discovered the true mammalium ovum in the previous year, showed how individuals within each of the main groups of living organisms undergo a distinct process of differentiation by which the specialized organs that characterize the group are formed. There is no single ladder of develop-

FIGURE 7.2 Anton Dohrn working at his microscope in 1889 in the Zoological Station he founded at Naples (reproduced with the permission of the archives, Stazione Zoologica "Anton Dohrn"). Microscopic examination of "primitive" creatures and their embryological development was routinely used at this time in an effort to reconstruct the history of life on earth, and marine biological stations allowed biologists to study live specimens with the best-available equipment, such as the microscope used here by Dorhn. Significantly, however, Dohrn fell out with Haeckel over the precise structure of the tree of life, and the evidence they produced could not resolve their differences.

ment—the history of the animal kingdom is best understood as a branching tree, just as Darwin would proclaim in his theory of evolution. Significantly, though, Haeckel subverted this vision by giving the tree a single main trunk running toward the human form. But in one respect, Haeckel's synthesis of embryology and evolutionism built on the latest developments in the study of living structure at the microscopic level. He was able to trace ontogeny (and hence by implication phylogeny) from a single cell, the fertilized ovum, through a complex process of differentiation in which that cell divided and subdivided, eventually forming a spherical body cavity as the foundation from which the embryo would be built (fig. 7.3). This focus on the fertilized ovum as the foundation for development would provide the basis for later work by August Weismann and others on the process by which the chromosomes of the cell nucleus transmit the information of heredity from parent to offspring (see chap. 8, "Genetics").

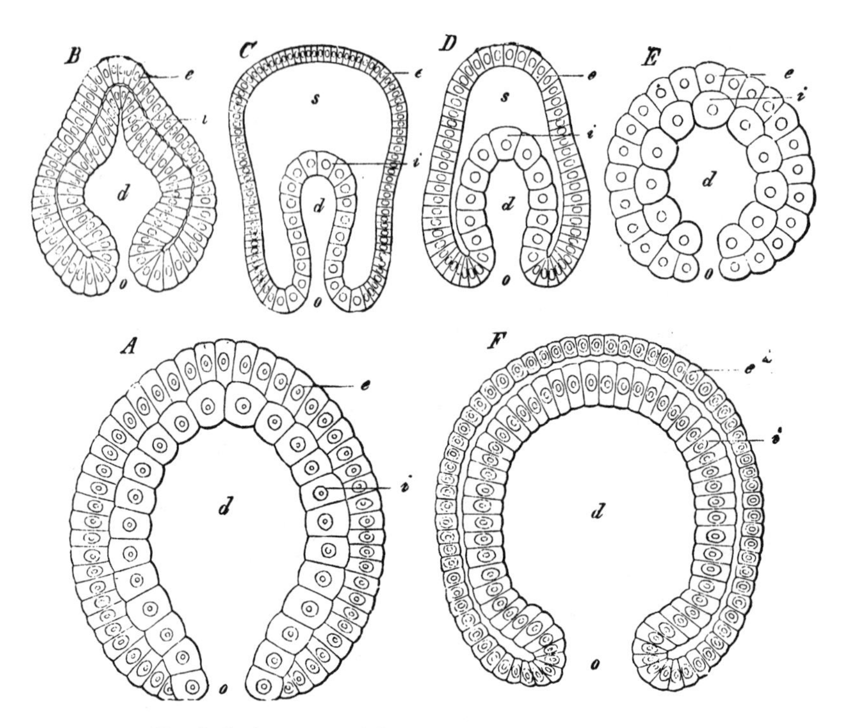

FIGURE 7.3 Haeckel's depiction of the very early "gastrula" stage of development in different organisms, from his *Evolution of Man* (London, 1879), 1:193. The bottom two are (*left*) a primitive zoophyte and (*right*) a human. Note how Haeckel shows the two layers of cells of which this stage of the embryo is built. He argued that the hollow gastrula represented an early common ancestor of the whole animal kingdom.

The idea that the cell was the fundamental unit of life, and that all larger organisms are thus composed of cells, had emerged in parallel with these developments in embryology. Cells had been observed in plant tissues by early microscopists such as Robert Hooke, but their nature and function remained a mystery until the improved microscopes of the nineteenth century allowed a more fine-grained analysis of tissue structures. In 1847 the German botanist Jakob Mathias Schleiden and the zoologist Theodor Schwann proclaimed their "cell theory," in which cells were the basic units from which all living tissues were constructed (fig. 7.4). They differed on how the cells were formed, however, Schleiden holding that new cells appeared within old ones by crystallization around a newly formed nucleus, while Schwann thought they formed from the featureless material surrounding the existing cells. At this point the theory could thus be understood in many different ways, but in 1855 another German, the embryol-

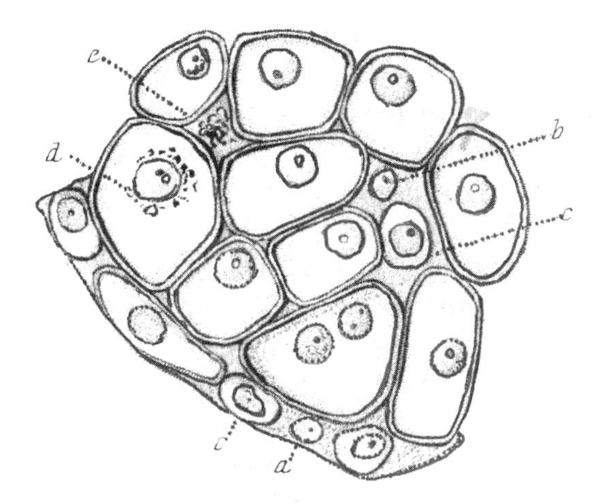

FIGURE 7.4 Microscopic study of the structure of a plant showing the cells and their nuclei, from Theodor Schwann's *Microscopical Researches* (London, 1847), facing p. 27. Schwann showed that all tissue, animal and plant, was composed of cells and argued that the cell was the basic unit of life.

ogist Robert Remak, showed that, in the early stages of growth, cells are formed by a process of division apparently initiated in the nucleus. In his *Die Cellularpathologie* of 1858, Rudolf Virchow proclaimed the final version of the cell theory: cells are the fundamental units of all life and new cells are formed only by the division of existing cells—"Omnis cellula e cellula." For Virchow, this latter point was a key factor in the defense of a vitalistic philosophy in which living things were driven by forces that somehow transcended those of the physical world. Only life could generate life, and theories of the spontaneous generation of living tissue from inorganic chemicals were necessarily false. Rejection of spontaneous generation was common among conservative thinkers, and Virchow was inclined to conservatism in both his philosophy and his politics. One historical study argues that Virchow's vision of the body as a unified and coherent assemblage of specialized cells was inspired by his preference for a political system in which all individuals find their true purpose in life within an ordered society (Ackerknecht 1953).

This vitalistic interpretation was not the only one, however. Other biologists had focused on the fluid material within the cells, largely ignoring the nucleus, whose function was little understood at the time. In the 1840s, Jan Purkinje and Hugo von Mohl defined this material as "protoplasm" and suggested that it was the basic material of life. On this model, the cell was important but only because its wall served to separate the protoplasm from the environment—it was the activity of the protoplasm itself that made life possible. Perhaps more important, this focus on the material substance of the protoplasm, rather than the ordered structure of the cell, encouraged a more materialistic view of life. If one could hold out the hope of chemistry eventually explaining the processes that the protoplasm performed to maintain life, then there was no need for special vital forces. This was the message proclaimed in T. H. Huxley's popular essay of 1868, "The Physical Basis of Life." Six years later, Huxley reinforced his essentially materialistic view in a talk titled "On the Hypothesis That Animals Are Automata, and Its History" in which he traced the continuity between nineteenth-century materialism and Descartes's original view that animals are nothing more than machines (both reprinted in Huxley 1893). At this level of debate there was a genuine interaction between the morphologists who studied how cells were assembled into larger organisms and the physiologists who were now applying the experimental method to understand the processes that maintained life.

The Functions of the Living Body

William Harvey's theory of the circulation of the blood, published in 1628, is sometimes presented as the foundation stone of modern physiology. The discovery undermined the traditional theory of how the body functioned, proposed by the Roman physician Galen, but it did not in itself explain the purpose of the blood being circulated through the lungs and then through the rest of the body. Perhaps for this reason it had little impact on the actual practice of medicine. Harvey's theory certainly prompted further research, including the microscopist Marcello Malpighi's discovery of the capillaries that joined the arteries to the veins in the muscles. But Descartes's suggestion that animals could be understood as just complex machines was incapable of forming a basis for a serious research tradition. Perhaps the heart was a pump, but what powered it and the other muscles in the body was unknown, as were the function of both digestion and respiration. The chemistry of the time had nothing to offer as a means of understanding these processes. Nevertheless, the study of physiology—the functioning of the animal and human body—came into existence as a recognizable discipline within the medical faculties of eighteenth-century universities. Most active was the Swiss biologist Albrecht von Haller, whose *First Lines in Physiology* (1747) offered an early survey. He was best known for defining the difference between those parts of the body that are irritable (contract when touched) and those that are sensible (transmit sensations through the nerves to the brain). But Haller's physiology was still only a somewhat more animated version of anatomy: it sought to establish the functions of the parts of the body more carefully, but it still offered no real explanation of how those functions operated (see chap. 19, "Science and Medicine"; for a broad survey of the history of physiology, see Hall [1969]).

Some historians would make the same point for the more sophisticated tissue doctrine of Marie-Francois-Xavier Bichat, expounded in his *Anatomie générale* of 1801. If there is a great divide separating the thought of the eighteenth and the nineteenth centuries, as Michel Foucault (1970) argues, then Bichat's efforts to classify the vital functions and associate each with the particular type of body tissue within which it was performed still fits into the eighteenth-century mold (Albury 1977). Traditionally, Bichat has been regarded as the archetypal vitalist; for him the vital functions were the sum total of the forces that resist the physical world's tendency to destroy life—which is why the body decays so rapidly after death. Each tissue had its own vital function, such as sensitivity or irritability, and the existence of these functions was a self-evident deduction from the facts of observation. The sheer variability of organic functions made it obvious that the vital forces were not governed by the mechanistic and predictable laws of the physical world. To make physiology scientific, these unique forces had to be identified, classified, and localized in the body, a technique that paralleled the eighteenth century's fascination with the classification of biological species. If this aspect of Bichat's thought is stressed, there is a clear gulf between his approach and that of the next generation, typified by François Magendie's relentlessly experimental technique of trying to understand how the functions operated. Yet Bichat was also a pioneer of vivisection and hence one of the founders of experimental physiology. Perhaps, as John E. Lesch (1984) suggests, his work had two dimensions, one relating to medicine and one to surgery. Physiology at this point was trying to establish itself within the new academic environment created by the French revolutionary government, and to some extent it sat uncomfortably among medicine, surgery, and natural science.

In another respect, Bichat was well aware of the latest developments in related areas of science. In 1777, the chemist Anton Lavoisier had suggested that his oxygen theory of combustion could be applied to explain the phenomenon of "animal heat" (Goodfield 1975). An animal's body is warm because a process equivalent to the burning of its food material is taking place in its lungs. In the 1780s Lavoisier worked with the physicist Pierre Simon Laplace, using an ice calorimeter, to show that the amount of heat generated was approximately the same in both combustion and respiration. Here was a direct application of a materialistic approach to physiology: a major vital function now seemed to be potentially explicable in purely physical terms. Bichat was well aware of this theory and supported the modification of it that supposed that the oxidation took place in the tissues of the body, not in the lungs, the blood being responsible for conveying both oxygen and food materials to the tissues. But he remained convinced that many other vital functions were not capable of being reduced to physical processes. In this sense, Lavoisier set the scene for the vitalist-mechanist debate of the following century, in which some would follow Bichat while others would argue for the eventual reduction of all vital processes to physical ones. But Bichat's own position warns us of the complexity of the issues involved: the vitalists cannot be dismissed as backward looking thinkers hoping to retain a role for a mystical or spiritual dimension in science.

This debate would be conducted primarily in the physiological laboratories of France and Germany, with Britain now lagging far behind the Continental developments. There is a long-standing assumption that early nineteenth-century German biology was deeply affected by the mystical values of the antimechanist and romantic *Naturphilosophie.* But as Lenoir (1982) insists, the influence of *Naturphilosophie* has been overstated. Much German biology is best described as teleomechanist: it assumed that the body obeys lawlike principles, but interpreted those principles as working for the goal of maintaining life. There was thus no barrier to the application of experiment on living things, on the assumption that physicochemical processes were involved. An important model for the new biology was provided by the research school established in chemistry by Justus von Liebig (Brock 1997). Liebig was appointed professor of chemistry at Giessen in 1824 and established an Institute of Chemistry there. This became a magnet for students from all over Europe, who came to imbibe Liebig's message of the importance of laboratory-based experiment for the study of organic and animal chemistry. The institute's motto was "God has ordered all His creation by weight and measure." In keeping with the new quantitative ethos in experimental philosophy, Liebig insisted on the importance of accurate measurement and analysis. He regarded biological functions as the results of chemical and physical processes going on in the body, invoking the modified form of Lavoisier's theory of respiration to explain animal heat. The aim of the quantitative program outlined in his *Animal Chemistry* of 1842 (reprint, 1964) was to examine carefully what went into the human or animal body at one end and what came out at the other, in effect seeking to use physiological processes like nutrition and respiration to explain the body's sources of energy. Liebig's belief that the degradation of proteins explained muscle activity while the oxidation of carbohydrates and fats generated only heat would soon be abandoned. His methodology was nevertheless an inspiration to later physiologists, although Liebig refused to abandon vitalist philosophy. Like Bichat, he seems to have thought that there were vital forces resisting decay. But he assumed that these forces were lawlike and worked in harmony with the laws of physics and chemistry. They were not inherently capricious, and there was no analogy with the soul or the mind. In effect, he was thinking of a vital energy that was interchangeable with other forms of energy.

One of the most influential departments promoting the new approach to biology was at Berlin under Johannes Müller. Originally influenced by the mysticism of *Naturphilosophie,* Müller turned to careful observation and experiment, working both in morphology and physiology. Some of his most important work was done on the sensory and motor nerves, building on the work of Charles Bell and François Magendie (discussed below). Müller articulated a law of specific nerve energies, which posited that however a sensory nerve is stimulated, it always gives rise only to a particular specific sensation. Despite his commitment to observation, though, Müller's early exposure to a more mystical approach ensured that he remained committed to a vitalism that was far more prescriptive than Liebig's. He was convinced that the living body is governed by a creative force that generates purposeful structures, the ensemble of different species reflecting the divine plan of the universe.

Three of Müller's students turned their backs on his vitalism and helped to found the most influential materialist school in nineteenth-century biology. They were Hermann von Helmholtz, Carl Ludwig, and Emil du Bois Reymond. There was a strong link with liberal political principles, the challenge to romanticism being seen as a challenge also to conservative ideology. It was no coincidence that the movement was founded in 1847, immediately before the year in which many European countries were convulsed by revolution. Their materialism was as much a reaction against the mysticism of *Naturphilosophie,* which they saw still at work in Müller's vitalism, as it was the outcome of demonstration through the new experimental techniques. They saw the advances being made in physics and chemistry and assumed that a program based on similar principles would have the same effect in biology. There were some important results, including du Bois Reymond's work on the electrical nature of nerve activity. Helmholtz also worked on the nerves and virtually founded the science of physiological optics, but he then moved to physics and became one of the founders of the law of the conservation of energy. In effect, the materialists viewed the animal body as a machine working in accordance with this law: there was no special vital form of energy associated only with life. This was equivalent to the program that T. H. Huxley proposed in his "Physical Basis of Life," although Huxley focused on the protoplasm within the cell as the prime locus for the crucial biochemical processes.

It has to be said that although the materialist-reductionist program played an important role in nineteenth-century debates in the philosophy of science, its implementation proved much more difficult than its early proponents imagined. At one time it was assumed that Friedrich Wöhler's synthesis of urea in 1828 drove the first nail in the coffin of vitalism. That a chemical known previously only as a byproduct of organic activity should be synthesized from material of a purely nonorganic origin must surely have convinced everyone that there was no need for a vital force. But further historical studies of the reception of Wöhler's work have shown that the synthesis was not perceived as having such far-reaching consequences at the time (Brooke 1968). The whole picture of a single classic experiment undermining the philosophy of vitalism turns out to be a myth, and vitalistic ideas continued to influence major biologists for at least another generation. Working out the details of how physiological processes functioned was not an easy task, even using experiments on living animals. It was the French experimentalists, using a less dogmatic approach to the study of living functions, who made perhaps the more substantial contributions to the founding of a scientific physiology.

The Experimental Method

Although the German school was founded on the use of systematic observation and experiment, there were some who could not bring themselves to experiment on living animals. Müller himself felt this way and later turned to comparative anatomy because he was aware that physiology could not be advanced without vivisection (Huxley remained an anatomist for the same reason). To study function, it was necessary to interfere in a controlled way with a living body and observe the results (fig. 7.5). We have already noted that in France Bichat was using vivisection from the start of the century, so his legacy can be traced as much through his contribution to experimental physiology as through his vitalism. His successor as the leading experimental physiologist of early nineteenth-century France was François Magendie, who gained a reputation as a brutal vivisectionist, indifferent to the suffering of the animals he used in his experiments. Ma-

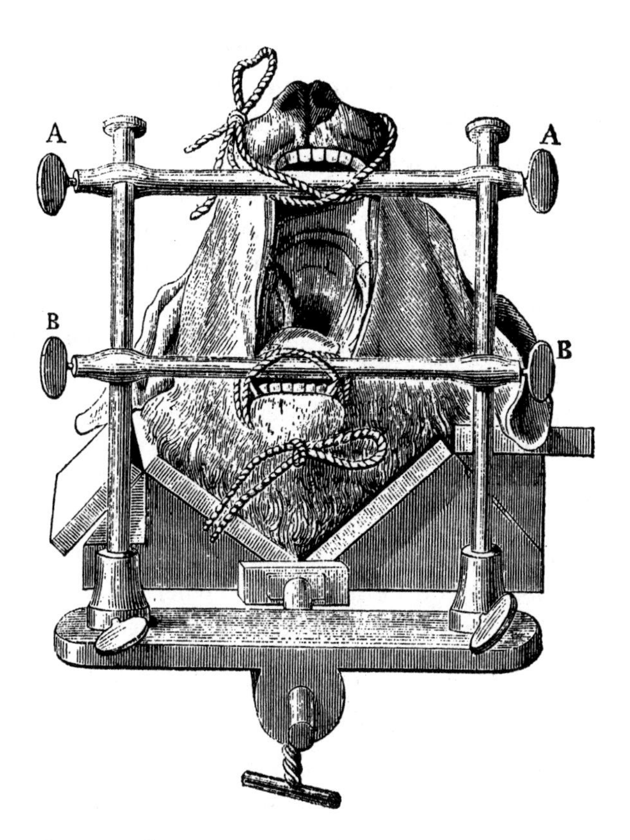

FIGURE 7.5 Apparatus for restraining the head of a dog during a vivisection experiment on the salivary glands or on the nerves of the neck, from Claude Bernard's *Leçons de physiologie opératoire* (1879), 137. Vivisection, or experimentation on live animals, was thought to be essential for understanding how the vital processes worked. But many nonscientists were outraged at the apparent indifference of the scientists to the suffering of their animal subjects, and the antivivisection movement became an early focus for popular opposition to science. This image was reproduced in an antivivisectionist pamphlet, *Light in Dark Places,* by Frances Power Cobbe, distributed in London (1883) by the Victoria Street Society for the Protection of Animals from Vivisection and the International Association for the Total Suppression of Vivisection.

gendie is remembered as the codiscoverer of the Bell-Magendie law that recognizes that the anterior (frontal) nerves running from the spinal cord govern the motion of the muscles, while the posterior (rear) nerves convey sensation to the brain. Significantly, the Scottish anatomist Sir Charles Bell hypothesized the law on the basis of a single experiment done in 1811—he would not follow up the discovery because he was reluctant to engage in further vivisection. When Magendie turned to the problem a decade later, he undertook a whole series of experiments with living animals that put the law on a firm foundation (Lesch 1984, 175–79).

Magendie's program for a scientific physiology rested on the application of experimental techniques, not on any philosophical commitment to materialism. He used experiment to develop explanation in terms of physical processes as far as possible and criticized Bichat for allowing vital forces to play an active role in his theories. Yet in the beginning of his career he seems to have accepted that there might be limits to how far the search for materialist explanations could go—perhaps the actual processes going on within the nerves might prove impossible to explain in purely physical terms. But the vital force could play no role in science so long as the physiologist was unable to postulate laws governing its behavior. This was what has been called a "vital materialism" as opposed to the rigid mechanistic materialism of the German school; it pushed materialism as far as possible without being dogmatic on the question of whether the body was governed solely by physical processes. In his later career, Magendie dismissed the vital force as a romance, a mere excuse to cover processes we do not yet understand, although he still refused to speculate explicitly about the complete elimination of such a force through future investigations. For Magendie, it was the experimental method that guaranteed that future work would be based on hard facts. Speculating about the ultimate nature of life was not part of the scientific process.

Magendie's best-known student at the Collège de France was Claude Bernard, who started as a laboratory demonstrator and went on to make a reputation for himself as a skilled and methodical experimenter. He was made professor of general physiology at the Sorbonne in 1854 and in the same year became a member of the Académie des Sciences. In 1855 he succeeded to Magendie's position at the Collège de France. Bernard's research focused on the role of the liver in maintaining blood glucose levels, on the digestive function of the pancreas, and on the action of poisons such as carbon monoxide and curare. He was admired for the simplicity of his experimental techniques and designs and for his skill at keeping his animals alive through to the end of his investigations (Holmes 1974). His *Introduction to the Study of Experimental Medicine* of 1865 (translated in 1957) became a classic statement of the role of experiment in biology.

Significantly, Bernard, like Magendie, sidestepped the mechanism-vitalism debate by focusing on the body as a system designed to maintain the *milieu interior,* or the internal environment, within which physiological functions could proceed. Even if all of those functions were purely physical in nature, it would be pointless to reduce physiology to physics because the living body was a self-regulating system that could not be accounted for in terms of those laws. In effect, the body is more than the sum of its parts; it operates as a unified whole that transcends its individual functions. This would later become known as the philosophy of holism or organicism and would form the most influential current of thought opposing mechanistic materialism in the twentieth century. The question of how such complex systems came to be constructed became a key problem for evolution theory, and it is significant that many physiologists and biochemists have remained suspicious of that theory's ability to explain the production of this degree of complexity in purely materialistic terms.

On the whole, however, physiology and the biomedical sciences tended to move ever more firmly into the mechanist camp, seeking to explain every function solely in terms of physics and chemistry. Further research continued to drive back the limits within which purely vital functions could be postulated, leaving most biologists convinced that the whole vitalist program had merely held back the development of their science. It has now become almost an article of faith that modern biology was founded on a program of explaining all physical functions in physicochemical terms. The emergence of biochemistry as an independent discipline in the early twentieth century also contributed to this process (Kohler 1982). Yet the refusal of so many early physiologists to dogmatize on the issue of materialism, and the continued efforts of later workers to defend a role for the body as an organized whole, warn us not to place too much emphasis on this philosophical debate. To a significant extent, the emergence of modern physiology rested on the application of the experimental method within an essentially pragmatic worldview that merely sought to extend natural explanations as far as possible.

Historical studies of the later developments in which mechanistic explanations became dominant have been hampered by the sheer complexity of the technical issues involved. But some important studies have driven home the point that the main driving force of theoretical innovation was not always the urge to promote reductionist materialism. Philip Pauly's study (1987) of the German-American physiologist Jacques Loeb—who achieved notoriety as an advocate of the mechanistic view of life—shows that he was an experimentalist who remained impressed by the complexity of the body's "engineering." It was Loeb's *The Mechanistic Basis of Life* of 1912 that caught the public's attention, but he also wrote *The Organism as a Whole* four year later. The eminent British physiologist J. S. Haldane, who made important advances in the study of respiration, openly repudiated mechanistic materialism, using the analogy of the dependence of the parts of the body on the whole to bolster an ideology in which the individual is subordinated to society (Sturdy 1988). In Germany, too, early twentieth-century biologists such as Hans Driesch resisted the overrigid application of mechanist principles. More generally, there was a reaction against the mechanistic view of the previous century, with a number of scientists exploiting a holistic view of nature (Harrington 1996). A detailed study by Frederick L Holmes (1991, 1993) of the process by which the biochemist Hans Krebs worked out the citric acid cycle in animal tissues (the Krebs cycle) shows that he was deeply influenced by the concept of the organism as a balanced whole. The experimentalist program has certainly helped to eliminate the concept of nonphysical forces from biology, thus realizing one aspiration of materialist philosophy. But some of its most eminent practitioners have retained the sense that the organism must be treated as a system whose structure is so complex and well-integrated that biology will never form a mere subdepartment of the physical sciences.

Institutionalizing the New Biology

Morphology had established a place for itself in the natural history museums founded in many European cities in the early nineteenth century. It gradually adapted itself to the university system but always tended to fall between the two stools of anatomy (in the medical faculties) and natural history. The shift into museums transformed natural history from a discipline devoted to the collecting and describing of species into a centrally located research enterprise where resident experts studied specimens sent to them by fieldworkers of a far lower professional standing (see chap. 14, "The Organization of Science"). It was physiology, however, that would decisively transform the educational system by helping to create the specialized and highly technical departments of what would become known as biology. In the process, natural history was marginalized—and eventually so was morphology, although to begin with, it had ridden into the new world on the coattails of the new experimentalism. But even physiology at first struggled to gain a professional locus for itself because its emphasis on a more scientific study of living processes offered both opportunities and threats to the established tradition of medical education. It was also seized on by popular writers arguing for a more materialistic perspective.

These problems were obvious in France, where even Magendie and Bernard struggled to create a professional focus for the new physiology. Magendie gained the support of both Cuvier and Laplace, but there was no section devoted to physiology in the Académie des Sciences. Both Magendie and Bernard taught at the Collège de France, and Bernard exploited links with the Société de Biologie, a group of physicians who favored the new scientific approach. It was in Germany that the rapidly expanding university system created a framework within which institutes and departments of promoting the new biology could be established. Building on the model provided by Liebig's laboratory at Geissen, Müller and others established programs that often linked physiology and morphology. One of the earliest applications of the sociological approach to the history of science was the suggestion that competition between the different German universities formed a particularly favorable environment for the establishment of new departments in fashionable subjects such as this.

Britain lagged behind, partly because physiology was associated with a more materialistic approach that seemed hostile to the academic elite's enthusiasm for natural theology. It was Darwin's bulldog, T. H. Huxley, who became the most outspoken proponent of systematic laboratory training as an integral part of medical education. As the older universities were modernized and new ones created, this program began to take effect, although it was dogged by a strong antivivisection movement based on a concern for animal rights (French 1975; see fig. 7.5). At Cambridge, Huxley's protégé Michael Foster was appointed prelector at Trinity College and in 1883 was appointed to a university chair with the resources to establish a physiology laboratory (Geison 1978). Foster's *Textbook of Physiology* (1877) played a key role in establishing laboratory-based training in medicine. Huxley introduced summer schools for high school teachers in London based on laboratory courses, with his young disciples as the demonstrators. Here morphology and physiology were presented as twin components of a truly scientific study of living things, form and function being seen as inseparable parts of what was increasingly being called "biology" (Caron 1988). In America, the rapid expansion of research-based universities in the later decades of the century created the opportunity for a similar expansion of the new biology (Rainger, Benson, and Maienschein 1988). Johns Hopkins became the model for the new breed of university within which experimental biology flourished, and its graduates fanned out across the country to found other departments.

The Revolt against Morphology

By the last decades of the nineteenth century, animal physiology had emerged as the paradigm for the new experimental biology. It was paralleled by developments in botany, as Julius Sachs and others began to focus on plant physiology, to some extent eclipsing the old focus on classification and the study of geographical distribution. William Thiselton-Dyer spread the new botany into Britain, just as Foster spread the new animal physiology. It was within this rapid expansion of experimentally based studies that what Allen (1975) has called the "revolt against morphology" took place, completing the transition to the modern framework within which the life sciences are studied. Although pioneering figures such as Müller and Huxley tried to associate a laboratory-based study of form (based on the new microscope techniques) with the experimental study of living functions, it became increasingly clear to many of the next generation that morphology was still essentially a descriptive science. It used the study of dead organisms to throw light on their evolutionary affinities, but it could offer no insights into how those structures functioned in the living body. Nor, despite the emphasis on comparative embryology, could it explain how the structures were actually created within the developing organism. More recent studies have questioned whether there was a sudden revolt or merely a gradual transformation, but the end result was the same: descriptive biology was eclipsed by the study of function (Maienschein 1991).

One consequence of this process was the rapid specialization of the life sciences into a number of distinct disciplines, which did not always communicate as well as they might because their founders were intent on carving out their own institutional framework. Embryologists abandoned the recapitulation theory as a guide to evolutionary relationships and followed Wilhelm Roux's proclamation of the need for an *Entwickelungsmechanik,* a science that sought to explain how the embryo develops in terms of physicochemical processes. This would lay the foundations of modern experimental embryology, although some of the pioneers (including Hans Driesch) found it hard to abandon the old idea that there were more purposeful directing forces involved. This work also focused attention onto the processes within the fertilized ovum that prepared the way for the development of the embryo, playing a key role in the emergence of the theory of the chromosome and hence the gene as the determinant of the future organism's characters (see chap. 8, "Genetics"). E. B. Wilson and others founded the science of cytology to focus on the processes governing life at the cellular level. At the same time, the new science of Mendelian genetics focused on the experimental study of how characters are transmitted from one generation to the next. Although T. H. Morgan's theory of the gene would unite chromosomal studies with the Mendelians' breeding experiments, genetics lost touch with embryology and paid little attention to the process by which the gene's information was expressed in the developing organism.

The experimental disciplines were, in general, hostile both to the morphological tradition and to the older form of natural history that morphology had marginalized earlier in the nineteenth century. Classification and the reconstruction of evolutionary genealogies were dismissed as old fashioned, and even the revived Darwinism based on the genetical theory of natural selection struggled to find a home within the new biology. In one important respect, however, the experimental approach reinvigorated a topic that had been studied within the older natural history tradition, leading to the emergence of the discipline of ecology. Naturalists had always been interested in the relationship between the organism and its environment, and Darwinism had kept this interest alive because adaptation was the driving force of natural selection. But now both plant and animal physiologists began to think in terms of relating the functions they studied within the body to the physical conditions of the surrounding environment, extending the experimental techniques already in use. Most influential were the plant physiologists, including Eugenius Warming in Denmark and Frederick Clements in America (see chap. 9, "Ecology and the Environmental Sciences"). Ecology remained a fragmented discipline, however, and it too remained quite distinct from many of the other specialized forms of biology that had established themselves within the early twentieth century. The drive to create a range of disciplines focused on the experimental study of different living functions thus ended up fragmenting the life sciences into a group of distinct and sometimes hostile professional groups.

Conclusions

The life sciences underwent major transformations in the course of the nineteenth century that established the field of biology in something like its modern form. Natural history was marginalized, although some field naturalists, including amateurs, continued to play a role in areas such as taxonomy and the study of geographical distribution. The emphasis switched to laboratory-based research in the great universities and museums, with the field naturalist demoted to the mere collector who transmitted new information for processing at the center. But the pressure to develop an intrusive, experimental science of life, emanating from the biomedical areas of the life sciences, allowed physiology gradually to emerge as the model for what a truly scientific biology should look like. Eventually even morphology found itself eclipsed as a purely descriptive discipline with no real explanatory power. The great museums were themselves marginalized as mere repositories of material to be described and classified, activities little better than stamp collecting as far as the experimentalists were concerned. University departments and medical schools became the focus for the most prestigious research. Topics such as evolutionism, which sought to straddle the old and the new techniques, found themselves in almost the same predicament as the old natural history. In the course of these developments, the old theory of a distinct vital force was gradually abandoned, with increasing attention focusing on the drive to work out explanations based on physics and chemistry. Not all the pioneers were dogmatic materialists, however, and many biologists remain convinced that the complex interactions that sustain life can only be understood if the organism is treated as a coordinated whole.

Expansion of the new biology had been funded by the public's ever-increasing demand for improved medical techniques, but some legacies of the new biology have now become a focus for concern. The massive specialization of research disciplines led to a fragmentation of knowledge and expertise that some biologists are struggling hard yet to overcome today. Bridges have to be built, often with great difficulty, between areas such as genetics and embryology—although any old-fashioned morphologist would have told you that it was pointless to study the transmission of characters between the generations without also taking an interest in how those characters were developed in the individual organism. Evolution theory has also had to take on board the fact that changes in the ways genes are expressed may have had profound effects on the emergence of novelties in the history of life on earth. More seriously, perhaps, the isolation of ecology from other specialized areas of biology has fragmented our response to the current environmental crisis. Even the old disciplines of taxonomy and biogeography, long neglected along with the research departments of the great museums, are being hailed as essential factors in our effort to salvage the biosphere. If we do not know how many species there are, or where they live, how can we save them? The new biology created a wealth of opportunities in the biomedical sciences that have transformed our lives through treatments based on discoveries about how the body operates. But a study of the social transformations within the scientific community that created the life sciences as we know them today reveals that specialization and the relentless urge to focus research in the laboratory have their downsides too. If biology is to have a role in dealing with the environmental crisis, as well as satisfying our demand for better medical facilities, some of the developments on which the new biology was based may have to be reconsidered.

REFERENCES

Ackerknecht, Erwin. 1953. *Rudolph Virchow: Doctor, Statesman, Anthropologist.* Madison: University of Wisconsin Press.

Albury, W. R. 1977. "Experiment and Explanation in the Physiology of Bichat and Magendie." *Studies in the History of Biology* 1:47–131.

Allen, Garland E. 1975. *Life Science in the Twentieth Century.* New York: Wiley.

Appel, Tobey A. 1987. *The Cuvier-Geoffroy Debate: French Biology in the Decades before Darwin.* Oxford: Oxford University Press.

Bernard, Claude. 1957. *An Introduction to the Study of Experimental Medicine.* New York: Dover.

Bowler, Peter J. 1996. *Life's Splendid Drama: Evolutionary Biology and the Reconstruction of Life's Ancestry, 1860–1940.* Chicago: University of Chicago Press.

Brock, William H. 1997. *Justus von Liebig: The Chemical Gatekeeper.* Cambridge: Cambridge University Press.

Brooke, John H. 1968. "Wöhler's Urea and Its Vital Force?—a Verdict from the Chemists." *Ambix* 15:84–113.

Caron, Joseph A. 1988. "'Biology' in the Life Sciences: A Historiographical Contribution." *History of Science* 26:223–68.

Coleman, William. 1964. *Georges Cuvier, Zoologist.* Cambridge, MA: Harvard University Press.

———. 1971. *Biology in the Nineteenth Century: Problems of Form, Function and Transformation.* New York: Willey.

Foucault, Michel. 1970. *The Order of Things.* New York: Pantheon Books.

French, Richard D. 1975. *Antivivisection and Medical Science in Victorian Society.* Princeton, NJ: Princeton University Press.

Geison, Gerald L. 1978. *Michael Foster and the Cambridge School of Physiology: The Scientific Enterprise in Late-Victorian Society.* Princeton, NJ: Princeton University Press.

Goodfield, G. J. 1975. *The Growth of Scientific Physiology: Physiological Method and the Mechanist-Vitalist Controversy, Illustrated by the Problems of Respiration and Animal Heat.* New York: Arno Press.

Hall, Thomas S. 1969. *History of General Physiology.* 2 vols. Chicago: University of Chicago Press.

Harrington, Anne 1996. *Re-enchanted Science: Holism in German Culture from Wilhelm II to Hitler.* Princeton, NJ: Princeton University Press.

Holmes, Frederick L. 1974. *Claude Bernard and Animal Chemistry: The Emergence of a Scientist.* Cambridge, MA: Harvard University Press.

———. 1991. *Hans Krebs: The Formation of a Scientific Life, 1900–1933.* New York: Oxford University Press.

———. 1993. *Hans Krebs: Architect of Intermediary Metabolism, 1933–1937.* New York: Oxford University Press.

Huxley, T. H. 1893. *Methods and Results.* Vol. 1 of *Collected Essays.* London: Macmillan.

Kohler, Robert E. 1982. *From Medical Chemistry to Biochemistry: The Making of a Biomedical Discipline.* Cambridge: Cambridge University Press.

Lenoir, Timothy. 1982. *The Strategy of Life: Teleology and Mechanics in Nineteenth-Century German Biology.* Dordrecht: D. Reidel.

Lesch, John E. 1984. *Science and Medicine in France: The Emergence of Experimental Physiology, 1790–1855.* Cambridge, MA: Harvard University Press.

Liebig, J. von. 1964. *Animal Chemistry: or Organic Chemistry in Its Application to Physiology and Pathology.* New York: Arno.

Maienschein, Jane. 1991. *Transforming Traditions in American Biology, 1880–1915.* Baltimore: Johns Hopkins University Press.

Nordenskiöld, Eric. 1946. *The History of Biology.* New York: Tudor Publishing.

Nyhart, Lynn K. 1995. *Biology Takes Form: Animal Morphology in the German Universities, 1800–1900.* Chicago: University of Chicago Press.

Pauly, Philip J. 1987. *Controlling Life: Jacques Loeb and the Engineering Ideal in Biology.* New York: Oxford University Press.

Rainger, Ron, Keith R. Benson, and Jane Maienschein, eds. 1988. *The American Development of Biology.* Philadelphia: University of Pennsylvania Press.

Russell, E. S. 1916. *Form and Function: A Contribution to the History of Animal Morphology.* London: John Murray.

Rupke, Nicolaas A. 1993. *Richard Owen: Victorian Naturalist.* New Haven, CT: Yale University Press.

Sturdy, Steve. 1988. "Biology as Social Theory: John Scott Haldane and Physiological Regulation." *British Journal for the History of Science* 21:315–40.

CHAPTER 8

GENETICS

THE SUCCESS OF THE HUMAN GENOME PROJECT has focused a great deal of public attention onto the prospect that a better understanding of our heredity may help eliminate many debilitating medical conditions. So great is the level of public expectation that many experts now worry about the massive oversimplification that has crept into popular understanding of the role played by heredity in individual development. People expect there to be a gene "for" every particular character, good or bad, and look to a time when "designer babies" can be produced with only the best aspects of their parents' characters. The critics are concerned that such a possibility, if widely implemented, would have dramatic and not necessarily beneficial consequences for society. They also point out that the whole program is based on a misunderstanding of how the genes work: damage to a single gene may generate a particular medical condition, but there is no single gene that will ensure a high IQ—or a predisposition to criminal behavior. And even if a genetic component could be identified for such complex characteristics, the results would depend on an interaction between the genes and the environment in which the organism develops. The expectation that every character is rigidly predetermined by heredity reflects a particular and highly controversial view of human nature that has manifested itself from time to time over the past century or more, often with consequences that many of us would find very distasteful. We risk the reemergence of a new and even more insidious form of eugenics—and history warns us how easily the ideology of genetic determinism can get out of control (see chap. 18, "Biology and Ideology").

In these circumstances it is important to understand how the modern science of genetics emerged, and how it can be misused to promote an exaggerated view of the extent to which genes determine character. To some extent, the history of genetics has itself been used to persuade us that scientific knowledge of heredity has only advanced through the discovery and exploitation of the idea that organisms have characteristics that are transmitted as whole units because they are predetermined by single genes. Everyone has heard the story of how Gregor Mendel brought clarity to an area that had been mired in confusion by spotting the unit characters that could be traced through successive generations of the peas grown in his monastery garden. It was by linking such characters to particular sections of the chromosomes in the cell nucleus that the classic notion of the gene was formulated by T. H. Morgan and his team (traditional histories of genetics include Carlson [1966]; Dunn [1965]; Sturtevant [1965]). More recently, the discovery of the double-helix structure of DNA by James Watson and Francis Crick in 1953 is seen as providing the key to how the "genetic code" operates, laying the foundation for the development of molecular biology and the high-tech biology represented by the Human Genome Project and its applications.

A closer study of the history of genetics suggests a far more complex picture (Bowler 1989; Keller 2000; Olby 1985). The pre-Mendelian "state of confusion" reflected in part the absence of conceptual distinctions that were clarified in the early twentieth century, and then only at the expense of an oversimplification of the complex relationships between the transmission of characters from parent to offspring and the development of those characters in the embryo. Mendel's status as the "precursor" or "forerunner" of twentieth-century genetics has turned out to be problematic, in part because he was probably not searching for a new theory of heredity—his famous experiments were more likely intended to throw light on the origin of new species by hybridization. The reformulation of ideas about heredity that resulted in the creation of modern genetics after the "rediscovery" of Mendel's work in 1900 reflected a complex of intellectual, professional, and cultural interests. New ideas in evolution theory and cell theory focused attention onto the possibility that characters might exist as units that bred true over the generations. But an emphasis on the hereditary determination of character was also fostered both by the need for a new way of controlling animal and plant breeding for agricultural purposes and by the emergence of a social program that insisted that some humans were predestined to be inferior by their genetic endowment. The theory of the unit gene completely uncontaminated by environmental influences was used to create an independent discipline of genetics within the scientific community—but only in the English-speaking world. In France and

Germany, genetics did not become established as a distinct field, and there was much less enthusiasm for the idea of rigid genetic determinism, at least among biologists.

For several decades in the early twentieth century, Anglo American geneticists explored the notion of the unit gene on the assumption that it corresponded to a discrete portion of the chromosome in the cell nucleus. They could investigate the behavior of the chromosomes and link this to the inherited characters, but they did not know how the genetic information was "coded" into the chemical structure of the nucleus, and they largely ignored the question of how that information was then decoded in the development of the embryo. This situation began to change with the emergence of molecular biology in the years after World War II. Eventually the chemical nature of the material responsible was identified (DNA), and in 1953 Watson and Crick proposed their inspired solution to the question of how a chemical molecule could both duplicate itself in transmission and code for the construction of proteins in the developing organism. The subsequent elaboration of molecular biology has extended our understanding of how genes operate to such an extent that the old notion of a unit gene has virtually disappeared—there are many different concepts of the gene, depending on which function is being investigated. Work has also proceeded on understanding how the information in the DNA is decoded, although we still lack a coherent program to link this with the later stages of embryological development. Critics warn that it is the failure to understand just how much is still left to be done that encourages overoptimistic assessments of the Human Genome Project's ability to revolutionize medicine and an overly simplistic view of how rigidly the genetic information predetermines adult character. These failures allow something like the old-fashioned notion of the unit gene to retain a hold on the popular imagination, which in turn encourages the reemergence of social effects resembling the eugenics program.

This chapter will analyze the key steps in the history of genetics in light of the revisionist positions mentioned above. But we begin with a survey of the pre-Mendelian period to show how it was possible for several generations of naturalists to think about the relevant issues without realizing that it would be possible to have a separate discipline dealing with the study of heredity. This was not so much a state of confusion (although some issues were confused, by later standards) as one in which it simply did not seem conceivable that one could study the transmission of characters without thinking about how those characters were developed in the embryo. Debates in embryology were used to define alternative positions on the role of preformation and environmental influence, while evolution theory was eventually used to provide an understanding of why the embryo's development followed a preordained course. Studies of how individual characters were transmitted from one generation to another were occasionally conducted within this tradition but were more likely to reflect the practical interests of animal and plant breeders trying to put together a systematic framework to throw light on phenomena they needed to control.

Preformation versus Epigenesis

The possibility that the characters of the adult organism were predetermined from the moment of conception (or even before) was formulated in the late seventeenth century in response to a crisis precipitated by the application of the "mechanical philosophy" to living things. If the organism was just a complex machine, how could it be produced by a process of development from undifferentiated matter—surely the laws of mechanics could not organize matter to produce a purposeful structure? In an age in which natural theology still reigned, there was a possible solution to the dilemma. Perhaps the laws of nature didn't need to construct order out of chaos because the structure of the organism already existed in miniature, needing only to be "filled in" with extra matter in order to exhibit the parts to the naturalist who studied the growing embryo. In the most extreme form of this "preformation theory," the embryos of successive generations of the species were stored up one within the other like a series of Russian dolls, each waiting its turn to be developed. The whole human race had been created directly by God, enclosed within the sperm of Adam or the ovary of Eve (fig. 8.1; Pinto-Correia 1997; Roe 1981; Roger 1998).

This theory has been ridiculed by later biologists, and it does indeed seem bizarre and contrary to observation. Surely, as microscope studies were already showing before 1700, the embryo grows from an undifferentiated piece of tissue by the sequential addition of parts, a process known as epigenesis. The fact that microscope studies were regularly interpreted in support of preformation—by arguing that minute rudiments could often be seen before the main development of a structure—shows how easily observation is structured by theoretical preconceptions. And yet the idea of preformation was not as silly as it sounds: indeed the very term "preformation" was used in the late nineteenth century to denote theories that presupposed that the future structure of the embryo is somehow preordained at conception by information coded into the fertilized ovum. We now

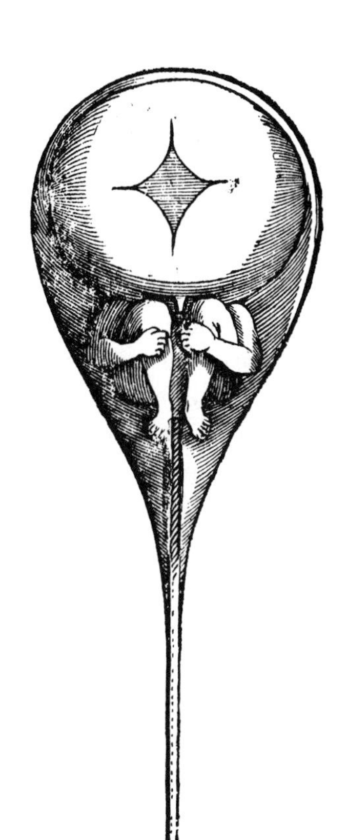

FIGURE 8.1 A male spermatozoon showing a hypothetical human figure preformed in the head, from Nicolas Hartsoeker's *Essai de dioptrique* (Paris, 1694). Hartsoeker did not claim to have observed such a figure but showed what would be expected if the whole organism did indeed preexist within the sperm. Most naturalists at the time thought that the miniature was more likely to be preformed within the female ovum, with the male semen serving as a stimulus to initiate its development (in which case the seminal fluid, not the sperm, was more likely to be the key agent in fertilization).

think of the information being inscribed in a chemical structure that is somehow "decoded" in the course of development. Such elaborations were not available to seventeenth- and eighteenth-century thinkers, so is it any surprise that they imagined actual miniatures waiting to be fleshed out? Far from being a silly theory, preformationism actually defined a crucial idea that had to be reformulated to create the classical conception of the gene.

There were, of course, problems with the preformation theory. To begin with, people argued over whether the miniatures were stored within the female ovum or the male sperm (it had to be one or the other). The ovum won out because otherwise vast number of fully formed miniature humans would be wasted in every male ejaculation. But how do we explain the transmission of characters derived from the parent who doesn't supply the miniature, such as the father's red hair transmitted to his children? An attack on preformationism by the French savant Pierre Louis de Maupertuis in 1745 included an early attempt to trace characters through a series of generations on both the male and female sides, sometimes portrayed as an anticipation of Mendel's work. The response was to argue that the male semen provides nourishment for the first steps in the ovum's growth, thus allowing some male characters to be transmitted. Maupertuis took a bolder course—like many of the more radical Enlightenment figures, he rejected the idea of a God who has designed everything. He argued that the laws of nature can indeed construct the embryo from a mixture of fluid semen provided by both parents (in this theory both the sperm and the ovum are irrelevant). But this brought him back face to face with the problem that had generated the preformation theory in the first place: How can the mere laws of mechanics control the movement of matter so precisely that they can construct an embryo out of a disorganized fluid?

Maupertuis evaded this problem by implying that matter itself has powers, such as memory and volition. Late eighteenth-century opponents of preformationism such as C. F. Wolff were openly vitalistic: to explain the gradual production of the embryonic parts by epigenesis they invoked purposeful, nonmaterial forces that imposed order on the material incorporated into the structure. By the start of the next century, preformationism was dead, and embryologists devoted themselves to the study of the gradual process by which the new organism was constructed. It was widely assumed that the pattern of development followed a more or less linear sequence or hierarchy similar to the old "chain of being." According to this theory, the human embryo was first an invertebrate, then successively a fish, a reptile, and a lower mammal before acquiring its distinctive human characters. It was still assumed that some nonphysical guiding force was in control. The situation was rendered more interesting when it became apparent that the sequence of development corresponded to the history of life on earth as revealed in the fossil record. Late nineteenth-century evolutionists such as Ernst Haeckel championed the "recapitulation theory" in which the development of the embryo (ontogeny) recapitulates the evolutionary history of its species (phylogeny). (See chap. 6, "The Darwinian Revolution," and Gould [1977].)

There was little room within this synthesis of evolutionism and embryology for the notion of rigidly predetermined characters or, indeed, for a separate study of how character differences would be inherited. The overall pattern of ontogeny was predetermined by the past history of the species, but like most recapitulationists, Haeckel accepted the Lamarckian theory of the inheritance of acquired characters. Ontogeny had to be flexible enough to allow the organism to adapt itself to changes in its environment—but Lamarckism assumed that such self-adaptations are pushed back into ontogeny so they can be inherited by future generations. Haeckel was no vitalist, but his philosophy of "monism" supposed that matter and mind are different aspects of the one underlying substance, and this allowed him to ascribe mental properties to even the most basic natural entities. For him and his followers, heredity was equivalent to memory—the developing embryo is, in effect, remembering the sequence of characters added in the course of its species' evolutionary ancestry. There was no prospect of anything resembling modern genetics emerging within such a worldview.

Haeckel actually called himself a Darwinist, although his evolutionism made little use of Darwin's theory of natural selection acting on individual variations. The selection theory did focus attention onto character differences between individuals, and it depended on the assumption that such differences are inherited. It has often been said that Darwin's theory cried out for the genetic model of inheritance, which would allow favored variations to be preserved as units for transmission to future generations. But Darwin explored a different viewpoint more in tune with the developmental model outlined above (Gayon 1998). His theory of "pangenesis," published in 1868, supposed that heredity works through the transmission, to the offspring, of minute particles or "gemmules" budded off from the various parts of the parents' bodies. He assumed that in most cases there would be a mixing of the parental gemmules for any structure, so that character differences would be blended together in the offspring. Most significant, the theory depended on the material structures responsible for heredity being formed by the parents' bodies—unlike the modern theory, there were no genetic units transmitted unchanged from one generation to the next.

Darwin himself accepted Lamarckism in addition to natural selection because changes acquired by the parents' bodies would be reflected in the gemmules they would bud off and could thus be inherited.

Mendel

The brief overview provided above provides an explanation of why Mendel's classic breeding experiments, published in 1865, fell on deaf ears: no one was thinking in terms of character units transmitted from one generation to the next. In the orthodox history of genetics, Mendel transformed the situation (at least potentially) by suggesting an entirely new model of heredity that would clear up all the confusions inherent in the earlier ideas. The problem was that it took time for the value of these insights to become apparent, so Mendel died in obscurity, leaving his model to be "rediscovered" in 1900 by the biologists who would go on to found modern genetics. The developments that made it possible to launch this new initiative form the topic of the following section, but first we must try to fit Mendel himself into the picture. Historians of science have become increasingly suspicious of precursors or forerunners who are supposed to have proposed new theories long before they were finally accepted. Given our prevailing assumption that scientific knowledge is context dependent, it seems intrinsically unlikely that an individual would be able to cut himself off from his own intellectual milieu and somehow anticipate that of a future generation. There was certainly something new in Mendel's approach, but recent historical studies suggest that the traditional image of him as a precursor of genetics was constructed in order to provide the new science with a creation myth based on a misunderstood founder. He certainly did not anticipate the whole conceptual system of early twentieth-century genetics, and in the words of one historian, Mendel was not himself a Mendelian (Olby 1979; 1985, app.).

The problem seems to have arisen because the rediscoverers read many of their own ideas into the text of Mendel's paper. They assumed that, like them, he was searching for a general law of heredity. They also seem to have assumed that his experiments did not make sense unless they were interpreted in terms of unit characters defined by some sort of material particle transmitted between the generations (the gene, as it became known). Recent historians have noted that Mendel's paper makes no mention of paired material particles—his discussion is solely in terms of character differences and offers no hypothesis about how those differences are maintained. More interesting, by looking for the context within which Mendel would have thought about the problem, we can see that he may not have been testing a law of heredity at all. The most radical reinterpretation suggests that he was really trying to substantiate an alternative to Darwin's theory of evolution—he did not anticipate that his results could be seen as the basis for a new way of thinking about heredity (Callendar 1988). This revisionist interpretation has the major advantage of rendering meaningless the whole problem of why no one understood his new "theory of heredity"—because there was no such theory.

Mendel developed his insights by hybridizing varieties of the garden pea that had distinctive characteristics and tracing those character differences through successive generations. There was an existing tradition of conducting such experiments, partly by horticulturalists seeking better control of plant breeding, but also by naturalists who were inspired by Carolus Linnaeus's suggestion in the previous century that new species might be generated by the hybridization of existing ones (Roberts 1929). Taking a renewed look at this idea would have been an obvious move for a Catholic priest who found Darwin's theory distasteful. By crossbreeding very distinctive varieties of peas, Mendel hoped to throw light on whether crosses between species might yield permanently new forms. This would explain why he was so alert to the possibility of tracing fixed characters through the hybrids and their progeny—but the real point was to establish the laws of hybridization, not the laws of heredity.

Mendel had gained a limited scientific education before he became a monk at the monastery at Brno in Moldavia where he conducted his experiments (Henig 2000; Iltis 1932; Orel 1995). He began with several varieties of the garden pea that had been artificially selected to breed true and picked out seven character differences to trace through the hybrid generations. Thus he crossed, for instance, a very tall variety with a short form and found that there was no blending: all the plants of the first hybrid generation were tall rather than of intermediate height. The short character had apparently disappeared. When he crossed the hybrids to produce the second hybrid generation he obtained his famous 3:1 ratio. The short character had reappeared, but in only a quarter of the plants; the other three-quarters were tall. This proved that the character states existed as discrete units and that one state was somehow "dominant" over the other (the recessive). The potential for the recessive character could exist within a hybrid form, but if the potential for the dominant was present too, the recessive would be completely masked in the adult organism. The experiments showed that we have to think of heredity in terms of paired character determinants, with each organism inheriting one determinant from each parent and transmitting one to each offspring. Mendel did not specify that the characters were determined by material particles transmitted from parent to offspring, and although most of the early geneticists assumed that he must have had such a situation in mind, there is no evidence that he did.

If we translate the experiments into the later terminology of the gene (which is how the paper was read after 1900), we have to assume that for a particular character such as the height of the pea plants, there are two genetic units capable of controlling that character (two "alleles"), in this case one for tall (*T*) and one for short (*S*). Each plant has a pair of genes inherited from its parents, and in the pure strains this must be *TT* for the tall plants and *SS* for the short. The first hybrid generation must derive one gene from each parent (*TS*), but here the dominant-recessive relationship takes over and only the tall gene manifests itself.

TT × *SS*
(tall) (short)
↓
TS
(tall)

The hybrids are physically identical to the tall parent, but are different genetically because each is carrying a hidden copy of the *S* gene. When the hybrids themselves are crossed, we get all four possible combinations of *T* and *S* in approximately equal numbers, and once again applying the dominant-recessive rule we get four states, of which three will produce tall plants while in one, the recessive short can again manifest itself:

TS × *TS*
↓
TT *TS* *ST* *SS*
(tall) (tall) (tall) (short)

Mendel also showed that the seven character differences he studied were all transmitted independently of each other. His posthumous followers assumed that this situation could be generalized to give a complete theory of heredity based on discrete unit characters transmitted unchanged through the generations and nonblending inheritance due to dominance.

Mendel's paper was read to his local natural history society in 1865 and published the following year (translation in Bateson 1902; Stern and Sherwood 1966). It was almost completely ignored. The one scientist who took him seriously, Carl von Nägeli, encouraged him to work on hawkweed, a plant whose complex genetics defied analysis by these techniques. Early histories of genetics tried to explain the long neglect of his paper by pointing to the obscurity of the journal in which it was published. We can now see more fundamental reasons why no one could take it seriously. Unit characters were incompatible with the whole theoretical framework within which most biologists thought about heredity and development. If Mendel himself saw the paper as a contribution to the debate on hybrid species, he would have had no interest in presenting it as the foundation for a theory of heredity. More practically, the clearly distinct character states he studied in his peas are not typical of most species, so his work would have seemed only an exception to the rule. Most characters in most species are controlled by a number of different genes that intermingle promiscuously in the population and give the superficial appearance of blending. More significant, they produce a continuous range of variation within the population, just as Darwin had observed. Human beings do not fall into the distinct categories of giants and midgets: most people are of around average height with only small numbers of tall and short individuals at either end of the range.

It would take a great leap of imagination to see that Mendel's laws could be used to make sense of the diverse phenomena of inheritance, and we must now ask what changes took place in the climate of scientific opinion between 1865 and 1900 to make the rediscovery of Mendel's work possible. There were major developments both in the understanding of the reproductive process and in evolution theory, focusing attention onto the idea of heredity as a force that predetermined adult character and on the possibility that those characters might be seen as discrete units. There was increased interest in the experimental control of phenomena such as heredity and development, especially when the recapitulation theory proved unreliable as a guide to evolution (Allen 1975). But this new emphasis within biology was in part a response to more general changes taking place in society at large. The growth of the eugenics movement focused public attention on heredity as a source of degenerate characters in the human population. Francis Galton's contributions to the debate on heredity were inspired by his belief that human character—good or bad—is predetermined at birth by heredity. The work of plant and animals breeders was also becoming crucial as agriculturalists sought for better ways of applying artificial selection to the production of useful new varieties. A niche was beginning to open up for a new science of heredity that could offer the information on which the control of both human and nonhuman populations could be based.

A number of developments were promoted by this focusing of attention onto the problem of how characters are transmitted. Cell theory dominated biology at that time (see chap. 7, "The New Biology"). In 1875 Oscar Hertwig showed that the embryo grows from the single cell of the female ovum fertilized by material from the nucleus of a single male sperm. Edouard van Beneden showed that the gametes (egg and sperm) received only a single strand of the normally paired chromosomes—these rodlike structures were so-called because they take up the color of the stains used to make specimens more visible under the microscope. Clearly the act of fertilization created a pair for the offspring composed of one from each parent (fig. 8.3 below). These discoveries would form the basis of the mechanism proposed by the early geneticists to explain the pairing of characters in Mendel's experiments. August Weismann insisted that the chromosomes were the site of what he called the "germplasm"—the material basis of heredity that somehow transmitted characters from parent to offspring. But Weismann insisted that the germplasm was isolated from the rest of the body and was thus transmitted unchanged from generation to generation. Lamarckism was impossible on this model of heredity, and there was no room for vague ideas of the embryo "remembering" its evolutionary past. Weismann did not think of the characters being predetermined as large-scale units, favoring a Darwinian model of natural selection based on minute germinal variations.

This gradualist model of evolution came under fire in the last decade of the century as biologists renewed interest in the old idea that evolution works through sudden leaps or saltations. In 1894, the British biologist William Bateson published his *Materials for the Study of Variation* in which he attacked the Darwinian theory and insisted that detailed studies of many species suggested that new characters were produced by saltations. If a flower, for instance, was transformed from a four-petaled to a five-petaled variety, the extra petal would be created not by the slow expansion of a minute rudiment but by a sudden switch in the developmental process. The Dutch botanist Hugo De Vries introduced his "mutation theory" in which evolution worked by saltations, producing new varieties or even new species instantaneously. This was supported by work on the evening primrose, although it was subsequently shown that what De Vries was observing were not genetic mutations but recombinations of characters due to hybridization. The mutation theory became widely popular at the turn of the century, and it stimulated a climate of opinion in which biologists were inclined to think that if new characters were created as units, then they might subsequently breed true as units. It is no accident that many of the founding fathers of genetics began from an interest in evolution by saltations—De Vries was one of the rediscoverers of Mendel's work, and Bateson became the leading British advocate of what he called "genetics."

Mendelism and Classical Genetics

The scene was now set for the rediscovery of Mendel's laws. In 1900, two biologists who had been conducting hybridization experiments reported the laws of transmission already noted by Mendel. One was De Vries, the other the German botanist Carl Correns (the claims of a third rediscoverer, Erich von Tschermak, are now largely rejected on the grounds that he did not really understand the laws). Mendel was soon being mentioned as a precursor who had already published the laws, and indeed it is probable that the clarity of his exposition may have helped the later workers, De Vries especially, to understand what was going on. Bateson, too, was impressed when he read Mendel's paper and soon produced the first English translation along with a powerful argument that it be taken as the basis for a new science of heredity (Bateson 1902). The willingness of everyone involved to acknowledge Mendel as the founder of the new science may have been prompted by the desire to head off a potentially acrimonious priority dispute among the rediscoverers. For Bateson, especially, the laws offered a model that would completely transform the study of heredity. Characters that did not fit this model were irrelevant—a claim that prolonged an already acrimonious dispute with the biometrical school of Darwinism under Karl Pearson, which insisted that all normal variation exhibits a continuous range (Gayon 1998; Provine 1971). Most of the early Mendelians supported the mutation theory and assumed that new characters were introduced suddenly through dramatic alterations in the Mendelian factors. Curiously, De Vries soon lost interest in Mendelism, holding that mutated characters did not necessarily follow the laws.

Bateson coined the term "genetics" in 1905 and launched it at an international congress in the following year. He attempted to promote the new science at Cambridge University, but the real interest came from the animal and plant breeders, and Bateson eventually moved to the John Innes Horti-

FIGURE 8.2 Hybrid corn cobs showing Mendelian segregation of differently colored kernels, from W. E. Castle et al., *Heredity and Eugenics* (Chicago: University of Chicago Press, 1922), 94. Many of the early studies to confirm the laws of genetics were performed with species of economic importance, in the hope that by understanding how characters were transmitted breeders would gain information on how to improve yields.

cultural Institute. His disciple R. C. Punnett got the first chair (professorship) in genetics at Cambridge in 1916. In America, the new science was also taken up with enthusiasm by agricultural interests, although here it proved easier to get it established as an academic discipline because the university system was expanding. Many of the early demonstrations of Mendelian effects were in species with commercial value (fig. 8.2). In the first few years the discipline was based on a theoretical model defined solely in terms of the way in which characters are transmitted. Neither Bateson nor Punnett was interested in the possibility that the characters might be preformed by information coded into the material structure of the chromosomes. Bateson was philosophically opposed to materialism and dismissed the chromosomal theory of the gene even after it became widely accepted a decade later. The Danish botanist Wilhelm Johannsen introduced the term "gene" and insisted that the organism's "genotype" (its genetic constitution) is the only factor relevant for working out its effect on future generations—thereby reconfirming Weismann's opposition to Lamarckism. Yet, like Bateson, Johannsen did not think of the gene as a material particle, preferring to see it as a stable energy state within the organism as a whole.

What became known as classical genetics emerged in the period 1910–15 through the efforts of the American biologist T. H. Morgan and his school to link the laws of inheritance with the behavior of chromosomes in the process of fertilization (Allen 1978). Morgan had initially rejected Mendelism, although he had attacked Darwinism in the name of mutation theory. He now became interested in the obvious parallel between the way the paired chromosomes are formed by the fusion of egg and sperm and the transmission of Mendelian characters (fig. 8.3). He focused on the fruit fly, *Drosophila,* whose chromosomes are usually large and hence could be studied more easily (Kohler 1994). He showed that the gene could best be understood as a section of the chromosome that was somehow coded to produce the corresponding character in the developing organism. He and his followers were even able to produce maps showing approximately where each gene was located on its chromosome. Their results were summed up in a book *The Mechanism of Mendelian Inheritance* (Morgan et al. 1915), which defined the classical theory of the gene.

Morgan and his school also studied the production of new genetic characters by mutation. They showed that there were occasional sudden transformations of an existing gene into something that coded for a new character, which was then transmitted unchanged to the next generation, in effect replacing the original gene. Whatever the material structure of the gene, it could obviously be changed so that it coded for something new. Mutations were produced by external forces, such as radiation, and many mutations were trivial or even harmful. But Morgan also noted that most were quite small and that their carriers seemed to breed normally with the rest of the population. This insight, along with a growing willingness to admit that many characters are influenced by more than one gene, paved the way for the eventual reconciliation between genetics and Darwinism. Mutation was the source of the random variation that Darwin postulated in every population, while Mendel's laws allowed the process of selection to work by reducing the frequency of a harmful gene and increasing the frequency of the occasional one that conferred an adaptive benefit.

Genetics was soon firmly established in the scientific communities of America and Britain and, along with it, the assumption that the chromosomal gene absolutely predetermines the character to be developed by the

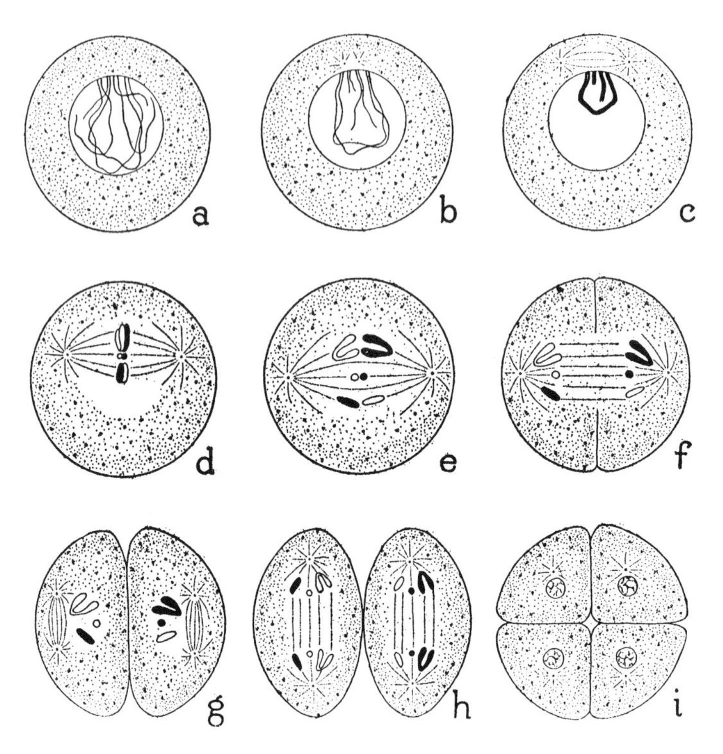

FIGURE 8.3 Behavior of the chromosomes during reduction division or meiosis, shown here in the formation of a sperm cell, from T. H. Morgan, *Evolution and Genetics* (Princeton, NJ: Princeton University Press, 1925), 80. The crucial part of the process for explaining Mendel's laws is shown in the middle row (stages *d–f*), where the chromosomes are divided and then separate into two cells, each of which contains only one of the original pair in the parent cell. The process is in fact very complex and involves a second division, ending at stage *i* with four sperm cells. This is an idealized representation of what is seen under the microscope and combines the results of many years of investigation by a number of biologists in the late nineteenth and early twentieth centuries.

organism to which it is transmitted (this is why the theory could be described as a revival of preformationism). But the situation was very different outside the English-speaking world, illustrating the extent to which even major scientific developments reflect the local context within which research is done. In France, hardly anyone took the theory seriously, and the one important geneticist there, Lucien Cuénot, was more interested in how the gene is expressed in the developing organism (Burian, Gayon, and Zallen 1988). Cuénot's work became known as physiological genetics, whereas that of the Morgan school came to be called transmission genetics. In Germany, the theory had more success, but it was not used to define a new biological discipline (Harwood 1993). Here, too, biologists were interested in physiological as well as transmission genetics, and there were many who doubted the rigid preformationism of the chromosome theory. Perhaps the surrounding material of the cell, the cytoplasm, also played a role in heredity, and this might not be so rigidly insulated from environmental effects (Sapp 1987).

These geographical differences tell us that the classical genetics of the Anglo American scientific community was not the inevitable expression of an unambiguous step forward in our understanding of nature. The chromosomal theory of the gene was immensely important, but it focused on a limited set of issues and excluded ideas and insights that would later prove to be crucial. Most obviously, the narrow focus on transmission alienated the geneticists from the biochemists and embryologists, leaving them with no handle on (and indeed no interest in) the question of how the gene is able to exert control over the growing embryo in so deterministic a way. All that mattered within the chromosomal theory of the gene was how it got from one generation to the next. This narrowing of the research program not only divided biology into rival areas, it also fueled the wider public perception that the gene was the determinant of character in the human individual. Many early geneticists supported the eugenics program and its policy of restricting the breeding of those with "unfit" genes (see chap. 18, "Biology and Ideology"). Although they soon began to realize the oversimplifications involved, they were slow to speak out against the policy in the years before the excesses of the Nazis in Germany highlighted the horrific consequences of applying it in a rigorous way. The problem was that it suited the transmission geneticists to pretend that nothing very interesting happened in the process of decoding the gene during embryological development. They were thus locked into an ideology that neglected the possibility that environmental factors might affect the way the gene is expressed and hence the character of the adult organism. To some extent we are still influenced by the conceptual blinkers this approach imposes on our way of thinking about the relationship between genes and organisms.

Molecular Biology

The weaknesses of classical genetics were made apparent by the fact that many of the key steps in working out the nature of the genetic code were inspired by research done outside its influence. Classical genetics said nothing about the nature of the genetic code—it simply assumed that somehow a section of the chromosome contained a chemical agent capable of predetermining embryological development in a certain way. Working out the nature of the code would require new ideas and new techniques and, hence, a revolution in the very foundations of genetics. Information was needed to determine how a chemical could duplicate itself so accurately that identical copies could be transmitted from one cell to another. But more important, a whole new area of research would be needed to link the biochemical processes operating in the genes to the early stages of embryological development. How did the chemical code not only copy itself but also, in different circumstances, trigger a cascade of complex chemical transformations that would influence the ways in which the cells of the embryo are formed? These were the questions that would be tackled by the new science of molecular biology that emerged in the middle decades of the twentieth century (Echols 2001; Judson 1979; Olby 1974). Historians are still debating whether the emergence of this new discipline constitutes a scientific revolution in the Kuhnian sense, or whether it is better understood as the application of a new layer of understanding, derived from studies as diverse as biochemistry and physics, to the traditional problems identified by genetics.

It had been discovered in the 1930s that viruses (which are essentially naked genes) have a structure composed of 90% protein and 10% nucleic acid. Not surprisingly, it was at first assumed that the protein carried the genetic message, and it was only in the 1940s that attention began to focus on the nucleic acid. By this time it was known that there were two types of nucleic acid, ribonucleic acid (RNA) and deoxyribonucleic acid (DNA), and new work on viruses then confirmed that it was the DNA that was the genetic messenger. The question then became: How could the structure of the DNA molecule both replicate itself and carry coded information that would trigger development in the organism? Erwin Chargaff showed that, of the four bases involved, the proportions of adenine and thiamine are the same, as are those of guanine and cytosine. X-ray diffraction studies of the molecule by Maurice Wilkins and Rosalind Franklin suggested a spiral arrangement, and it was this that allowed James Watson and Francis Crick to make their pioneering announcement in 1953 that the molecule was a double helix with the information carried in the arrangement of the bases that make up the arms of the spiral (figs. 8.4–8.6; for a highly personal account of the discovery see Watson [1968]). If adenine can only bond with thiamine, and guanine with cytosine, then when the spiral is unwound, each strand can re-create the other because the bases can only add on in a predetermined way. The genetic code can thus be copied indefinitely. Much of the early work on understanding the processes involved was done on the simplest

FIGURE 8.4 James Watson and Francis Crick in the Cavendish Laboratory at Cambridge in 1952, showing off their model of the double helix structure of DNA.

possible organisms, bacterial viruses or bacteriophages, which are in effect naked genes. The "phage group," founded by Max Delbrück, Salvador Luria, and Alfred Hershey, led the way in these early studies.

The breakthrough leading to an understanding of the genetic code still did not explain how the information carried by the sequence of bases was decoded to shape the development of cells and hence of the embryo. George Beadle and Edward Tatum proposed the "one gene–one protein" hypothesis, in which each section of DNA would somehow control the production of a single protein. Working from information theory, George Gamow argued that the bases must act in threes, or triplets, in order to specify the amino acids of which proteins are composed. Francis Crick suspected that the RNA served as an intermediate by which the information in the DNA triplets was used to manufacture the amino acids. Eventually it

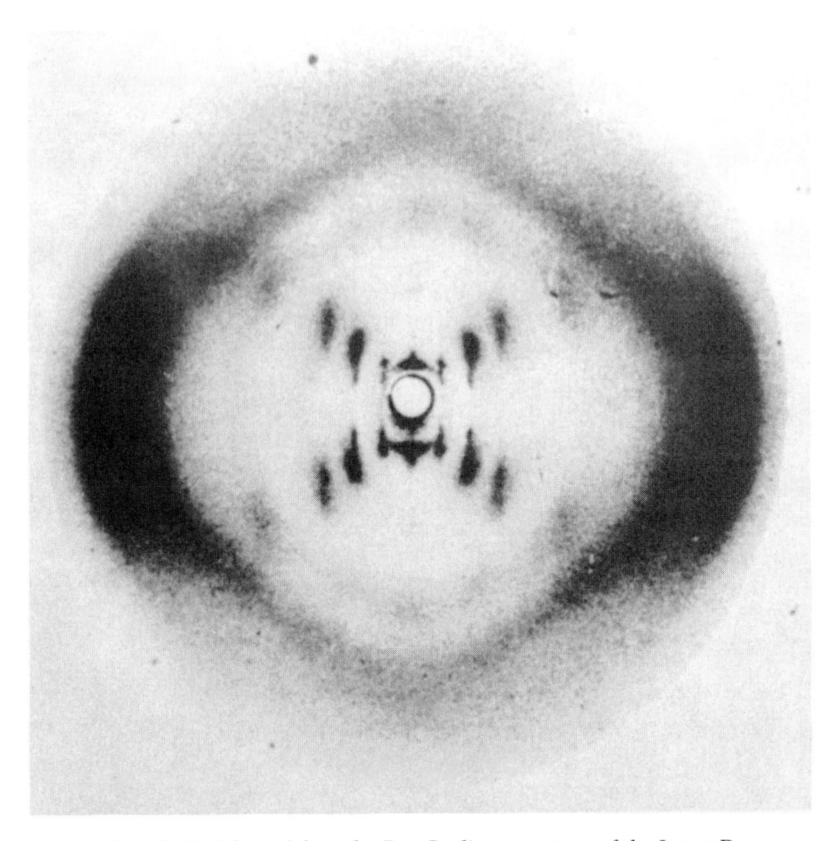

FIGURE 8.5 DNA B-form (photo by Ray Gosling, courtesy of the James D. Watson Collection, Cold Spring Harbor Laboratory Archives, Cold Spring Harbor, NY). X-ray spectroscopy photograph obtained from DNA. The substance is subjected to X-rays, which are scattered in particular ways by molecules of particular structure. The characteristic cross pattern was known to indicate a spiral structure in the molecule. It was photographs similar to this obtained by Rosalind Franklin that provided Watson and Crick with a vital clue to DNA's structure.

was shown that there are two kinds of RNA: François Jacob and Jacques Monod proposed that the soluble form acted as a "messenger" (transfer RNA) to convey the information to the insoluble (ribosomal) RNA on which the amino acids are assembled. They went on to show that the one gene–one protein model was inadequate because some genes act solely to regulate others by switching their activity on or off.

These discoveries have gone a long way toward explaining how the genetic code operates. They have established the "central dogma" of molecular biology, which is essentially a confirmation of preformationism and Weismann's claim that the germplasm cannot be affected by changes in

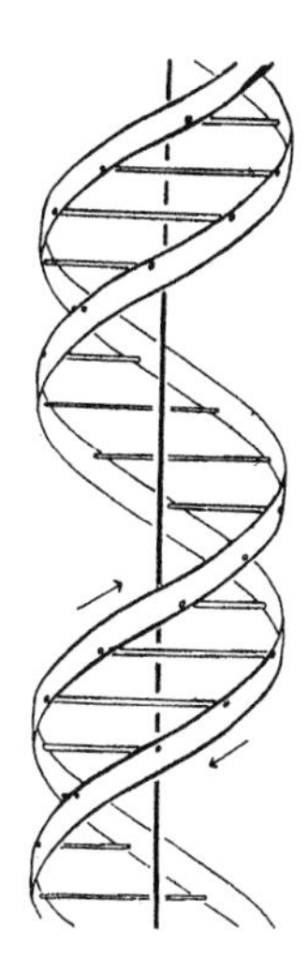

FIGURE 8.6 The spiral structure of the DNA molecule as depicted in Watson and Crick's classic paper, "Molecular Structure of Nucleic Acids," *Nature* (April 25 1953), 737. The two spiral ribbons represent phosphate sugar chains, while the horizontal rods are the base pairs that hold the molecule together.

the developing organism. DNA makes RNA, and RNA makes proteins, and there is no way in which changes to the protein makeup of cells can be transferred back into the coding in the arrangement of base pairs in the DNA. In this sense, the advent of molecular biology has refined rather than revolutionized the concepts of traditional genetics. Yet in another sense, everything has changed. Molecular biology is essentially a reductionist research program—it seeks to explain the phenomena of life (heredity and development) in terms of the behavior of chemical molecules. Some of the most successful biologists now argue that the way forward is to reduce everything to the laws of physics. Those who seek to understand the way living organisms function in the natural world, including ecologists and evolutionists, are frustrated by the molecular biologists' tendency to dismiss their work as just old-fashioned natural history. The jury is still out on the extent to which biology will continue to be dominated by the molecular approach in the twenty-first century.

Conclusions

The reductionism of molecular biology represents the most aggressive phase of a tradition that has existed since Descartes first declared that animals are only complex machines. The limitations of this approach can only be displayed by focusing on the importance of levels of analysis that it would be meaningless to express in molecular terms. Trying to describe the colonization of a territory by a newly arrived immigrant species in molecular terms is simply pointless and would loose sight of genuine problems that ecologists and evolutionists need to tackle. But there is a more serious implication that has arisen from the power that the new genetics has put into our hands. The modern project to work out the entire genome of the human species (and an increasing number of others) shows how we can now hope to specify with some completeness the whole sequence of information in the genome. It is this work, coupled with the central dogma of genetic determinism, that fuels the public's expectation that soon every character of every organism (including a human being) will be shown to be rigidly predetermined by a single gene. The genetic determinism of the age of classical genetics and the eugenics movement has been given a new lease on life by the popular interest in the medical implications of molecular biology. To be fair to the scientists, the significance of what they were doing was apparent to all from the very start of the project (Kevles and Hood 1992)

The potential dangers can be seen by recognizing that there is still a long way to go before we understand how the genome of any organism actually operates, except in a few clear-cut cases where the loss of a vital function results when a gene is damaged by mutation. Although we know in principle how the information in the gene is decoded, in practice there is still much work to be done before we can trace how complex organs and functions, which may be affected by many genes, are developed. So complex has the investigation become that the whole idea of "the gene" has become difficult to define. Many different functions are involved; some DNA seems to have more than one function, and some has no purpose whatsoever (junk DNA). Different areas of molecular biology have to work with different definitions of what constitutes a gene—yet to the layperson the gene represents an unambiguous piece of biological hardware.

More seriously, there is still much to be found out about the interaction of the genetic information with the environment within which the organism develops. Critics of the ideology of genetic determinism point out that there is little to justify the claim that every gene has an unambiguous function that will be automatically expressed in any environment. In many cases, the way in which the genetic information is expressed will depend on the circumstances provided by the environment. The developmental process of the embryo is immensely flexible and can often respond in a purposeful manner when things are disturbed by outside forces. The more we become aware of these factors, the less easy it is to have faith in the simpleminded assumption that every character has a genetic foundation. The organism is a complex whose structure is shaped by the interaction of the genes and the environment, and in this situation it is wrong to pretend that every character is predetermined. In the old debate about preformation versus epigenesis, we should not allow the apparent success of genetic determinism to obscure the fact that epigenesis still has a vital role to play. History shows that there have been several episodes when preformationism seemed to gain the upper hand, but this was always at the expense of oversimplification. To be fair, oversimplification is sometimes necessary to make a start on clarifying a complex phenomenon, and the trend toward specialization in modern science often encourages such initiatives. But the pendulum has usually had to swing the other way once the initial burst of narrowly focused exploration has run out of steam. This may well happen again as the current focus on genetic preformationism becomes bogged down in the complexities of trying to explain epigenesis.

References and Further Reading

Allen, Garland E. 1975. *Life Science in the Twentieth Century.* New York: Wiley.

———. 1978. *Thomas Hunt Morgan: The Man and His Science.* Princeton, NJ: Princeton University Press.

Bateson, William. 1894. *Materials for the Study of Variation, Treated with Especial Regard to Discontinuity in the Origin of Species.* London: Macmillan.

———. 1902. *Mendel's Principles of Heredity: A Defence.* Cambridge: Cambridge University Press.

Bowler, Peter J. 1989. *The Mendelian Revolution: The Emergence of Hereditarian Concepts in Modern Science and Society.* London: Athlone; Baltimore: Johns Hopkins University Press.

Burian, R. M., J. Gayon, and D. Zallen. 1988. "The Singular Fate of Genetics in the History of French Biology." *Journal of the History of Biology* 21:357–402.

Callendar, L. A. 1988. "Gregor Mendel—an Opponent of Descent with Modification." *History of Science* 26:41–75.

Carlson, Elof A. 1966. *The Gene: A Critical History.* Philadelphia: Saunders.

Dunn, L. C. 1965. *A Short History of Genetics.* New York: McGraw Hill.

Echols, Harrison. 2001. *Operators and Promoters: The Story of Molecular Biology and Its Creators.* Berkeley: University of California Press.

Gayon, Jean, 1998. *Darwinism's Struggle for Survival: Heredity and the Hypothesis of Natural Selection.* Cambridge: Cambridge University Press.

Gould, Stephen Jay. 1977. *Ontogeny and Phylogeny.* Cambridge, MA: Harvard University Press.

Harwood, Jonathan. 1993. *Styles of Scientific Thought: The German Genetics Community, 1900–1933.* Chicago: University of Chicago Press.

Henig, Robin Marantz. 2000. *A Monk and Two Peas: The Story of Gregor Mendel and the Discovery of Genetics.* London: Weidenfeld & Nicolson.

Iltis, Hugo. 1932. *Life of Mendel.* Reprint, New York: Hafner, 1966.

Judson, H. F. 1979. *The Eighth Day of Creation: Makers of the Revolution in Biology.* London: Jonanthan Cape.
Keller, Evelyn Fox. 2000. *The Century of the Gene.* Cambridge, MA: Harvard University Press.
Kevles, Daniel J., and Leroy Hood, eds. 1992. *The Code of Codes: Scientific and Social Issues in the Human Genome Project.* Cambridge, MA: Harvard University Press.
Kohler, Robert E. 1994. *Lords of the Fly: "Drosophila" Genetics and the Experimental Life.* Chicago: University of Chicago Press.
Morgan, T. H., A. H. Sturtevant, H. J. Muller, and C. B. Bridges. 1915. *The Mechanism of Mendelian Inheritance.* New York: Henry Holt.
Olby, Robert C. 1974. *The Path to the Double Helix.* London: Macmillan.
———. 1979. "Mendel No Mendelian?" *History of Science* 17:53–72.
———. 1985. *The Origins of Mendelism.* Rev. ed. Chicago: University of Chicago Press.
Orel, Vitezslav. 1995. *Gregor Mendel: The First Geneticist.* Oxford: Oxford University Press.
Pinto-Correia, Clara. 1997. *The Ovary of Eve: Egg and Sperm and Preformation.* Chicago: University of Chicago Press,
Provine, William B. 1971. *The Origins of Theoretical Population Genetics.* Chicago: University of Chicago Press.
Roberts, H. F. 1929. *Plant Hybridization before Mendel.* Princeton, NJ: Princeton University Press.
Roe, Shirley A. 1981. *Matter, Life, and Generation: Eighteenth-Century Embryology and the Haller-Wolff Debate.* Cambridge: Cambridge University Press.
Roger, Jacques. 1998. *The Life Sciences in Eighteenth-Century French Thought.* Edited by K. R. Benson. Translated by Robert Ellrich. Stanford, CA: Stanford University Press.
Sapp, Jan. 1987. *Beyond the Gene: Cytoplasmic Inheritance and the Struggle for Authority in Genetics.* New York: Oxford University Press.
Stern, Kurt, and E. R. Sherwood. 1966. *The Origins of Genetics: A Mendel Sourcebook.* San Francisco: W. H. Freeman.
Sturtevant, A. H. 1965. *A History of Genetics.* New York: Harper & Row.
Watson, J. D. 1968. *The Double Helix.* New York: Athenaeum.

CHAPTER 9

ECOLOGY AND ENVIRONMENTALISM

AT FIRST SIGHT IT MIGHT SEEM OBVIOUS that the two topics listed in the title above should be linked together. The environmentalist movement has sought to warn of the dangers posed by humanity's ever-more powerful efforts to exploit the world and its inhabitants through industry and intensive agriculture. It points to the increasingly common catastrophes that can be attributed to the uncontrolled exploitation of the world's resources and notes that we are now witnessing a mass extinction of geological proportions caused by the destruction of species' natural habitats. If we are not careful, the environmentalists warn, we shall wipe ourselves out by rendering the whole world uninhabitable. To make this point they sometimes call on the science of ecology, which seeks to describe and understand the relationships between organisms and their environment. Indeed the term "ecological" is often taken to mean "environmentally beneficial," as though the science went hand in hand with the social philosophy that seeks to defend the natural world (see the title of Bramwell's 1989 book, which is actually about environmentalism). Many assume that ecology is a science created by environmentalists to provide them with the information they need about the balance of nature and the ways in which disturbing influences such as human exploitation upset and ultimately destroy that balance. Such an interpretation of the origins of ecology would take it for granted that the science is based on a holistic worldview that seeks to understand how everything in nature interacts to produce a harmonious and self-sustaining whole. Ecology is the science behind James Lovelock's image of the earth as "Gaia"—a sustaining mother to all living things who will not hesitate to discipline one of her children if it gets out of line and threatens the whole.

One pioneering study by Donald Worster (1985) sought to present such a unified picture of the origins of both environmentalist thought and scientific ecology. But subsequent work has uncovered a more complex and far less coherent pattern of relationships. To a large extent, the environmentalist movement has opposed modern science as the handmaiden of industrialization, seeking its image of nature in a romantic impressionism rather than in scientific analysis. To the extent that it has had an impact on science, it has done so by encouraging a holistic methodology that openly challenges the materialistic and reductionist approach favored by the majority of scientists. There are thus some forms of scientific ecology that do draw inspiration from environmentalist concerns—but there are others that owe their origins to the reductionist viewpoint that is anathema to the romantic vision of natural harmony. Many of the first professional ecologists used physiology as a model, arguing that just as the physiologists saw the body as a machine, so they should apply a purely naturalistic methodology to studying how the body interacted with its environment. Some schools of ecology have remained resolutely materialistic, depicting natural relationships in terms more of a Darwinian struggle for existence than of harmony. Ecologists from these backgrounds are among the leading critics of Lovelock's efforts to depict nature as a purposeful whole that seeks to maintain the earth as an abode for life.

Modern historical studies force us to see ecology as a complex science with many historical roots. Indeed, it is not really a unified branch of science at all, since its various schools of thought have such different origins that they still find it hard to communicate with one another. Providing hard evidence for the environmentalist campaign is certainly not on most scientific ecologists' agenda. As in so many other areas, a historical study forces us to contextualize the rise of science, breaking down the more obvious links, such as those assumed to exist between ecology, holism, and environmentalism. Instead, we see the science emerging from a number of different research programs instituted in different places and times and for different purposes, some of them designed more to encourage the exploitation of the environment than to promote its protection. Far from originating as a unified response to a single philosophical message, ecology is a composite of many rival approaches that even today have not coalesced into a single discipline with a coherent methodology.

We begin with an overview of how science became associated with the drive to exploit the world's resources, then move on to an account of how the environmentalist movement emerged to counter this program. The second half of this chapter then outlines the emergence of scientific ecology from the late nineteenth century onward, showing how different research problems and different philosophical an ideological agendas promoted theoretical disagreement almost from the beginning.

Science and the Exploitation of Resources

From the Scientific Revolution of the seventeenth century onward, the rise of science has been linked to the hope that better knowledge of the world would allow a more effective use of natural resources. The ideology promoted by Francis Bacon stressed the use of observation and experimentation to build up practical knowledge that could be applied through improvements to industry and agriculture. The world was depicted as a passive source of raw materials to be exploited by humanity for its own benefit. Even the methodology of science stressed the dominance of humanity and the passivity of the natural world: the experimenter sought to isolate particular phenomena so they could be manipulated at will. There was no expectation that everything might interact in a way that would negate the insights gained from the study of the particular. If the whole universe was just a machine, there was no reason why humanity should not tinker with individual parts for its own benefit. Carolyn Merchant (1980) sees this attitude as characteristic of an increasingly "masculine" attitude toward nature (see chap. 21, "Science and Gender"). By the end of the eighteenth century, this attitude was already bearing fruit as the Industrial Revolution got underway, and in the course of the following century the role that science could play in promoting technological development became obvious to all (see chap. 17, "Science and Technology").

At the same time, science was increasingly involved in the effort to locate and exploit natural resources around the world (fig. 9.1). The voyages of discovery undertaken by navigators such as Captain James Cook were intended to bring back information on the plants and animals of remote regions for Europeans to study and classify, but they were also intended to locate new territories that might be colonized. Sir Joseph Banks accompanied Cook on his first voyage to the South Seas (1768–71) as a naturalist. In his later capacity as president of the Royal Society, he helped to coordinate the British navy's efforts to explore and map the world, often with a view to discovering useful natural resources (MacKay 1985). The voyage of H.M.S. *Beagle,* which provided Darwin with crucial insights, was undertaken to map the coast of South America, a region vital to British trade. In the 1870s, the British navy provided a vessel, H.M.S. *Challenger,* for the first deep-sea oceanographic expedition (fig. 9.2). Although much information

FIGURE 9.1 A European naturalist in the tropics, from Pierre Sonnerat, *Voyage à la Nouvelle Guinée* of 1776. The naturalist describes the exotic creatures brought to him by the people of the region—an idealized relationship that was seldom maintained when European traders and colonists began to exploit the resources of these distant lands.

FIGURE 9.2 Deep-sea dredging equipment carried by H.M.S. *Challenger* on her pioneering oceanographic voyage from 1872 to 1876, from *Report of the Scientific Results of the Voyage of H.M.S. Challenger: Zoology* (London, 1880), 1:9. *Challenger* was equipped as a specialist survey vessel with on-board laboratories. The expedition scientists discovered a wealth of new marine species and disproved the widely held theory that the depths of the ocean were devoid of life. They also discovered manganese nodules on the deep-sea bed that are now seen as a potential source of minerals.

of scientific interest was generated, funding for marine science was increasingly provided in the expectation that there would be benefits for navigation, fisheries, and other practical concerns.

On land, too, there were many expeditions designed to explore remote regions to satisfy curiosity about the world (see below), but there were also explicit signs of science's growing involvement with imperialism. Many European nations established botanical gardens at home and in their colonies with the deliberate intention of identifying commercially useful plant species and studying how foreign species could be imported as new cash crops. Kew Gardens in London was the center of the British effort, under the direction of botanists such as Joseph Dalton Hooker, a leading supporter of Darwin (Brockway 1979). The cinchona plant, source of the antimalarial drug quinine and hence vital to European efforts to colonize the tropics, was transported via Kew from its home in South America to found commercial plantations in India. The rubber plant was smuggled out of Brazil despite a government prohibition to create the worldwide rubber-production industry. North America was transformed as European farming methods were adapted to its wide range of different environments. By the early twentieth century, the Bureau of Biological Survey under C. Hart Merriam was coordinating deliberate attempts to eradicate native "pests" such as the prairie dog that destroyed the farmers' crops. Europeans and Americans were now interfering on an unprecedented scale with natural ecosystems, destroying native habitats and importing alien species as cash crops (for a survey of these developments, see Bowler [1992]).

The Rise of Environmentalism

These developments were not without their critics, and gradually an articulate movement evolved to criticize the unrestricted exploitation—and often the consequent destruction—of the natural environment (McCormick 1989). The Romantic thinkers of the early nineteenth century celebrated the wilderness as a source of spiritual renewal and hated the industrialists who destroyed it for profit. Significantly, writers such as William Blake saw mechanistic science as a key component of the unrestrained exploitation of the natural world. A later generation of writers such as Henry Thoreau also celebrated the recuperative value of wilderness for a humanity increasingly alienated by an urban and industrialized lifestyle. In 1864, the American diplomat George Perkins Marsh wrote his *Man and Nature* to protest against the destruction of the natural environment. He warned that, contrary to early optimistic expectations, there was a degree of human destructiveness that nature might never be able to repair: "The earth is fast becoming an unfit home for its noblest inhabitant, and another era of equal human crime and human improvidence . . . would reduce it to such a condition of impoverished productiveness, of shattered surface, of climatic excess, as to threaten the depravation, barbarism and perhaps even extinction of the species" (Marsh 1965, 43). Marsh was not calling for a halt to all human interference but for better management that would allow the earth to retain its self-sustaining capacities. Partly as a result of his efforts, the U.S. government set up the Forestry Commission to manage the nation's resources, and eventually areas of woodland were set aside to be protected from logging. Public concern also led to the designation of areas of outstanding natural beauty as national parks, Yosemite Valley in California in 1864 and Yellowstone in Wyoming in 1872. The Sierra Club, founded in 1892 by John Muir, was dedicated to the protection of wilderness areas. In Europe, where there was little true wilderness left to protect, efforts were nevertheless made to create nature reserves where stable environments that had existed for centuries could be conserved (on nature reserves in Britain, see Sheal [1976]).

There was considerable tension between those who called for a more careful management of nature in order to allow resources to be renewed and an increasingly vocal movement that depicted all human interference as evil and potentially damaging to the earth as a whole. The former group was willing to call in science, in the form of the newly developed ecology, to help better understand the ways in which natural ecosystems would respond to human interference. But a more extreme form of environmentalism developed out of an alternative, more romantic vision of nature that, if it had any use for science at all, insisted that it must be a science based on holistic rather than mechanistic principles. This movement cut across all traditional political divisions and was by no means always sympathetic to a democratic approach to government. After all, the common people may well vote for more industrialization out of a short-sighted desire for more material goods. In Germany, a "religion of nature" often linked to the philosophy of the evolutionist Ernst Haeckel, became part of Nazi ideology—and the Nazis created nature reserves on ground cleared of Jews and Poles sent to the death camps. Soviet Russia had a strong environmentalist policy until Stalin's drive for industrialization led to unrestricted exploitation of the country's resources (on European environmentalism, see Bramwell [1989]).

In America, there were debates between those who saw the "dust bowl" on the Great Plains in the 1930s as part of a natural climatic cycle and those who insisted that it was a consequence of the unsuitability of the prairies for farming. The latter position was increasingly typical of the more active environmentalist movement, which allied itself with those who saw the preservation of wilderness as essential for human psychological health, to say nothing of the health of the planet as a whole. In America, Aldo Leopold's *Sand County Almanac,* published posthumously in 1949, recorded the transition of a Wisconsin game manager into an environmentalist with an emotional and aesthetic attachment to wilderness. For Leopold, scientific ecology was not enough because it needed to be supplemented by an ethical commitment that recognized that all species have a right to exist, a right that should not be compromised by human expediency: "Conservation is getting nowhere because it is incompatible with our Abrahamic concept of land. We abuse land because we regard it as a commodity belonging to us. When we see land as a community to which we belong, we may begin to use it with love and respect. There is no other way for land to survive the impact of mechanized man, nor for us to reap from it the esthetic harvest it is capable, under science, of contributing to culture" (Leopold 1966, x). Leopold's environmentalism did not rule out a role for the scientific study of nature, but that had to take place within a framework in which humanity was part of nature, not dominant over it.

Such an attitude has grown in influence, as more people have become aware of the dangers of the unrestricted exploitation of the environment. Rachel Carson's *Silent Spring* of 1962 highlighted the damage done to many species by the use of insecticides. Numerous environmental catastrophes have driven home the same message, although there are still significant differences between the ways in which different communities have responded. In America, despite the activities of those who cherish the wilderness, the public seems content to let corporate agriculture manipulate nature in the interests of producing cheaper food. In Europe, by contrast, the use of chemical fertilizers and insecticides has become unpopular, while genetic manipulation of food crops is restricted. In the Third World, however, genetic engineering is seen as perhaps the lesser of the two evils, since it might increase yields without leaving farmers dependent on expensive and potentially dangerous chemicals.

The Origins of Ecology

A distinct science of ecology only began to emerge at the end of the nineteenth century, although concepts we associate with the discipline had long been recognized. The Swedish naturalist Linnaeus wrote of the "balance of nature" in the mid-eighteenth century, noting that if one species increased its numbers due to favorable conditions, its predators would also increase and tend to restore the equilibrium. For Linnaeus, this was all part of God's plan of creation, and the natural theologians routinely described the adaptation of species to their physical and biological environment as an illustration of divine benevolence.

Systematic study of such relationships was also part of Alexander von Humboldt's project for a coordinated science of the natural world, which focused especially on the geographical factors that shaped different environments. Humboldt was impressed by the Romantic movement popular in the arts around 1800, with its emphasis on the ability of wilderness to inspire human emotions, but he insisted that a serious study of the natural world must use the scientific techniques of measurement and rational coordination. His aim was a science that focused on material interactions but interpreted them as parts of a coordinated whole in which each natural phenomenon was interlinked with all the others. He spent the years 1799–1804 exploring South and Central America, taking numerous scientific measurements in a variety of environments that were used to throw light on the interactions between their geological structure, physical conditions, and biological inhabitants. Humboldt made important contributions to geology—he was a follower of A. G. Werner and named the Jurassic system of rocks after the Jura Mountains of Switzerland (see chap. 5, "The Age of the Earth"). He also produced maps showing the variations of temperature and other climatic factors on a worldwide scale and others showing cross sections of mountainous regions illustrating how the characteristic vegetation changed with altitude (fig. 9.3). Humboldt's accounts of his South American voyage inspired many European scientists, including Darwin, and his emphasis on the earth as an integrated whole encouraged a whole generation to undertake systematic surveys of a variety of physical and biological phenomena. Under the influence of "Humboldtian science" biologists were taught to think in what we would now call ecological terms, looking for the ways in which the distribution of animals and plants was determined by the character of the soil and underlying rocks, the local climate, and the other native inhabitants of the region.

In the next generation, Darwinism, too, stressed the adaptation of the species to its environment but encouraged a more materialistic view of each population in competition not just with its predators but also with rivals seeking to exploit the same resources (see chap. 6, "The Darwinian Revolution"). Darwin also focused attention on biogeography, which illustrated how species adapted to new environments. It was the German Dar-

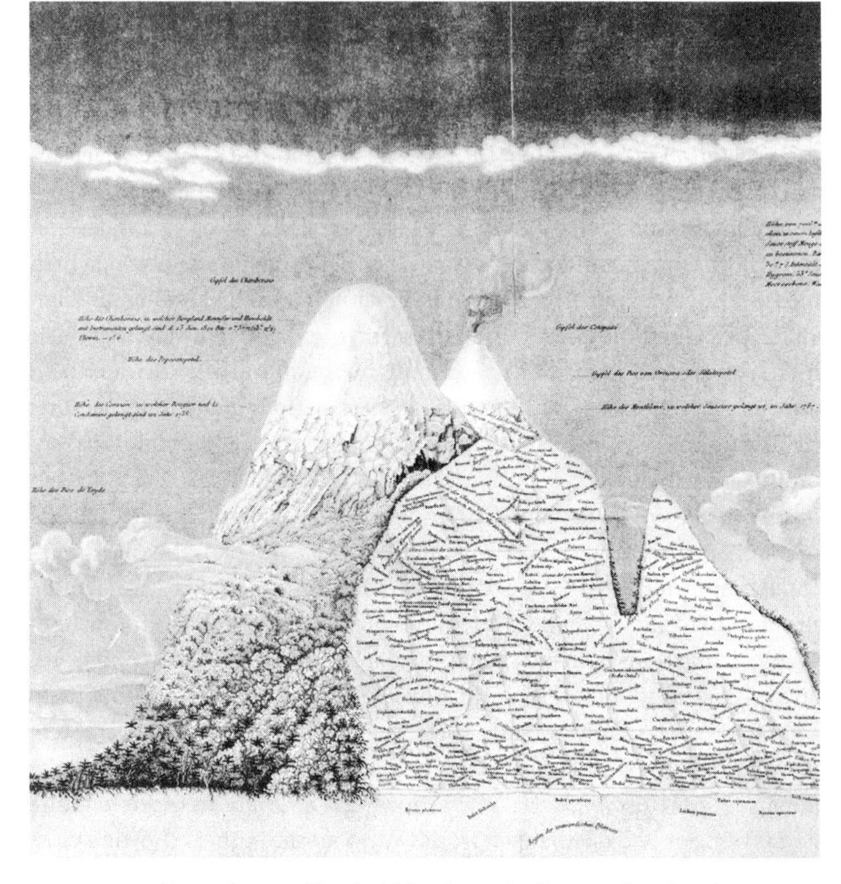

FIGURE 9.3 Alexander von Humboldt's schematic diagram showing the zones of vegetation at different levels on the South American mountain Chimborazo, from his *Essai sur la geographie des plantes* (1805). Humboldt's work helped to lay the foundations of ecology by showing how variations in the physical environment correlated with different forms of animals and plants.

winist Ernst Haeckel who coined the term "oecology" in 1866 from the Greek *oikos,* referring to the operations of the family household—the ecology of a region showed how the species there interacted to exploit its natural resources. But unlike Darwin, Haeckel adopted a nonmaterialistic view of nature in which living things were active agents within a unified and progressive world. The tension between the materialistic and holistic worldviews ensured that the science of ecology would be driven by theoretical disagreements from its inception. There were a number of different research programs, each trying to tackle the complex relationships between species and their environment in a different way. Because they began from different origins, they often adopted different theoretical outlooks.

The stimulus for the creation of the new biological discipline that would adopt the name ecology came from the breakdown of the descriptive or morphological approach to nature at the end of the nineteenth century. At that juncture, the emphasis was on experimentation, with physiology as the model, and a number of new biological disciplines arose in response to this challenge, including genetics. It was much harder to apply the experimental method to the study of how species relate to their environment, but there were several avenues that pointed the way to a more scientific approach to this topic. One was the increasing refinement of Humboldt's biogeographical techniques. In America, C. Hart Merriam of the Bureau of Biological Survey developed detailed maps showing the various "life zones" or habitats stretching from east to west across the continent. In 1896, Oscar Drude of the Dresden botanical garden published a fine-grained plant geography of Germany that showed how local factors such as rivers and hills shaped the vegetation of each region.

Plant physiology provided the model for other pioneers of plant ecology. Experimental studies had produced a much better understanding of how the internal functions of a plant operate, but by the end of the century a number of botanists began to realize that it would also be necessary to look at how the plant's physical environment affected these functions. This insight was especially obvious to those who worked in botanical gardens established in the tropics and other extreme environments, where the role of adaptation was crucial (Cittadino 1991). The founder of plant ecology, botanist Eugenius Warming, was trained in plant physiology in Denmark and had worked for a time in Brazil. He developed his approach as an alternative both to pure physiology and to the traditional focus of most botanists on classification (Coleman 1986). His *Plantesamfund,* published in 1895, was translated into German the following year and into English as *Oecology of Plants* in 1909. Warming could see how the physical conditions of an area determined which plants could live there, but he also realized that there was a network of interactions between the plants that were characteristic of a particular environment. These typical plants formed a natural community, each dependent in various ways on the others. The concept of a natural community had already been described by naturalists such as Stephen A. Forbes of Illinois, whose 1887 address to the Peoria Scientific Association, "The Lake as a Microcosm," had stressed that all the species inhabiting a lake were dependent on one another. It was a concept that was all too easily taken up by the opponents of materialism to argue that the community formed a kind of superorganism with a life and purpose of its own. But Warming resolutely opposed this almost mystical view of the community; for him the relationships were just a natural consequence of evolution adapting species to the biological as well as the physical environment. He acknowledged that all the species were competing with each other in a constant struggle for existence and that when the original community was disturbed (as by human interference) there was no guarantee that the original collection of species would reestablish itself. If we cut down a forest, the trees may never get a chance to grow again because the soil has been modified in a way that prevents them from reseeding themselves. This view was also characteristic of one of the first American schools of ecology founded at the University of Chicago by Henry C. Cowles.

There was another American research tradition, however, that developed around a very different viewpoint. At the state university in Nebraska, Frederic E. Clements sought to put the study of grassland ecology on a more scientific footing (Tobey 1981). The European techniques were not suited to the vast uniform areas of the prairies, and Clements realized that in these conditions the only way to get really accurate information about the plant population was literally to count every single plant growing in a series of sample areas. He marked out measured squares or quadrats spread over a wide region and compounded the information to give a much more precise assessment of the overall population (fig. 9.4). By clearing quadrats of all vegetation, he was able to see how the natural plant community reestablished itself and became convinced that in these circumstances there was a definite sequence by which the natural or "climax" population was built up. Clements's *Research Methods in Ecology* (1905) publicized the new techniques, and the school of grassland ecology established itself, especially in institutions dealing with the practical problems of the farmers whose activities inevitably destroyed the natural climax grassland of the prairies. Clements was an influential writer and he promoted a philosophy of ecology that was very different from the materialistic approach of Warming and Cowles. He saw the natural climax population of a region in almost mystical terms: nature was predestined to move toward this community whenever it was disturbed, and the community had a reality of its own that required it to be seen as something more than a collection of competing species. Here was an ecology that seemed to derive from the romantic image of nature as a purposeful whole that resisted human interference, yet it was being used to give advice to the farmers whose activities had destroyed the natural environment of the plains.

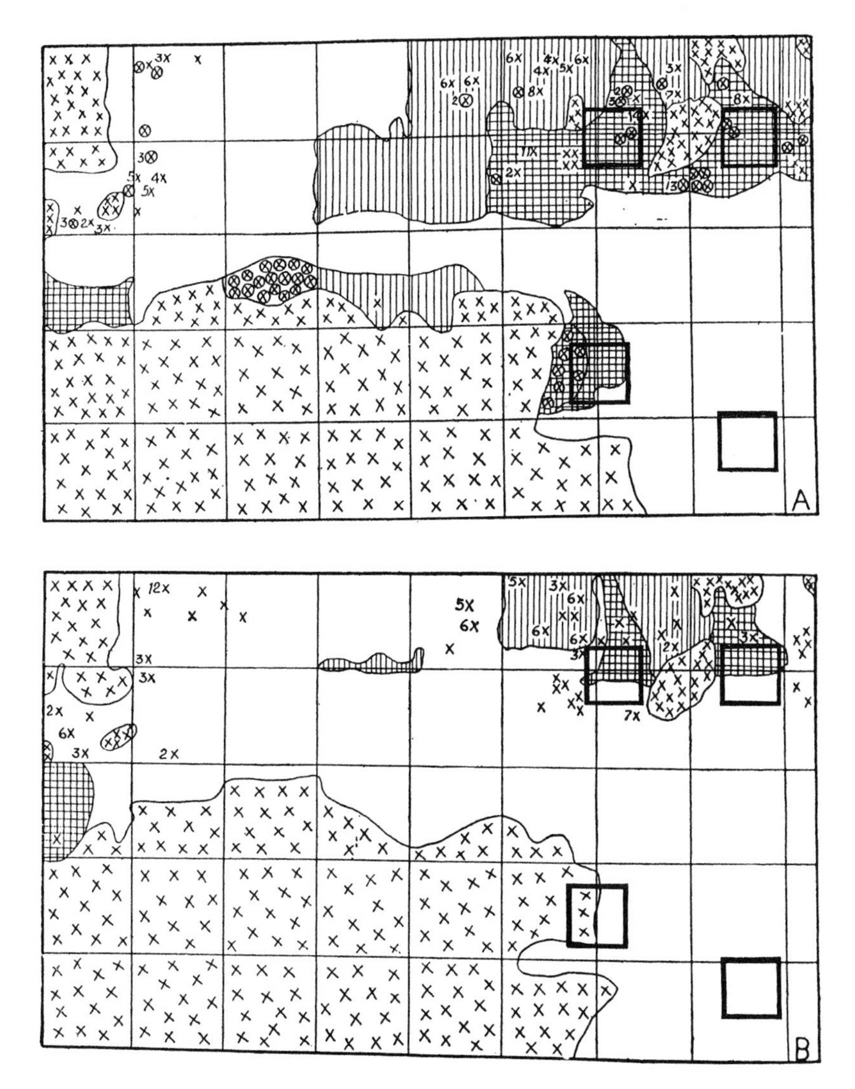

FIGURE 9.4 Typical plant ecology survey, from John E. Weaver and Frederic E. Clements, *Plant Ecology* (New York: McGraw-Hill, 1929), 41. An area of overgrazed pasture at Lincoln, Nebraska, has been marked off in five-foot squares and the position of different types of vegetation noted: individual wolfberry shrubs are marked with an *X*, areas covered by bluegrass with *vertical hatch,* buffalo grass with *cross hatch,* and wheat grass left *blank*. The upper survey was taken in 1924, the lower in 1926, showing an expansion of the shrubs and a decrease in the area covered by bluegrass and buffalo grass. The small *squares* indicated in bold are quadrats marked off for a more detailed survey in which each individual plant would be counted.

Consolidation and Conflict

In the early decades of the twentieth century, the rival approaches to ecology pioneered by Warming and Clements gained enough attention for the area as a whole to become recognized as an important branch of science. But new developments continued the original tensions, and there was competition among the different research schools for control of its journals and societies and for access to government and university departments where it might flourish. In fact, despite a promising start, expansion was slow until after World War II. The British Ecological Society was the first ecological society to be founded, in 1913 (Sheal 1987), followed two years later by the Ecological Society of America (whose journal, *Ecology,* first appeared in 1920). But the new discipline's bid to establish itself in academic departments was slow, except in America, and even here the membership of the Ecological Society remained static through the interwar years. In Britain, pioneer ecologists such as Arthur G. Tansley had to struggle for academic recognition; Tansley spent some time as a Freudian psychologist and blamed the slow growth of ecology in part on the loss of promising young scientists in World War I.

In America, Clements's school of grassland ecology continued to flourish into the 1930s, when it provided support for the claim that the prairies should be returned to their natural climax of grassland to recover from the erosion of the Dust Bowl. The idealist notion of the climax community as a superorganism with a life of its own was linked by his student John Phillips to the holistic philosophy being popularized by the South African statesman Jan Christiaan Smuts, whose *Holism and Evolution* appeared in 1926. Smuts made an emotional appeal to a vision of nature as a creative process with inbuilt spiritual values and depicted evolution as a process designed to bring about complex entities whose properties were of a higher level than anything visible in their individual parts. In Britain, Tansley had to compete with South African ecologists wedded to Smuts's philosophy who were threatening to dominate ecology throughout the British Empire (Anker 2001).

Although Clements and his supporters tried to explain the Dust Bowl, the fact that the soil had disappeared effectively undermined their claim that the natural climax vegetation could reestablish itself. Other schools of ecology developed, especially in university departments that did not have to deal with the problems of the prairie farmers. Henry Allan Gleason and James C. Malin both challenged Clements's ideas by arguing that changes could take place in the vegetation of a region due to fluctuations in the climate and the natural invasion of species from other regions. In Britain, Tansley—who eventually gained a chair at Oxford—argued strenuously against Phillips's use of the superorganism concept, openly dismissing it as little more than mysticism. Yet Tansley used research methods very similar to those of the Clements's school, and it was he who coined the term "ecosystem" in 1935 to denote the system of interactions holding the species of a particular area together. For any European biologist, it seemed obvious that most apparently "natural" communities were to some extent the product of human activity, perhaps extended over centuries, so there was little point in trying to claim that a particular ecosystem had some sort of prior claim to be recognized as the only one appropriate for a certain area. Tansley and other critics also worried that promoting the idea of a superorganism would play into the hands of mystics who wanted to block any scientific study of the natural world. In continental Europe, an entirely different form of ecology based on the precise classification of all the plants in an area was developed, and in this the notion of a superorganism was simply irrelevant.

A clear indication of the fragmentary origins of ecology can be seen in the fact that it was not until the 1920s that systematic study of animal ecology began. But here, too, the tensions between the materialistic and holistic viewpoints immediately asserted themselves. At the University of Chicago, Victor E. Shelford applied Clements's approach to the study of animal communities and their dependence on the local vegetation. Also at Chicago, Warder Clyde Allee began to study animal communities on the assumption that cooperation between the members of the population is an integral part of how a species deals with its environment (fig. 9.5). Allee dismissed the Darwinian view of individual competition as the driving force of behavior and of evolution—he explicitly rejected the notion of a "pecking order" determining individuals' rank within the group. For him, evolution promoted cooperation, not competition, a view closely allied with the holistic philosophy characteristic of Clements's group. Allee and his followers also developed the political implications of their vision of natural relationships as an alternative to the "social Darwinism" that presented individual competition as natural and inevitable (Mitman 1992).

A very different approach was developed in Britain by Charles Elton, who worked at the Bureau of Animal Populations at Oxford from 1932 (Crowcroft 1991). His book *Animal Ecology* (1927) established itself as a textbook for the field and popularized the term "niche" to denote the particular way in which a species interacted with its environment. Elton had worked with the records of the Hudson's Bay Company that gave details of

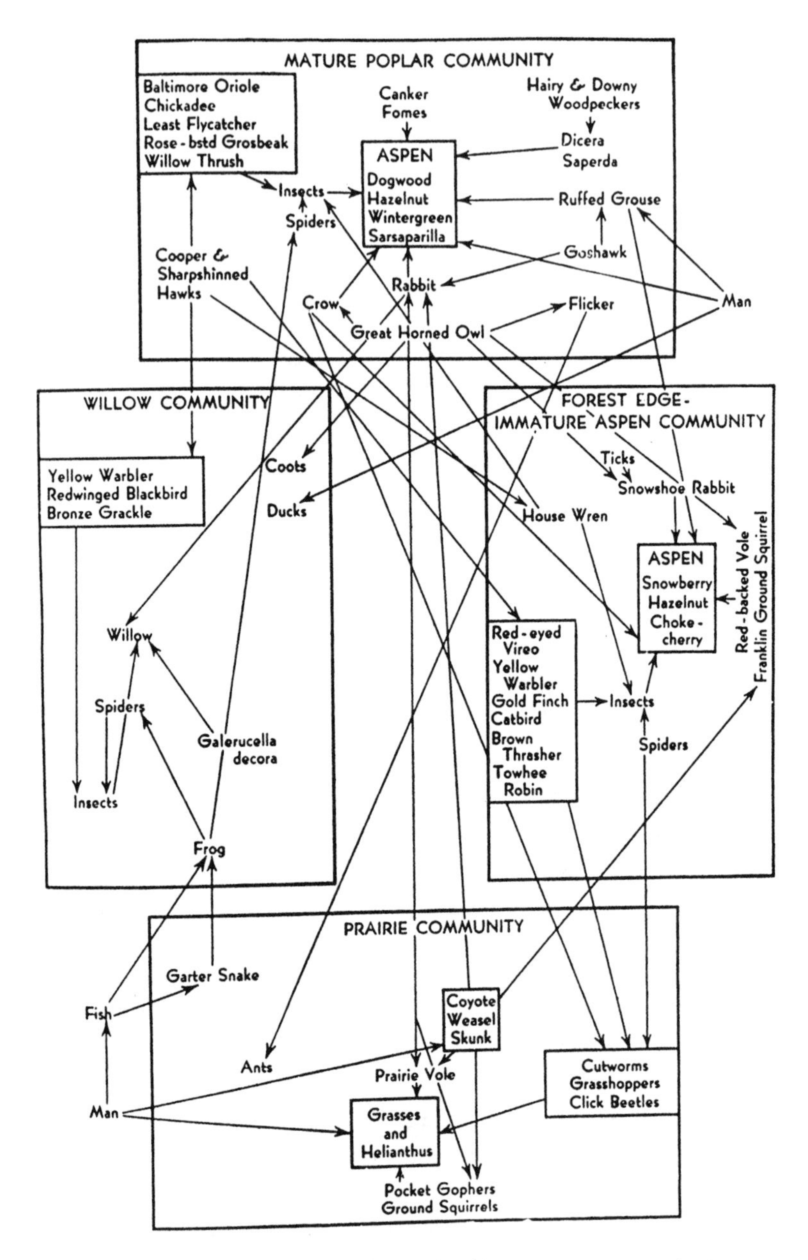

FIGURE 9.5 Scheme of ecological relationships between species in the aspen parkland of Canada, from W. C. Allee et al., *Principles of Animal Ecology* (Philadelphia: W. B. Saunders, 1949), 513. Allee and his colleagues in the Chicago school of ecology stressed the harmonious interactions between individuals and species in order to minimize the role of the struggle for existence in both nature and human society. Their textbook was colloquially known as the “great AEPPS book” after the initials of its authors’ names (Allee, A. E. Emerson, Orlando Park, Thomas Park, and K. P. Schmidt).

fluctuations in the numbers of fur-bearing animals trapped over many years. These revealed occasional massive increases in numbers (plagues of lemmings are the classic example) caused when rapidly reproducing species outstrip their natural predators in a time of plentiful resources. The occurrence of such episodes made nonsense out of the old idea of a "balance of nature" and confirmed Darwin's Malthusian image of populations constantly tending to expand to the limit of the available resources.

Elton made common cause with Tansley and with the young Julian Huxley to promote their vision of ecology, which Huxley was also concerned to link with the new Darwinism emerging in evolution theory. By denying the existence of a natural ecosystem characteristic of any environment, their approach made it easier to see the natural world as something that could be adjusted to human activity through scientific planning. Such a vision had clear social implications and was popularized in the science fiction novels being written by H. G. Wells (who also collaborated with Huxley on a major popular work, *The Science of Life,* in 1931). At this point, however, they did not envisage ecology as a subject that could be analyzed using mathematical models, partly because the rapid fluctuations in population density observed by Elton seemed unpredictable. But others were becoming more interested in the possibility of using mathematics, perhaps by seeing an analogy between the behavior of individual molecules in a gas and of individual animals interacting with their environment. The American physical chemist Alfred J. Lotka published a book on this topic in 1925, and this approach was subsequently taken up by the Italian mathematical physicist Vico Volterra, who had become interested in predicting the fluctuations in commercial fish populations. In the late 1930s, the Russian biologist G. F. Gause performed experiments on protozoa to test the "Lotka-Volterra equations," and his efforts to substantiate the mathematical techniques would play a vital role in stimulating the expansion of ecology after World War II (Kingsland 1985). For the time being, however, there were many who shared Elton's suspicions, feeling that the unpredictable dynamics of natural population changes were an unsuitable field for the application of abstract mathematical models.

Modern Ecology

Ecology expanded rapidly in the 1950s and 1960s as the world became more aware of the pressing environmental problems created by human activity. But the pressure was not necessarily coming from environmentalist groups. Those who sought to control and exploit nature also wanted information that would help them manage the ever more complex problems that they were confronting (Bocking 1997). The ecologists exploited the new image of a more "scientific" approach made possible by the mathematical techniques developed by Lotka and Volterra before the war. They were also able to make common cause with the Darwinian synthesis now beginning to dominate evolutionary biology following the emergence of the genetical theory of natural selection (itself based on mathematical modeling of populations). A school of population ecology emerged based on the exploitation of the Darwinian idea that competition was the driving force of natural relationships. There was no overall theoretical consensus, however, because at the same time a rival school of systems ecology emerged, exploiting analogies between ecological relationships and the stable economic structures existing in human society. Here there was a renewed focus on the harmonious nature of communities, drawing not on the old vitalistic philosophy but on the models of purposeful natural systems created in cybernetics. When James Lovelock's Gaia theory extended this approach into something that looked like the old mysticism, he was violently criticized by most biologists for abandoning the materialist ethos of science and pandering to the romanticized image of nature favored by the extreme environmentalists.

The Lotka-Volterra equations reinforced the lessons of Darwinism by implying that in a world dominated by competition, the best-adapted species in any environment would drive all rivals to extinction. This became known as the "principle of competitive exclusion," which states that there can be only one species occupying a particular niche in a particular location. This principle was tested by David Lack, a student of Julian Huxley, in the case of "Darwin's finches" on the Galapagos Islands. Although Darwin had used these birds as a classic example of specialization, later studies had shown that there were often several different species feeding in apparently the same way on the same island. Lack showed that this was not the case because each species was actually exploiting a different way of feeding—just because they were all mingling together did not mean they were taking the same food in the same way. His book *Darwin's Finches* (1947) helped to establish the new Darwinian synthesis in evolutionism and the principle of competitive exclusion in ecology, while at the same time renewing interest in Darwin's role as the founder of the selection theory.

The British-trained ecologist G. Evelyn Hutchinson, who had moved to America in 1928, launched an attack on Elton's refusal to use mathematical models in animal ecology. He argued that where there were difficulties in applying the Lotka-Volterra equations, the best approach was to modify the mathematical models, not reject the technique altogether. Hutchinson wanted to use the mathematical models to unify ecology and evolution theory, as proclaimed in the title of his 1965 book *The Ecological Theatre and the Evolutionary Play.* His student Robert MacArthur went on to found a new science of community ecology based on Darwinian principles of struggle and competitive exclusion (Collins 1986; Palladino 1991). MacArthur used mathematical models to address questions such as how close the niches could be in a particular environment and whether the niches evolved along with the species. Like Lack, MacArthur became interested in the problems posed by the structure of populations on isolated islands. He teamed up with Edward O. Wilson to develop a theory that predicted that the diversity of species on an oceanic island was directly proportional to its area. The number of species was maintained by a balance between immigration and extinction, the latter always a threat to small isolated populations. Wilson became interested in the way in which different reproductive strategies would help or hinder a species trying to establish itself on a new island and subsequently went on to develop the science of sociobiology.

Hutchinson had other interests, however, and these helped to create a rival school of systems ecology based on very different theoretical principles. He wanted to study communities using not an organismic analogy but an economic one, which traced the flow of energy and resources through the system and sought to identify feedback loops that maintained the stability of the whole. This was an approach pioneered by the Russian earth scientist V. Vernadskii, who had coined the term "biosphere" earlier in the century. The concept of feedback loops was central to the new science of cybernetics founded by Norbert Weiner to explain the activity of self-regulating machines. Hutchinson imagined such feedback loops working on a global scale to maintain the various ecosystems in a stable state. He also saw an analogy between this model of nature and the economists' attempts to depict human society as a stable system based on the cooperative use of resources. Hutchinson's student Raymond Lindemann wrote an influential paper in 1942 analyzing the flow of energy derived from the sun through the ecosystem of Cedar Bog Lake in Minnesota. This model of energy flow was then built on by the brothers Howard and Eugene Odum, the founders of systems ecology. The Odums studied the energy and resource circulations in a wide variety of environments, basing their work on the assumption that large-scale ecosystems would have a substantial robustness in the face of external threats. Some of their studies were funded by the U.S. Atomic Energy Commission, anxious about the potential damage that might be caused by nuclear war or accident. Systems ecology saw the human economy as just one aspect of a global network of energy and resource consumption and presented models suggesting that all levels of the process could be managed successfully if the flow patterns could be understood. Howard Odum's *Environment, Power and Society* (1971) presented a technocrat's dream of a society carefully structured and managed so that it could maintain itself even in the face of the more restricted levels of resources that will be available to humanity in the future (Taylor 1988).

Community ecology and systems ecology thus represented rival visions of how to construct a model of the ecosystem, the one based on the Darwinian principle of competition, the other on a more holistic vision of apparently purposeful feedback loops. Philosophically and politically, they invoked very different implications about nature and human society. The result was a deep level of conflict in which each side dismissed the other as philosophically naive and scientifically incompetent. The later twentieth century thus did not witness a unification of ecology around a coherent paradigm. There were still different schools with different research programs, methodologies, and philosophies. The one thing they all seemed to agree on was that scientific ecology had to present itself as essentially materialistic, offering no opening for communication with the kind of nature mysticism favored by the extreme environmentalist movement. Although systems ecology retained a holistic approach reminiscent of Clements's vision of the ecosystem as an organism in its own right, the advent of cybernetics and the link to economics allowed even this school to distance itself from the old idealism.

It is in this context that we can judge the reaction to James Lovelock's Gaia hypothesis of 1979, in which the whole earth is seen as a self-regulating system designed to maintain life. Gaia is the name of the ancient Greek earth goddess and was chosen to imply that the earth is mother to all living things, humans included. Lovelock made no secret of his support for environmentalism, criticizing those who advocate unrestricted exploitation of nature by implying that Gaia will, if necessary, take steps to eliminate humanity if it becomes a threat to the whole biosphere. Lovelock had impeccable scientific credentials, having worked in the space program developing systems to monitor the earth's surface from satellites, but the rhetoric with which he presented his theory clearly touched a raw nerve with many scientists. Although apparently similar to the systems approach, Gaia seemed to go beyond the cybernetic analogy and return to the older organicism in which ecosystems (in this case the biosphere as a whole) have a real existence and can act on their own behalf to achieve their own purposes. Critics were not slow to point out these implications, dismissing the whole theory as a perversion of science that pandered to the romanticism of the environmentalist movement. For Lovelock, it was as though a dogmatic scientific establishment had closed its ranks in defense of materialism: "I had a faint hope that Gaia might be denounced from the pulpit; instead I was asked to deliver a sermon on Gaia at the Cathedral of St. John the Divine in New York. By contrast Gaia was condemned by my peers and the journals, *Nature* and *Science,* would not publish papers on the subject. No satisfactory reasons for rejection were given; it was as if the establishment, like the theological establishment of Galileo's time, would no longer tolerate radical or eccentric notions" (Lovelock 1987, vii–viii). Nothing could more clearly indicate the gulf that still existed between scientific ecology (in all its forms) and radical environmentalism.

Conclusions

Although many people associate the term "ecology" with the environmentalist movement, we have seen that scientific ecology has a variety of origins, most of which were not linked to the defense of the natural environment. Science has more often been associated with efforts to exploit natural resources, and historical studies show that ecology emerged more from a desire to manage that process than to block it. At best, the majority of biologists have been concerned to ensure that humanity's engagement with the natural world does not do too much damage: sustainable yields are preferable to the wholesale destruction of a resource. Even those ecologists who imagined the ecosystem as a purposeful entity with a life of its own were willing to offer advice to farmers and others whose activities necessarily interfered with the untouched state of nature. In Europe, the whole idea of a purely natural landscape seemed meaningless, so ancient and so pervasive was the human role in shaping the environment. Although the more radical environmentalists can draw comfort from theories such as Lovelock's Gaia, they cannot lay claim to ecology as a science that inevitably lends support to their view that nature should be left untouched.

Equally interesting for the historian of science is the diversity of origins and theoretical perspectives from which the various branches of ecology emerged. Here was no single discipline shaped by a common research program and methodology. On the contrary, the movement toward what became known as ecology occurred in different places and at different times. The various locations of the scientists who became involved shaped the problems they sought to answer and hence the methodologies they thought appropriate. A technique that made sense on the open prairie of the American Midwest would have been inappropriate for the much-tilled landscape of Europe or the tundra of Hudson's Bay. Into these diverse environments came scientists with different backgrounds and interests; some were plant physiologists seeking to extend the experimental method to the interaction between plant and environment, some were biogeographers or taxonomists. All were driven by a determination to make the study of the interactions between organisms and their environment more scientific, but what they defined as "scientific" depended on their background and the problems they confronted. There was much suspicion to begin with over the application of mathematical techniques to modeling ecosystems. The majority of ecologists wanted to portray their science as materialistic, and this eventually led to a link with the revived Darwinism of the evolutionary synthesis. But there has been a persistent current of philosophical opposition to this movement, paralleling similar doubts in other areas of biology. Smuts's holism was by no means uncharacteristic of a nonmaterialist current of thought in early twentieth-century science. It certainly appealed to some of the early ecologists, and although that way of thought became less fashionable in the late twentieth century, its revival in the form of the Gaia hypothesis ignited a new level of debate. This debate reminds us of the gulf that still exists between the majority of scientists and the almost mystical vision of nature that has sustained the more radical environmentalist movement.

References and Further Reading

Anker, Peder. 2001. *Imperial Ecology: Environmental Order in the British Empire, 1895–1945*. Cambridge, MA: Harvard University Press.

Bocking, Stephen. 1997. *Ecologists and Environmental Politics: A History of Contemporary Ecology*. New Haven, CT: Yale University Press.

Bowler, Peter J. 1992. *The Fontana/Norton History of the Environmental Sciences*. London: Fontana; New York: Norton. Norton ed. subsequently retitled *The Earth Encompassed*.

Bramwell, Anna. 1989. *Ecology in the Twentieth Century: A History*. New Haven, CT: Yale University Press.

Brockway, Lucille. 1979. *Science and Colonial Expansion: The Role of the British Royal Botanical Gardens*. New York: Academic Press.

Cittadino, Eugene. 1991. *Nature as the Laboratory: Darwinian Plant Ecology in the German Empire, 1880–1900*. Cambridge: Cambridge University Press.

Coleman, William. 1986. "'Evolution into Ecology?' The Strategy of Warming's Ecological Plant Geography." *Journal of the History of Biology* 19:181–96.

Collins, James P. 1986. "Evolutionary Ecology and the Use of Natural Selection in Ecological Theory." *Journal of the History of Biology* 19:257–88.

Crowcroft, Peter. 1991. *Elton's Ecologists: A History of the Bureau of Animal Population*. Chicago: University of Chicago Press.

Kingsland, Sharon E. 1985. *Modeling Nature: Episodes in the History of Population Ecology.* Chicago: University of Chicago Press.
Lovelock, James. 1987. *Gaia: A New Look at Life on Earth.* New ed. Oxford: Oxford University Press.
Leopold, Aldo. 1966. *A Sand County Almanac: With Other Essays on Conservation from Round River.* Reprint, New York: Oxford University Press.
Marsh, George Perkins. 1965. *Man and Nature.* Edited by David Lowenthal. Reprint, Cambridge, MA: Harvard University Press.
Mackay, David. 1985. *In the Wake of Cook: Exploration, Science and Empire, 1780–1801.* London: Croom Helm.
McCormick, John. 1989. *The Global Environment Movement: Reclaiming Paradise.* Bloomington: Indiana University Press; London: Belhaven.
Merchant, Carolyn. 1980. *The Death of Nature: Women, Ecology and the Scientific Revolution.* London: Wildwood House.
Mitman, Greg. 1992. *The State of Nature: Ecology, Community, and American Social Thought, 1900–1950.* Chicago: University of Chicago Press.
Palladino, Paolo. 1991. "Defining Ecology: Ecological Theories, Mathematical Models, and Applied Biology in the 1960s and 1970s." *Journal of the History of Biology* 24:223–43.
Sheal, John. 1976. *Nature in Trust: The History of Nature Conservancy in Britain.* Glasgow: Blackie.
———. 1987. *Seventy-five Years in Ecology: The British Ecological Society.* Oxford: Blackwell.
Taylor, Peter J. 1988. "Technocratic Optimism, H. T. Odum, and the Partial Transformation of Ecological Metaphor after World War II." *Journal of the History of Biology* 21:213–44.
Tobey, Ronald C. 1981. *Saving the Prairies: The Life Cycle of the Founding School of American Plant Ecology.* Berkeley: University of California Press.
Worster, Donald. 1985. *Nature's Economy: A History of Ecological Ideas.* Reprint, Cambridge: Cambridge University Press.

CHAPTER 10

CONTINENTAL DRIFT

The 1960s witnessed a dramatic revolution in the earth sciences. Within a decade or so, principles that had been accepted since the "heroic age" of geology in the nineteenth century were overthrown and replaced by a new model of the earth's interior. The surface was now seen to be composed of interlocking but mobile plates that were constantly being renewed by volcanic action at one edge and destroyed by subduction into the interior at another. As a consequence of this new theory of "plate tectonics," the idea that the continents may drift horizontally across the face of the earth—which had been rejected or ridiculed for decades—now seemed perfectly plausible. The continents are like rafts of lighter rock carried along by the motion of the underlying plates on which they rest.

Not surprisingly, historians and philosophers of science have sought to use this episode as a case study to test theories of scientific change (Frankel 1978, 1985; Le Grand 1988; Stewart 1990). Was this a "revolution" in T. S. Kuhn's sense, in which a long-established paradigm entered a crisis state and was then replaced by another? Many of the participants certainly saw it in this light. Or was something more complex going on, perhaps requiring explanation in sociological terms related to the formation of research groups and new disciplines? According to Robert Muir Wood (1985), the revolution was actually a successful takeover bid for the earth sciences in which the newer discipline of geophysics displaced the more traditional science of geology. Much of the knowledge established by the geologists was retained, but the underlying principles were reformulated in the light of the new understanding of the earth's interior provided by geophysics. The sequence of geological formations established by nineteenth-century geologists (see chap. 5, "The Age of the Earth") was still valid, but their explanations of mountain building were abandoned. At the same time, one of the most controversial axioms of earlier geology, Charles Lyell's principle of uniformity, was triumphantly vindicated. The motions postulated by plate tectonics were slow and gradual and are still going on today. In part, the theoretical transformation had been made possible by new technologies allowing exploration of the deep-sea bed, revealing geological agencies that Lyell's generation had been unable to observe.

The situation is complicated by the fact that the idea of continental drift was suggested by Alfred Wegener as early as 1912 but was largely repudiated until the revolution of the 1960s. Was Wegener a pioneer of the theory that would later be accepted, and if so, why did a whole generation of geologists resist his arguments so vehemently? Or was his insight only a superficial anticipation of plate tectonics, a lucky guess that just happened to hit on one key aspect of the later theory while totally failing to anticipate the more fundamental revolution in our understanding of the earth? Wegener did not foresee the reformulation of ideas about the mechanisms going on within the crust that are integral to plate tectonics. Yet even when similar mechanisms were proposed in the 1920s, as a consequence of the new understanding of radioactive heating, the majority of geologists remained skeptical. Perhaps the fact that Wegener was himself a geophysicist, not a geologist, helps us to understand why his ideas were not taken seriously by those trained in the older way of thinking. In this case, we may want to think carefully about Wood's suggestion that the revolution was a consequence of the belated triumph of geophysics, prompted by the emergence of new techniques for studying the earth's crust.

The Crisis in Geology

Alfred Wegener was not the first to notice that the apparent "fit" between the coastlines of Africa and South America makes it look as though the Atlantic Ocean had been created by the continents being pulled apart. But he was the first to build this insight into a whole theory that sought to explain a wide range of geological phenomena in terms of continental drift. His theory was greeted with widespread skepticism, in part because he suggested no plausible mechanism by which continents could move horizontally across the earth's surface. Yet he did articulate a number of serious objections that had begun to plague the existing theories of geological change and hinted that a "mobilist" alternative might resolve these problems. In this sense, Wegener can be taken seriously as an architect of the downfall of the previous paradigm in the earth sciences, even if his anticipation of the new theory was limited in its scope. It is worth remembering that neither Copernicus nor Kepler was able to foresee the explanation of planetary motions offered by Newton, and Wegener himself saw his drift theory as a preliminary outline that would await future vindication by a generation that would reformulate ideas about the earth's underlying structure.

To understand the crisis to which Wegener was responding, we need to go back to the theories of the earth proposed during the nineteenth century (Greene 1982). As we saw in the chapter on the age of the earth (chap. 5), the predominant theory was that the earth is cooling down, with a consequent diminution in the rate of geological activity such as earth movements. Charles Lyell's uniformitarian alternative had been resisted largely because it implied that the earth had been in a "steady state" for an uncountable period of time. Lyell had some success in persuading the catastrophists to scale down the upheavals they postulated in earlier periods, but very few abandoned the basic claim that the earth was a more violent place in the distant past. Nor was Lyell able to explain away the evidence for dramatic, if not actually catastrophic, events in the geological record. The divisions between the geological periods did indeed seem to mark punctuation marks between periods of relative calm and episodes of massive mountain building and mass extinction caused by the resulting climatic transformation. By the later part of the century, most geologists believed that these episodes were caused by relatively sudden crumplings of the crust needed to relieve the pressure built up as the interior as the earth cooled and hence reduced in volume. Even the continents themselves were formed by such large-scale warping of the crust, so even they were relatively impermanent—any part of the earth's surface might be pushed down to form ocean bed or pushed up to form continents and mountains, depending on the precise location of the weaknesses that gave way to the pressure caused by contraction. The timescale of the whole sequence was defined by how long it had taken the earth to cool from an initially molten state.

By the end of the century, many aspects of this theory had been called into question, in part by the emergence of a new approach to the study of the earth that came to be known as geophysics. This new breed of earth scientists was not interested in the geologists' efforts to provide a relative dating for the sequence of events in earth history: they wanted to understand the actual physical processes that drove the activities going on deep in the planet's interior. Lord Kelvin's efforts to work out a timescale for the earth's cooling were part of this initiative, and Kelvin was certainly interested in the processes by which heat would be conducted up to the surface. One consequence of his work was the realization that the amount of heat reaching the surface from the interior was insignificant compared to that received from the sun. So even an advocate of the cooling earth would not expect the climate to cool down, at least in the later phases.

But some of the calculations performed by the geophysicists were more serious for the prevailing theory. Most important, it turned out that even if the earth were cooling and hence contracting, the amount of contraction was not enough to produce the enormous amounts of folding and faulting observed in the crust. By the early twentieth century, the cooling-earth model itself had come under fire, as the theory of radioactive heating suggested that the internal temperature could be maintained over thousands of millions of years. The contraction mechanism of mountain building was dead, and to Wegener it seemed obvious that horizontal movements of the continents would provide an alternative explanation.

Equally suggestive was evidence coming from new studies of the actual nature of the rocks making up continents and oceans. In his *Physics of the Earth's Crust* of 1881, the British geophysicist Osmond Fisher collected evidence suggesting that the continental rocks were composed of lighter material than those from the deep ocean bed. The continents were composed mainly of silicates of aluminum (later abbreviated to "sial") while the ocean floor was mostly silicates of magnesium ("sima"). The implication was obvious: the continents are not formed by uplift from the ocean but are better visualized as rafts of the lighter sial floating on an underlying global crust of sima. This concept was built into the theory of "isostasy" proposed in 1889 by the American geophysicist Clarence Dutton. On this model, the continents floated in hydrostatic equilibrium, rising and falling as material was eroded or deposited at one point or another.

By this time, the majority of geologists had accepted that the continents were extremely ancient, but many still believed that areas of land had been sunk beneath the sea at certain points in geological time. The present continents had once been linked by "land bridges" or even more extensive areas of land, now vanished beneath the waves. These land bridges explained certain anomalies in the fossil record, including the fact that the populations of Africa and South America seemed to have been identical up to the Mesozoic era, after which they steadily diverged. The assumption was that a land bridge linking the continents had been submerged at that point. But in the model proposed by Fisher and Dutton, such land bridges were implausible—it would be physically impossible for lighter continental rock to be forced down to a level where it could form the bed of the South Atlantic or any other ocean. Continents might occasionally be invaded by very shallow seas, but they could never form deep ocean bed. Here again

Wegener was able to seize on a weakness in the existing theory that, he claimed, could be overcome by postulating a horizontal motion of the continental rafts themselves.

Wegener and the First Theory of Drift

Wegener's theory was thus an attempt to provide an alternative to a paradigm that, he was able to argue, was already defunct. The problem was that most of his contemporaries thought the new idea was even more implausible than the old. There were certainly some important lines of evidence pointing to the possibility that the continents had moved, including some that had once been used to justify the postulation of land bridges. But Wegener did not move forward to a complete reformulation of ideas about the earth's internal structure, and his theory thus lacked any plausible explanation of how the continents could be dragged across the face of the earth against the enormous frictional force that would resist any such movement. Equally serious, Wegener himself was an outsider to the community of traditional geologists. He was a meteorologist whose original interests lay in paleoclimatology (Schwarzbach [1989]; for more general discussion, see Hallam [1973]). Along with his father-in-law Wladimir Köppen he supported the theory that the onset of ice ages was triggered by fluctuations in the amount of heat received from the sun. This interest in ice ages led him to do research in Greenland, where he eventually died on an expedition in 1930. His work on continental drift was thus, in a sense, peripheral to his main career in the meteorological aspects of geophysics. Historians have argued that Wegener's lack of training in orthodox geology may have given him the flexibility of mind needed to invent a completely new idea about earth movements, but it also alienated him from the professional community of geologists, who saw him as an outsider and a dilettante.

Wegener conceived his theory in 1910 when he noticed the relationship between the coastlines of Africa and South America, and he immediately began a search of the geological literature looking for arguments that would support the idea. Two years later, he began lecturing on the topic, and his book *The Origin of Continents and Oceans* appeared in 1915 (not translated into English until 1966). The book presented an effective summary of all the evidence that had built up against the old theory of mountain building and then went on to make the case for drift as an alternative. Few now doubted that the continents could be seen as rafts of lighter material resting on a denser layer of crust exposed on the ocean bed. Wegener's point was that if the continents were somehow pushed horizontally across the surface, friction would cause the leading edge of the continental plate to crumple, thus generating mountain ranges. If America were moving away from Africa and Eurasia, this would explain the ranges of mountains that run down the western edges of both North and South America. Wegener argued that all the continents had once been joined in a single great landmass he called Pangaea, which had begun to split up in the Mesozoic (fig. 10.1). This explained why the inhabitants of South America and Africa had only begun to diverge after that point. It also explained why the early geological structure of the two areas was also very similar. The argument from the fit of the coastlines was based on more than mere geography—the actual geological formations would also be continuous if one imagined them joined together. Wegener used an effective analogy: "It is just as if we were to refit the torn pieces of a newspaper by matching their edges and then check whether the lines of print run smoothly across. If they do, there is nothing left but to conclude that the pieces were in fact joined this way" (Wegener 1966, 77). In his eyes, the evidence for a splitting apart of the continents in the Mesozoic was inescapable.

Wegener also used his knowledge of paleoclimatology to provide other lines of evidence. The fossil record suggested that many continental areas had experienced an ice age during the Permian period. This was hard to explain if the continents had then been positioned as they are today but would make sense if they had once been united to form a larger landmass located near the South Pole. The warm conditions enjoyed by other regions at the same time could be explained if they had been located in the tropics. Much less reasonably, Wegener tried to argue, in addition, that Europe and North America had also been linked in the last ice age. Since this was very recent in geological terms, this theory would imply a very rapid opening of the North Atlantic. Wegener even cited some very dubious measurements suggesting that Greenland and Europe are currently separating at the rate of ten meters per year.

Moreover, Wegener had to explain how the continents were moved across the surface, and here his efforts proved much less convincing. He still thought of the underlying crust of sima as static, so the continental rafts would have to be pushed across this surface against a tremendous frictional resistance. To make the idea seem more plausible, he argued that the crust was not absolutely rigid. Like pitch, it resisted a sudden blow but would flow gradually when subjected to a continuous pressure. But even so, the resistance to a moving continent would be enormous, and to supply the necessary pressure Wegener had only two suggestions. One was a hypothetical "flight from the poles" caused by centrifugal force stemming

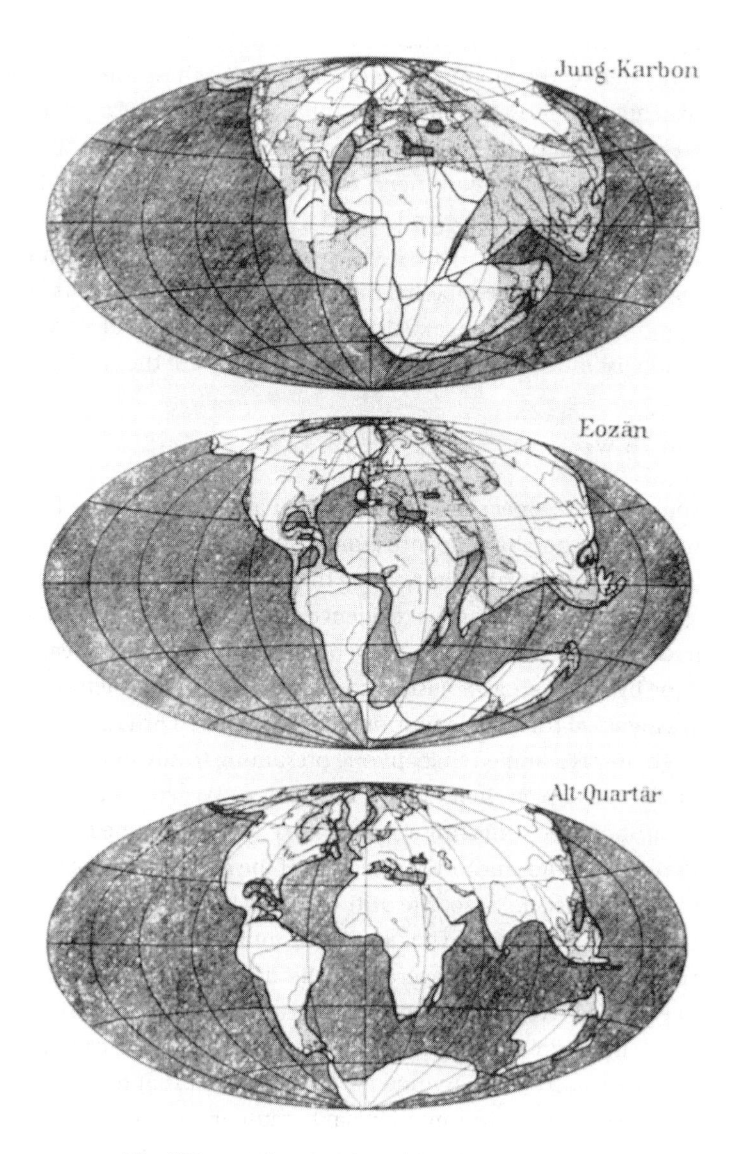

FIGURE 10.1 Alfred Wegener's maps showing continental drift, from his *Die Entstehung der Kontinente und Ozeane,* 3d ed. (1922), 4. The *upper* map shows the earth in the late Carboniferous period, with most of the land united in a single supercontinent, Pangaea. The *lower* maps show the fragmentation in the Eocene and finally in the early Quaternary period, by which time the modern distribution is becoming apparent.

from the earth's rotation. The other was a westward pressure caused by tidal forces generated by the moon. The problem was that these forces not only seem inadequate to most geophysicists but also failed to explain why Pangaea had broken up in the Mesozoic. Presumably the flight from the poles had been in effect since the continents were formed, so they should all have moved steadily to the equator and stayed there. And if the tidal force was pushing America westward, why was it having no effect on Eurasia and Africa? Wegener had seen the superficial evidence for continental drift, but he had not appreciated that to make this theory work one would have to develop a mobilist model for the whole underlying crust of the earth.

Reaction to Wegener

The response to Wegener's theory was muted at first, but in the English-speaking world it soon built up into almost universal hostility. German geologists were more sympathetic, treating the idea as potentially interesting, although in need of much further evidence if it were to be taken really seriously. In Germany, there was a tradition of theoretical work in the earth sciences done by armchair geologists who did no fieldwork of their own but, instead, assembled their evidence from the literature. In Britain and America, though, it was assumed that anyone presuming to advance a new theory must first have paid their dues in the field, so Wegener was seen very much as an outsider venturing into territory already claimed by others (Oreskes 1999). At a now notorious meeting of the American Association of Petroleum Geologists in 1926, the drift theory was widely rejected and in some cases openly ridiculed. The old idea of sunken land bridges was still used to explain the fossil evidence, despite its incompatibility with the geophysical evidence. Wegener was depicted as an uncritical enthusiast who combed the literature looking for evidence favorable to his cause, while ignoring a mass of contrary arguments. It was also felt that the theory undermined the logic of uniformitarianism because it seemed to imply that there was an arbitrary starting point for the whole process of drift in the Mesozoic.

Even the geophysicists proved hard to convince, and here the weakness of the actual mechanisms suggested by Wegener proved crucial. In his influential textbook *The Earth,* first published in 1924, the British geophysicist Harold Jeffreys argued that the forces postulated by Wegener were many orders of magnitude too small to overcome the friction that must occur if the continent were to be pushed across an underlying static crust.

A few geologists did take the theory seriously, although for several decades they were voices crying in the wilderness. The Harvard University geologist R. A. Daly postulated a mechanism for drift based on the continents sliding down from a polar "bulge" in the earth's surface. Most enthusiastic of all was the South African geologist Alexander Du Toit, who appreciated the similarities between the structure of his homeland and of South America. In his 1937 *Our Wandering Continents,* he toned down some of the more excessive claims that Wegener had made about the rapidity of drift and postulated two ancient supercontinents, Laurasia and Gondwana, instead of one.

For those historians seeking to understand why a theory so close to the modern one was rejected at the time, the most interesting line of support came from the geophysicist Arthur Holmes, who had a substantial reputation based on his work on radioactive dating of the earth (Frankel 1978). Holmes calculated that the amount of heat produced by radioactivity deep in the earth was so great that some mechanism in addition to conduction was needed to bring it to the surface. Extensive vulcanism was an obvious possibility. In 1927, Holmes argued that there might be convection currents in the earth's crust, in which hot material rose to the surface while cool material was subducted into the interior elsewhere. In effect, new crust was created from molten rock over a "hot spot" and old crust destroyed by subduction, and in between the crust would move horizontally. Holmes soon realized that such convection currents would provide a mechanism for continental drift because if the continental raft floated on an area of crust in motion, it would move with it. The arguments against Wegener based on the level of friction between continent and underlying crust were undermined by this new model of what was going on within the crust itself.

Holmes suspected that hot spots would tend to build up under continents and thus fragment them by drift. He did not realize that the implications of this are that most hot spots will now be found beneath the oceans created by the breakup of the original continent. In this respect his idea did not anticipate the notion of seafloor spreading that became central to the theory of plate tectonics, yet the theory of convection currents in the earth's crust was an uncanny anticipation of later developments. Even so, no one paid any attention, and Holmes's suggestions did nothing to boost the fortunes of Wegener's theory. Historians are thus led to ask why a theory that, by this stage, had come very close to that which would be accepted in the 1960s, continued to be rejected for another generation. One suggestion is that Holmes's early version of the theory was untestable and hence could not be used as the basis for a viable research program. Even if he had realized that the place to search for hot spots was in the middle of the oceans, there were no techniques available for studying the deep-sea bed at the time. More serious, though was the continuing influence of the old geological community, which was still not willing to allow the upstart geophysicists to dictate its worldview.

Plate Tectonics

The developments that revolutionized the earth sciences in the 1950s and 1960s were in part a spin-off from military technology developed during World War II and the Cold War. The threat posed by submarines made it vital for the world's navies to know more about the deep-sea bed, and it was to the geophysicists that they turned for information. Improved instruments were developed for mapping the magnetic structure of the ocean floor, and from this emerged new insights that would transform scientists' theoretical models of the earth's crust. This would allow the idea of continental drift to enjoy a belated triumph as it rode in on the coattails of the new theory of plate tectonics. But it was not just an existing paradigm that was replaced. As a result of its new level of funding and influence, the younger science of geophysics was able to overturn the power balance that had so far kept it subordinate to traditional geology. The triumph of the new order was proclaimed by the International Geophysical Year (actually July 1957–December 1959) that brought it wide publicity even outside the scientific community. Over the next decade or more, university geology departments began to rename themselves as departments of "earth sciences," acknowledging that the subject was no longer dominated by old-style geology. The revolution that created the theory of plate tectonics was not a transformation occurring within a single discipline; it was a by-product of a newer research community's bid for control of an area that had hitherto been dominated by the older geological tradition. According to one recent study, what changed—at least for American scientists—was the definition of what counted as good science in this area (Oreskes 1999).

The most important additions to the technology available to geophysicists were those that allowed detailed study of the earth's magnetic field. There were major controversies among the physicists over the nature of magnetism and hence over the constancy of the earth's field. The British physicist P. M. S. Blackett had helped to produce an extremely sensitive magnetometer to detect magnetic mines during World War II and he now used these skills to trace minute magnetic fields locked into the rocks of the earth's crust. It was assumed that these fields were imprinted onto the rocks when they were formed, in effect providing a record of the earth's magnetic field throughout geological time. To everyone's surprise, when details of the remnant magnetism from rocks in different areas were compared, it was clear that they were not all aligned with the current state of the earth's field or with each other. Either the rocks had moved since they were formed, or the magnetic poles had shifted. Since the remnant fields were different in rocks from different parts of the world, the most likely explanation was that the continents were no longer in the position they had occupied in earlier geological periods.

Equally puzzling was the fact that in many rocks the remnant magnetism had the reverse polarity to that now observed. Geophysicists began to suspect that the earth's magnetic field must reverse from time to time, the north and south magnetic poles swapping positions. By putting a mass of observations together it was possible to build up a timetable of these geomagnetic reversals. At the same time, more refined techniques of radiometric dating were allowing the construction of a more fine-grained timetable of rock formation through the Pleistocene era. By putting the two lines of evidence together, a team at Berkeley led by Richard Doell, Alan Cox, and G. Brent Dalrymple were able to work out a sequence for the magnetic reversals correlated with the existing geological timescale. The last reversal was pinned down from tests on rocks at Jaramillo, New Mexico, and published in 1966 (Glen 1982). It would soon play a vital role in the case for continental drift.

A parallel line of development took place in oceanography. During World War II and then the Cold War, detection of enemy submarines became of primary importance to the military. Better information about the nature of the ocean floor was crucial if concealed submarines were to be detected, and efforts were made to extend the range of the new, more sensitive magnetometers so they could produce detailed magnetic maps of the seabed. This research completely overturned expectations based on the idea of a static earth because the rocks of the seafloor turned out to be remarkably uniform and extremely recent in geological terms. Research with sonar and other techniques revealed a pattern of mid-ocean ridges, underwater mountain ranges stretching down the middle of otherwise flat seabeds. The ridges were sites of extensive seismic and volcanic activity. When rocks from the ridges were dredged up, they were found to be younger than any of the others, only recently solidified from a molten state. Here, in a totally unexpected location, were Holmes's predicted hot spots.

A leading figure in this transformation of ideas about the ocean bed was the American geophysicist Harry Hess. He had commanded a naval vessel in the Pacific war against Japan and had used its sonar system to map the

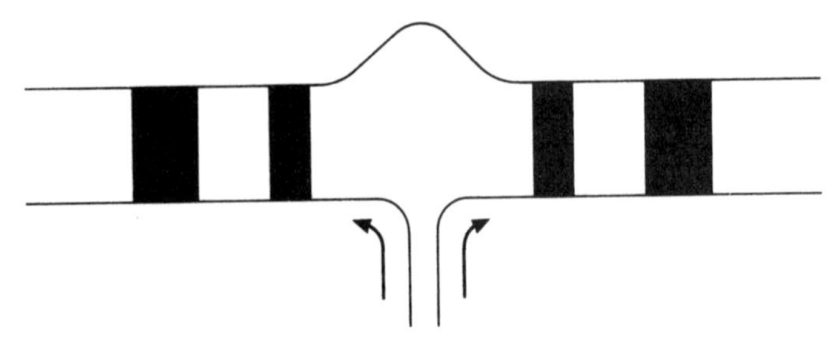

FIGURE 10.2 Cross section of the deep-ocean bed at a mid-ocean ridge, showing the effect of seafloor spreading. Hot material upwelling at the ridge spreads out equally on either side. The light and dark bands represent the magnetism of the earth's magnetic field imposed on the rock as it cools, either normal (*white*) or reversed (*black*). The effect is to produce parallel bands of normal and reverse magnetism on either side of the ridge, as shown in fig. 10.3. The continents form slabs of lighter rock lying on top of the denser, deep-ocean crust. As the crust spreads outward from the mid-ocean ridge, the continents are pushed apart.

ocean floor. In the mid-1950s he began to suggest that the mid-ocean ridges were the sites at which hot rock welled up from the interior of the earth. Here was where new crust was being produced, and the deep oceanic trenches were where old crust was being thrust down into the depths. The seabed was young because it was constantly being renewed—only the continents, riding high because of their lighter density, preserved evidence of the distant past. Holmes's theory of convection currents in the crust was right, but all the activity was taking place on the seafloor, where no one had been able to observe it before. The term "seafloor spreading" was coined by Robert Dietz in 1961.

At first Hess's ideas were greeted with skepticism, but he fired the enthusiasm of Fred Vine and Drummond Matthews of Cambridge University. They were trying to make sense of the patterns of magnetism being revealed on the seabed and were puzzled by the existence of parallel stripes of normal and reversed magnetism alongside the mid-ocean ridges. In 1963 they published a paper arguing that this pattern was exactly what would be expected if new seafloor were constantly being produced at the ridge and then forced away from it in either direction. As new rock upwelled, it would be imprinted with the current direction of the earth's magnetic field, but when the field reversed a new strip of reverse-magnetized rock would begin to form, steadily pushing the original strip further away from the ridge. The ridge should thus be surrounded on either side by a pattern of normal and reversed magnetic strips (figs. 10.2 and 10.3).

Vine and Matthews already had some evidence for this striping effect, but it was too indistinct to convince most of their fellow geophysicists. Workers at the Lamont Geological Observatory were skeptical, and it was their survey ship *Eltanin* that was producing the best magnetic maps of the seabed. In 1965 they were surveying the region of the Juan de Fuca ridge off the west coast of North America (the notorious San Andreas Fault in California is linked to this ridge). One magnetic sweep, *Eltanin* 19, demonstrated the parallel stripes so clearly that opinion now began to shift in favor of seafloor spreading (fig. 10.4). Vine was able to show that the sharper timescale of magnetic reversals provided by the Jaramillo event fit the pattern of magnetic stripes perfectly. At the same time, the Canadian geophysicist J. Tuzo Wilson developed the concept of "transform faults," which explained why the mid-ocean ridges and their associated magnetic patterns were occasionally shifted bodily to one side or another, creating an apparent zigzag effect.

The final version of the theory of plate tectonics was worked out in the mid-1960s by Jason Morgan, Dan McKenzie, and Xavier Le Pichon. They realized that the earth's spherical shape imposed constraints on the shape of the plates defined by mid-ocean ridges and associated subduction zones, explaining many effects that were confusing when viewed on a two-dimensional map. Le Pinchon produced a simplified version of the theory in which there were six major plates, each in constant motion because it was defined by the horizontal section of a convection cell in the underlying

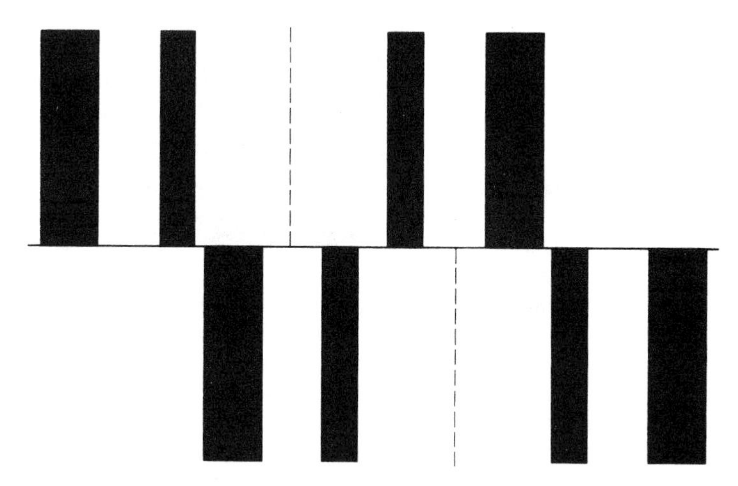

FIGURE 10.3 The parallel bands of normal and reverse magnetism produced by the process shown in fig. 10.2. The horizontal split in the middle of the pattern is a transform fault, where the whole ridge and its associated pattern of rocks are displaced at right angles to the ridge.

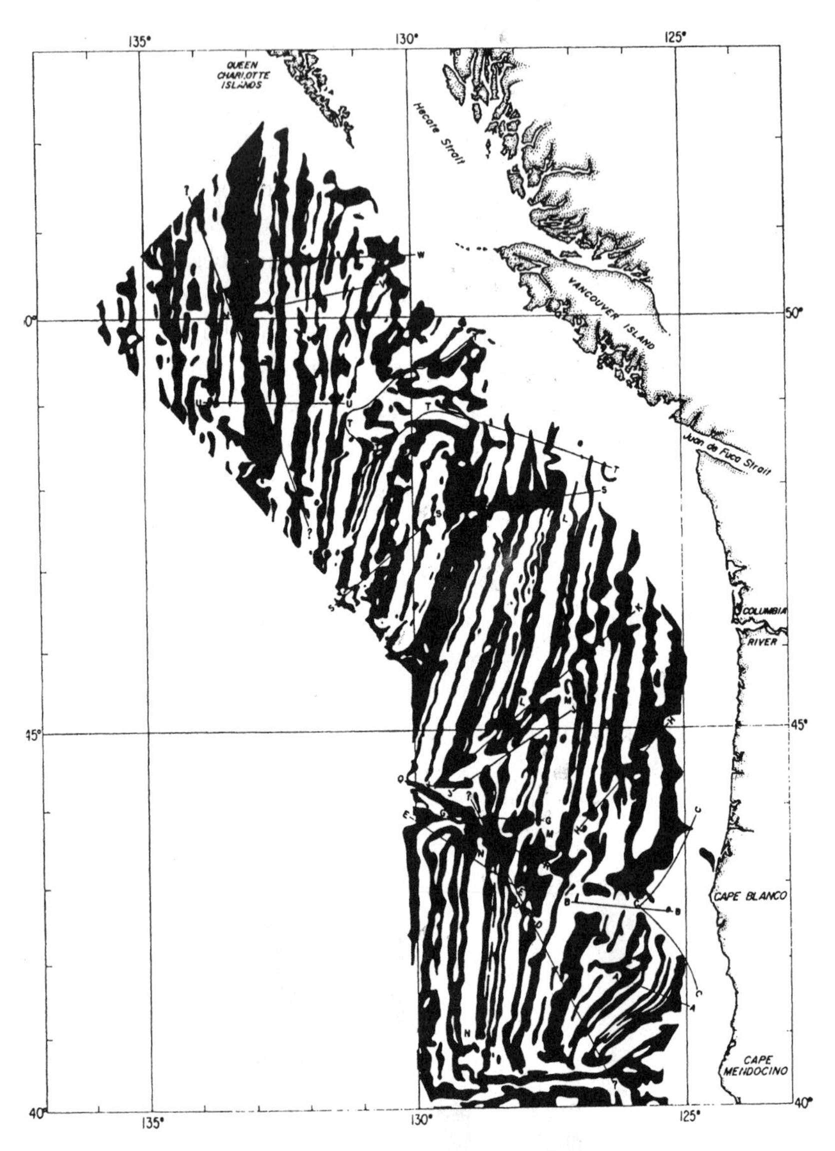

FIGURE 10.4 Map showing the magnetic anomalies on the ocean floor around the Juan de Fuca Ridge off the coast of Vancouver Island, produced by the survey vessel *Eltanin* in 1961, from R. Masson and A. Raff, in *Bulletin of the Geological Society of America* 72 (1961): 1267–70. Compare this with the idealized patterns shown in figs. 10.2 and 10.3. It was this survey that convinced many geophysicists that the hypothesis of seafloor spreading, coupled with the discovery of magnetic reversals, provided an explanation of continental drift.

crust. The continents, as in Holmes's theory, were simply carried along by the motion of the plates—America is separating from Eurasia and Africa because the Atlantic Ocean is expanding as the activity of the mid-Atlantic ridge continues to produce new crust. Mountains are formed either where a continent is riding up over a subduction zone, as in the case of the Rockies and the Andes, or where two continental masses are being forced together by the motion of separate plates, as in the Himalayas.

Conclusions

The widespread acceptance of the theory of plate tectonics in the late 1960s certainly marked a revolution in the earth sciences. Wegener's long-ridiculed idea of continental drift now made perfect sense, thanks to a complete reformulation of ideas about what was going on deep underneath the earth's crust. But this was not a paradigm shift within an established science. Orthodox geologists had focused on reconstructing the history of the earth but had not been very adventurous in seeking to explain the earth movements on which their theories relied to explain phenomena such as mountain building. It was the geophysicists who began to ask new kinds of questions about the structure of the earth and to seek new lines of evidence that would answer those questions. Although seen as junior partners by the established geological community of the late nineteenth and early twentieth centuries, they began to undermine the logic on which much of the older theorizing was based. To begin with, however, they had no serious alternative to offer, and even when the first hints of such an alternative were provided by Wegener, the geologists remained unwilling to admit that their existing ideas were vulnerable. To be fair, even some geophysicists were unimpressed, because without a much more radical rethinking of ideas about the earth's interior Wegener's idea was implausible. The revolution occurred when geophysics gained a new lease on life thanks to the oceanographic technology made available in the 1950s and 1960s. Simultaneously, the new evidence both precipitated a theoretical revolution and reduced the influence of the older community that would have been least likely to accept it.

In one sense, however, the revolution helped to reinstate a once-controversial principle of geological methodology. In the nineteenth century, Charles Lyell's uniformitarianism had gained only limited influence because few were prepared to believe that the earth was not cooling down. The massive expansion of the geological timescale made possible by the theory of radioactive heating rendered the idea of a steady state earth plausible at last. Plate tectonics reinforced this message by showing that the forces that drove the continents apart were still at work in the mid-ocean ridges today. All earth movements were slow and gradual, exactly equivalent to those we still observe. It is against this background that we must assess the later revolution of the 1980s, outside the scope of our study here, in which uniformitarianism was challenged once again by the advocates of mass extinctions caused by asteroid impacts (Glen 1994). Even if the earth's internal processes are slow and uniform, there is clear evidence of catastrophes caused by external, astronomical events. In addition, there are growing indications that vulcanism was so active at certain periods in the past that it generated environmental traumas as great as anything attributed to impacts. Modern science has been forced to take seriously some of the more alarming ideas promoted in the earliest days of catastrophism.

References and Further Reading

Glen, W., ed. 1994. *Mass Extinction Debates: How Science Works in a Crisis.* Stanford, CA: Stanford University Press.

Greene, Mott T. 1982. *Geology in the Nineteenth Century: Changing Views of a Changing World.* Ithaca, NY: Cornell University Press.

Frankel, Henry. 1978. "Arthur Holmes and Continental Drift." *British Journal for the History of Science* 11:130–50.

———. 1979. "The Career of Continental Drift Theory: An Application of Imre Lakatos' Analysis of Scientific Growth to the Rise of Drift Theory." *Studies in the History and Philosophy of Science* 10:10–66.

———. 1985. "The Continental Drift Debate." In *Resolution of Scientific Controversies: Theoretical Perspectives on Closure,* edited by A. Caplan and H. T. Englehart. Cambridge: Cambridge University Press, 312–73.

Glen, William. 1982. *The Road to Jaramillo: Critical Years of the Revolution in Earth Science.* Stanford, CA: Stanford University Press.

Hallam, Anthony. 1973. *A Revolution in the Earth Sciences.* Oxford: Oxford University Press.

———. 1983. *Great Geological Controversies.* Oxford: Oxford University Press.

Le Grand, Homer. 1988. *Drifting Continents and Shifting Theories.* Cambridge: Cambridge University Press.

Oreskes, Naomi. 1999. *The Rejection of Continental Drift: Theory and Method in American Earth Science.* New York: Oxford University Press.

Schwarzbach, Martin. 1989. *Alfred Wegener, the Father of Continental Drift.* Madison, WI: Science Tech.

Stewart, James A. 1990. *Drifting Continents and Colliding Paradigms: Perspectives on the Geoscience Revolution.* Bloomington: Indiana University Press.

Wegener, Alfred. 1966. *The Origin of Continents and Oceans.* Translated from the 4th rev. German ed. (1929) by John Biram. New York: Dover.

Wood, Robert Muir. 1985. *The Dark Side of the Earth.* London: Allen & Unwin.

CHAPTER 11

TWENTIETH-CENTURY PHYSICS

What happened to physics at the beginning of the twentieth century? In many ways this seems a fairly straightforward example of a revolutionary change in science. The way of looking at the world now usually referred to as "classical physics" was superseded by the new theories of relativity and quantum mechanics. These new theories did not just suggest novel mathematical techniques for understanding nature or new ways of carrying out and interpreting experiments. They inaugurated completely new philosophical perspectives. The theories of special and general relativity required a wholesale rethinking of the relationship between space and time. Quantum mechanics called for a systematic reconsideration of the relationship between cause and effect, as well as a reassessment of just what it might be possible to know about the fundamental structure of matter. It is certainly the case that by the middle of the twentieth century, physicists were asking themselves questions about the ultimate nature of matter that would have been considered unthinkable—if not completely illegitimate—less than a century previously. The luminiferous ether—the focus of so much late nineteenth-century physical inquiry—was dead and buried. Nevertheless, as we shall see in this chapter, it is as easy to chart continuities as discontinuities between late nineteenth-century physicists and their concerns and those of their successors (see chap. 4, "The Conservation of Energy").

It is also certainly the case that massive institutional changes took place in physics during the course of the past century (see chap. 14, "The Organization of Science"). These institutional changes were very closely related to the new ways in which physicists started understanding the world around them, so much so that it is difficult to consider either aspect entirely separately. If the professionalization of physics (like other sciences) can be said to have started during the nineteenth century, then the process certainly accelerated during the twentieth century. At the same time, the process of specialization that started in the nineteenth century continued to the extent that by the middle of the twentieth century it was increasingly difficult to see physics even as a self-contained discipline. Theoretical and experimental physics (let alone subdisciplines, such as relativity theory, quantum mechanics, or particle physics) were becoming increasingly distinct from each other. This had important consequences for the practice and the content of physics. Physics and its subdisciplines were becoming increasingly esoteric, to the extent that physicists working in adjacent laboratories in the same institute might not fully understand what the other was doing. Physics also became a science that was more and more dependent on massive resources. Experiments at the end of the nineteenth century—and even up until the 1930s—could fit onto a tabletop. By the 1950s and 1960s the scale of things had changed completely, with physicists talking about the size of their apparatuses in kilometers rather than meters.

We will start this chapter back in the 1890s when J. J. Thomson carried out the experiments that would later be hailed as the "discovery of the electron." Those experiments, as well as those that led to the discoveries of X-rays and radioactivity, opened up a whole new set of problems for physicists. At the same time, they provided them with the tools with which to set about solving them. The result was a new understanding of the structure of the atom. The publication of Albert Einstein's theory of special relativity, closely followed a few years later by his theory of general relativity, provided another set of powerful tools and concepts for rethinking the structure of the universe. Again though, as we shall see, the significance of Einstein's insights took a while to sink in. It was not as clear to his contemporaries that his theories were as revolutionary as they might appear with the benefit of hindsight to us. Niels Bohr's quantum theory of the structure of the atom, incorporating the idea that energy was exchanged at the atomic level in discrete packages (or quanta) was a breakthrough too. Nevertheless it was dissatisfaction with this model (not least on Bohr's own part) that led to the development of quantum mechanics during the 1920s. After the Second World War, attention turned to probing ever more deeply into the structure of matter, with a resulting proliferation of elementary particles. Discovering and tracking these new particles required massive resources, consequently turning particle physics into the ultimate big science.

Inside the Atom

For much of the nineteenth century, atomic theory—the idea that matter should be considered as being made up of discrete, fundamental atoms—was very much a theory. As far as many physicists were concerned, atoms were at best a useful hypothesis, not to be taken as real existing objects. They provided chemists with a convenient way of balancing the books in chemical reactions but that was all (see chap. 3, "The Chemical Revolution"). It also seemed to many that inquiry into the fundamental structure of matter—to find out of it was made up of discrete units like atoms or was continuous and indefinitely divisible, for example—was beyond the scope of experiment. Theories about the structure of matter could in the end be nothing more than just theories. From the late 1850s, however, it seemed to some investigators, such as the German Julius Plücker, or William Robert Grove and John Peter Gassiot in England, that their experiments with discharge tubes provided new insights or, at least, new tools for investigating, the ultimate structure of matter. In experiments like these, in which currents of electricity were passed through attenuated gases in sealed tubes (a little like modern neon light tubes), strange glows appeared. The experimental physicist William Crookes, during the 1870s, argued that these cathode rays, as he called them, provided a new way of understanding the basic makeup of matter (fig. 11.1). By the 1880s, cathode ray experiments were part of the standard repertoire of physicists' experimental research.

One place where cathode ray experiments were taken up with enthusiasm was Cambridge's Cavendish Laboratory under the directorship of the physicist J. J. Thomson (fig. 11.2). Starting in the mid-1880s, Thomson himself began experimenting with gaseous discharges, looking for ways of investigating the relationship among matter, electric fields, and the ether. He also wanted empirical evidence for his model of matter as being made up of interlocking vortices in the ether. In 1897, Thomson announced that his latest experiments of cathode rays showed that they were made up of a stream of small negatively charged particles, each of which had a mass of about a thousand times smaller than a hydrogen atom—usually regarded at the time as the smallest unit of matter. He did this by means of measuring the ratio of electric charge to mass by deflecting the cathode rays in a magnetic field and, in later experiments, in an electrostatic field as well. He also suggested that his particles, or corpuscles, were the components from which atoms of matter were made. Such ether theorists as Joseph Larmor and George FitzGerald suggested that the corpuscles that Thomson had

FIGURE 11.1 A cartoon of William Crookes holding a cathode ray tube, from *Vanity Fair* (image courtesy of the Science and Society Picture Library, London).

FIGURE 11.2 J. J. Thomson at the Cavendish Laboratory in Cambridge, working with the apparatus he used in the discovery of the electron in 1897. Photo courtesy of the Department of Physics/Cavendish Laboratory, University of Cambridge).

identified were "electrons"—a word that Larmor had coined some years earlier to describe packets of pure electrical energy in the ether. One reason that they suggested this was because they were unhappy with Thomson's suggestion that his corpuscles, rather than atoms, were the ultimate constituents of matter.

A year before Thomson's announcement, the German physicist Wilhelm Röntgen had claimed the discovery of an entirely new kind of ray—soon dubbed X-rays. Like Thomson, he had made his discovery while experimenting with cathode rays from discharge tubes; in fact, it was as a result of Röntgen's work that Thomson commenced his own cathode ray experiments. The new X-rays appeared to have some amazing properties. They seemed to pass through solid objects as if they were sheets of transparent glass. Röntgen himself quickly discovered their use in taking photographs of the inside of the human body, publishing a photograph of the skeletal structure of a hand. Investigators were soon experimenting to understand the properties of the new rays. They could be reflected and refracted like beams of light but not, it seemed at first, diffracted. One of these experimenters, Henri Becquerel, soon came up with another new kind of ray, seemingly emanating from uranium salts. Inspired by Becquerel's discoveries, the Sorbonne student Marie Curie and her husband Pierre turned to study these new radiations as well. In 1898, they announced the existence of two new "radioactive" elements, polonium and radium, which gave off these new kinds of rays in copious quantities. The Curies argued that the source of the radioactivity seemed to be inside the atoms of their newly discovered elements.

As with X-rays, experimenters set out to investigate the properties of this mysterious radiation. Becquerel succeeded in deflecting it in a magnetic field, suggesting that it had negative charge. Thomson succeeded in measuring its charge to mass ratio, suggesting it was close to that of cathode rays. Thomson's student at the Cavendish, the New Zealander Ernest Rutherford, soon found that there was more than one kind of this radiation. Different kinds of radiation were stopped by different thicknesses of aluminum foil. Alpha rays were relatively easily stopped; beta rays were more persistent. The Frenchman Paul Villard showed in 1900 that there was an even more penetrating kind of ray—gamma rays—that seemed to pass through everything. By the early 1900s, Rutherford and his colleague Frederick Soddy were arguing that radioactivity seemed to emanate from inside the atom and—even more controversially—that in the process elements changed into other elements. Radioactivity appeared to be a source of energy from inside matter itself. It was soon suggested that it was the ultimate source of the sun's energy. It was established that beta rays were streams of Thomson's electrons. Rutherford suggested in 1905 that alpha rays were streams of positive ions of helium. Now based in Manchester, Rutherford used scintillation screens to count individual particles of radiation and worked at measuring their deflections in different magnetic and electric fields. Increasingly, it looked as if studying the new particles could unlock the secrets of the atom's interior.

In 1911 Rutherford announced his model of the atom, based on his latest experiments. He had been investigating the ways that alpha particles were scattered when passed through thin metal foil by looking at scintillations on a phosphorescent screen. These were difficult and sensitive experiments involving long hours of observing tiny flashes of light through a microscope in a darkened room. They also depended on access to the difficult-to-get radioactive sources. Only those with a secure supply of the precious radium could engage in such an enterprise. In the course of Rutherford's experiments it seemed as if some of the alpha particles bounced back off the metal foil. Rutherford was sure that each individual deflection was the result of a single interaction between an alpha particle and an atom. The alpha particles bouncing back from the foil must have done so as a result of having encountered a large and concentrated positive charge. This was the

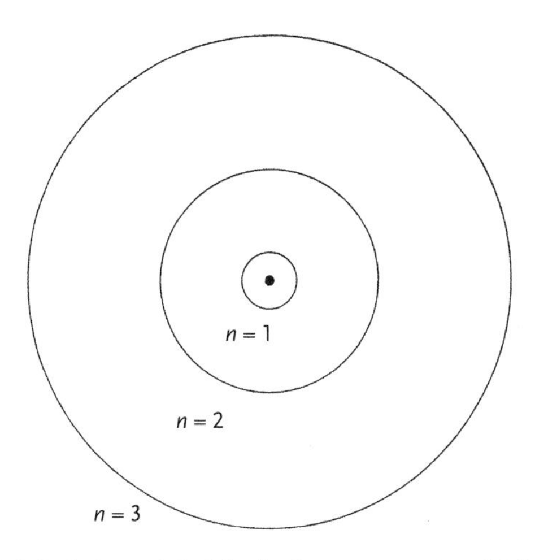

FIGURE 11.3 Niels Bohr's model of the hydrogen atom in which an electron could only orbit the central nucleus in orbits defined by Planck's Constant, *h*.

evidence on which he based his new model of atomic structure. He suggested that atoms were made up of a relatively large, positively charged core—the nucleus—surrounded by a number of relatively small orbiting electrons, just like the planets orbiting around the sun. Though seemingly simple, the model was not without its problems. In particular, Rutherford's model seemed to be unstable. According to physicists' understanding, the electrons orbiting the central nucleus should radiate energy as they did so. However, as they radiated energy they should also lose momentum and quickly end up spiraling into the central core. In other words, according to Rutherford's model, atoms should not exist—at least not for very long.

A young Danish physicist, Niels Bohr, suggested a solution to this problem. Bohr had worked with Thomson at the Cavendish and with Rutherford at Manchester. In 1913, Bohr suggested a model of atomic structure very similar to that proposed by Rutherford, but with one important difference. Bohr suggested that the electrons orbiting the central nucleus could only release their energy in distinct packets of energy, each with a distinctive frequency (fig. 11.3). This was how he solved the problem of atomic stability. The electrons orbiting the nucleus were not radiating continuously, they only did so at particular frequencies. Bohr was picking up on an idea first formulated by the German physicist Max Planck (of whom more later in this chapter) that energy could be released in quanta (i.e., discrete packets) defined by a constant factor—called Planck's constant (*h*) after its inventor. Albert Einstein had already made use of Planck's constant to argue that light could be treated as particles, each with an energy defined by the light frequency multiplied by *h*. What Bohr suggested was that atoms could exist in a number of stable states, each defined as a multiple of *h*. They only released energy when they changed from one state to another, and the energy they released in that process was a multiple of *h* and their change of frequency.

One crucial feature of Bohr's model of atomic structure was that it provided an explanation for the distinctive emission and absorption spectra of the different elements. It had been known for decades that the elements had distinctive spectra like this—different elements showed distinct dark lines in particular parts of the spectrum. This was how physicists used spectroscopy to identify the elements making up different substances: by comparing a sample to known elements and comparing their spectra they could use the spectral lines to identify the unknown elements. According to Bohr's model, this was because the individual atoms making up an element only vibrated at particular frequencies, corresponding to the spectral lines. In particular, Bohr's model explained Balmer's formula—an empirically derived formula worked out by the Swiss mathematician Johann Balmer showing that the position of these lines in the spectrum followed a regular pattern. Bohr managed to show that his equations fit the Balmer formula as well. The Rydberg Constant governing the relationship between the spectral lines was shown by Bohr to be itself a derivative of Planck's constant. Bohr had succeeded in bringing together the theory of discontinuous radiation pioneered by Planck and Rutherford's model of atomic structure. There was only one problem with the theory. It violated most of the then accepted laws of physics. British physicists such as Lord Rayleigh—J. J. Thomson's predecessor at the Cavendish—were unhappy with the introduction of the mysterious quantum. German theoretical physicists who had accepted Planck's views on the quantum of energy were unhappy with the idea that the atom was a real entity, let alone one whose physical structure could be discovered (Pais 1991).

Redefining Space and Time

One of the outstanding questions of late nineteenth-century physics was the issue of the earth's movement relative to the luminiferous ether. According to some theories it should be possible to detect the earth's movement through the ether by measuring differences in the velocity of light. To put it simply, when the earth was moving toward the light source, light should appear to be moving slower; when the earth was moving away

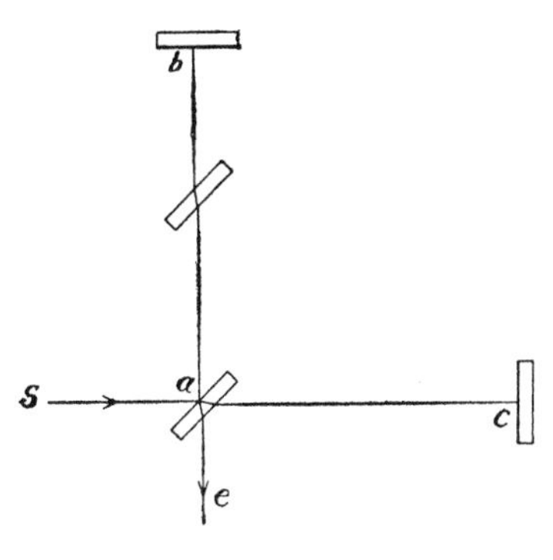

FIGURE 11.4 A diagram of the Michelson-Morley apparatus used in an attempt to measure the earth's movement through the ether. If the earth (and therefore the apparatuses) moved through the either, then the two beams of light aiming at the detector should arrive there slightly out of phase, causing an interference pattern, since one beam would have traveled slightly faster than the other. Michelson and Morley failed to detect any interference.

through the ether, light should appear to be faster. In 1888, two American physicists, Albert Michelson and Edward Morley, published the results of their experiments showing that they could detect no such deviation in the speed of light (fig. 11.4). Historians, philosophers, and physicists have often presented the experiment as a decisive refutation of the ether's existence. We will return to this point later. For the moment it should be enough to say that no physicist at the time—including the experimenters themselves—took it to be any such thing. At worst it was a problem to be solved, at best, to some it was even a potential confirmation of their own versions of ether theory. The extent to which the Michelson-Morley experiment played any role in the theoretical ruminations of the young Albert Einstein also remains a matter of considerable dispute, to which we will return again.

In 1905, when Albert Einstein published his paper on "The Electrodynamics of Moving Bodies" in the *Annalen der Physik,* he was an obscure patent examiner in Zurich, having graduated from the Zurich Polytechnic a few years previously. He had a handful of publications under his belt, but nothing to indicate that he was about to turn the world of physics on its head. In his 1905 paper, Einstein introduced two new principles into physics that led eventually to a completely new understanding of the nature of space and time. According to his principle of relativity, there was no privileged, absolute perspective from which to view events in the universe. All movement could only be measured relative to some particular frame of reference. Everything was relative—except for the velocity of light, which remained the same in all frames of reference. This was the second principle—the constancy in all frames of reference of the velocity of light. There was no such thing as Newtonian absolute space or absolute time according to this model. It turned out from Einstein's calculations that time itself was relative within this framework. Time as experienced within one frame of reference proceeded at a different rate from time as experienced from a frame of reference moving at a different velocity. In other words everything in Einstein's universe was relative.

Einstein's theory did not appear entirely from nowhere. The Dutch physicist Hendrik Antoon Lorentz had proposed the existence of a contraction effect in electrical charges moving at high velocities as a way of accounting for slight variations in the forces they exerted on one another. The Irish physicist George FitzGerald had made a similar suggestion. FitzGerald also suggested that this contraction effect accounted for Michelson and Morley's failure to measure the motion of the earth relative to the ether. According to FitzGerald, the contraction effect neatly counterbalanced the expected difference in the measured velocity of light. The mathematical equations translating the apparent dimensions of an object moving at one velocity from the perspective of someone at rest (or moving at a different velocity) were known as the Lorentz-FitzGerald transformations. In fact, questions like these to do with the electrodynamics of moving bodies (the title of Einstein's paper) were very much at the forefront of theoretical work on the properties of the ether, particularly by Cambridge-trained mathematical physicists such as FitzGerald or Joseph Larmor. What turned out to be different about Einstein's work, however, was the way in which he used electrodynamical calculations to propose a radical break not only with the ether but also with the Newtonian perspective that space was absolute.

Reactions to Einstein's new theory were mixed and slow in coming. To some commentators there appeared to be relatively little new in his formulation. It was certainly very easy for British-trained mathematical physicists to regard Einstein's contribution as just another paper on the electrodynamics of moving bodies, albeit possibly one written in unnecessarily obscure language. The science magazine *Nature,* for example, mentioned Einstein's views on relativity in the same breath as those of Larmor and the ether theory's foremost champion, Sir Oliver Lodge. German-trained theoretical physicists, more sympathetic to the tradition of research in which Einstein had been taught, were more receptive to the possibilities that his theory of relativity opened up. Einstein himself published a series of papers over the next few years, expanding and refining his theory. One of these supplementary papers contained his first proofs of the famous equation linking mass and energy, stating that the energy of a body is equal to its mass multiplied by the square of the speed of light. One of the first to respond positively to Einstein's theory was Max Planck, who presented a seminar on Einstein's theory at Berlin in 1905–6. In 1908, Hermann Minkowski, a former teacher of Einstein's at Zurich, gave a lecture at Göttingen in which he started to develop a simplified mathematical approach to relativity and introduced the possibility of expressing the relationship between space and time in terms of non-Euclidean geometry.

In 1907, Einstein published a review paper outlining work on the theory of relativity over the previous two years. In this review paper, he first raised the possibility that the scope of relativity theory might be expanded to consider systems undergoing relative acceleration as well as systems moving at constant velocities with respect to each other. He also suggested that relativity might be expanded into a theory of gravitation. It took him and others until 1915 to work out fully the implications of these suggestions and to produce what is now known as Einstein's general theory of relativity. According to Einstein's fully fledged theory, the principle of relativity did indeed apply to systems that were accelerating relative to each other. With the help of Marcel Grossman, a professor colleague at the Zurich Polytechnic, Einstein also developed a mathematical way to apply Minkowski's suggestions concerning non-Euclidean geometries of space and time to the theory of gravitation. They found a way to describe gravitation in terms of the curvature of space-time. Einstein's theory also suggested that the spectrum of light should shift toward the red end of the spectrum under the influence of a gravitational field. Another suggestion famously predicted that light rays would curve under the influence of gravity. In Minkowskian terms, light would continue to follow the shortest route between two points, but under the influence of gravity, space itself would be curved and so the shortest route that light could follow would be curved, too. General relativity also suggested that an observer would experience time differently in gravitational fields of different intensities.

One virtue that Einstein as well as other physicists saw in the general theory of relativity was that it appeared to be open to straightforward empirical confirmation. Einstein himself had already demonstrated that the theory could be used to account for anomalies in the orbit of Mercury that could not be explained using Newtonian gravitational theory. The real breakthrough came, however, when the British astronomer and enthusiast for general relativity Arthur Eddington announced his intention to test Einstein's prediction of light bending in a gravitational field during the forthcoming solar eclipse of 1919. Eddington aimed to use the opportunity of the eclipse to photograph the positions of stars around the sun's corona that would normally be obscured by the sun's light. By comparing their positions with those they appeared to occupy when the sun was not in their portion of the sky he could then determine whether light bending occurred as a result of the sun's gravitational field. The result was trumpeted as a stunning success for Einstein and general relativity theory. It was this decisive-seeming confirmation of his theory that made Einstein into a household name as newspapers across Europe and America splashed reports of the joint meeting of the Royal Astronomical Society and the Royal Society at which the announcement was made across their leading pages.

A great deal of ink had been spilled by historians, philosophers, and physicists alike over the issue of the relationship between Einstein's theories and their apparent empirical confirmations. One important focus of controversy has been the role of the Michelson-Morley experiment in Einstein's thinking leading up to the announcement of his special relativity paper. That paper makes no mention of the experiment, and Einstein in later years gave contradictory accounts of whether he had been aware of Michelson-Morley at the time. The experiment is nevertheless frequently cited as a decisive factor in the formulation and reception of relativity theory. It is also cited as a decisive refutation of the ether, with the efforts of ether theorists to accommodate it into their theoretical frameworks derided as clumsy post hoc rationalizations. Another focus of controversy is the role of Eddington's eclipse observations. Historians and philosophers have argued that the data Eddington and others provided as, in point of fact, ambiguous. They could have been interpreted differently so as to support classical Newtonian theory (which also predicts some light bending) rather than general relativity (Earman and Glymour 1980). What matters for the historian in cases like these is how the relevant information was used at the time, rather than how it might (or should) have been used—in which case, the Michelson-Morley experiment was clearly not decisive, while the Eddington observations were.

The relatively rapid acceptance of Einstein's theories—in some circles at least—is frequently described in terms of a decisive refutation of ether theory. As we just mentioned, the Michelson-Morley experiment is usually described as having struck the first blow, while Einstein's theory delivered the coup de grâce. As we have seen, however, the reality was rather more complex. Some ether theorists positively welcomed the Michelson-Morley results as confirmation of their versions of ether theory. This was how some contemporaries understood Einstein's theory initially as well. It was another theory that seemed to support the view of some theorists that the earth's motion through the ether could not be measured. What were more decisive in the reception of Einstein's theories were the changing institutions of physics itself. The tradition of mathematical physics, as taught at Cambridge, for example, was dying out. The newer German tradition of theoretical physics was in ascendancy (Jungnickel and McCormmach 1986). To the increasing numbers of physicists turning to new German theoretical practices and techniques, Einstein's theories looked more promising than the antiquated approaches of the previous generation. New physics research institutes—again predominantly in Germany and in countries that had adopted the German approach to physics—were also producing a new generation of physicists trained in the highly sophisticated and difficult to master mathematical techniques that Einstein adopted. To this new generation, Einstein's approach and that of others like him seemed more familiar, more powerful, and more promising.

The Uncertainty Principle

In the same year that Einstein published his paper on special relativity he also published another ground-shaking contribution, this time on the anomalous behavior of light. It was known that shining a beam of light onto certain substances caused some kind of electric emission. Hertz had noticed the phenomenon in 1887 during the course of the experiments that would lead him to electromagnetic waves (see chap. 4, "The Conservation of Energy"). In 1899, J. J. Thomson suggested that this photoelectric effect was the result of a stream of electrons being emitted from the substance. One peculiar feature of this photoelectric effect was that it seemed to depend on the frequency rather than the intensity of the beams of light. Hertz had noted that the phenomenon seemed to be a property of ultraviolet light, in particular. What Einstein suggested in his 1905 paper was that this phenomenon could be understood by assuming that under these circumstances light acted like a particle rather than a wave. He could then show that the energy required to make one electron leave the surface of the metal was given by the frequency of the light multiplied by a constant. It was as if light traveled in packets, each carrying just that amount of energy. When these light quanta, or photons, struck an electron, that energy was transferred to it.

The constant in Einstein's equation was Planck's constant, which we first encountered a few paragraphs ago. The physicist Max Planck had invented the number in the course of his investigations of the phenomenon of black body radiation. A black body was a theoretical construct that absorbs and emits radiation at all frequencies. The physicist Wilhelm Wien, during the 1890s, worked out the equations that dealt with this hypothetical situation by treating the radiation as an example of thermal equilibrium and applying the laws of thermodynamics, particularly those relating to entropy. As experimenters began to produce experimental set-ups that approximated a perfect black body; however, it started to become clear that the experimental data did not fit. Lord Rayleigh and James Jeans developed an alternative formulation that worked well for low frequencies of radiation but at higher frequencies was prone to the "ultraviolet catastrophe": the energy released was a function of the square of the frequency, which meant that at higher frequencies (like that for ultraviolet light) it veered toward infinity. Planck succeeded in producing his own solution to the problem that avoided the ultraviolet catastrophe at the expense of what looked to many like a deeply unsatisfactory fudge. He had to assume that the energy was released in packets depending on the frequency of the radiation multiplied by a constant factor. That factor was Planck's constant—what he called the quantum of action.

As we have seen already, Niels Bohr made good use of Planck's quantum of action when he was putting together his model of atomic structure. Bohr used Planck's constant to help define the different energy states in which the electrons orbiting the central nucleus of an atom could remain stable. Despite the model's success in explaining the empirical data derived from experiments such as those carried out by Rutherford at Manchester, as well as its heuristic value in suggesting new theoretical developments, many physicists—Bohr himself included—felt deeply unsatisfied with it. The problem was simple. It seemed that the Bohr model—and the quantum theory built around it—was a halfway house between classical physics and something else. The Bohr model was "classical" in that it largely followed the rules and assumptions of Newtonian mechanics. The atom consisted of discrete particles—electrons—orbiting a central core—the nucleus—in well-defined orbits. The only difference was that they could change orbits, indeed could change orbits only, according to principles that violated fundamental mechanical principles. By the 1920s, Bohr and other physicists were actively trying to find new and foundational physical principles that would allow them to make sense of quantum theory. Their problem was not with the physics of the Bohr model; it was with its metaphysics.

One of the first efforts to move toward an alternative formulation was the work of the young German physicist Werner Heisenberg. In 1924, Heisenberg spent six months in Copenhagen, carrying out research at the Institute for Theoretical Physics that Bohr had established. This kind of close collaborative working was to be crucial in the events that followed, as the key players met and worked together at colloquia, conferences, and research institutes. Frustrated by the ad hoc appearance of quantum theory, Heisenberg wanted to return to first principles and produce a completely new mathematical technology to deal with the phenomena. He wanted to do away with such theoretical concepts as atomic orbitals that had, in principle, no observable attributes. In his quantum mechanics (as he called it), Heisenberg replaced the notion of atomic orbitals with the assumption that atoms exist in different quantum states that can be mathematically defined. Following the suggestion of his mentor Max Born, Heisenberg used the mathematical notation of matrix calculus to express the different possible quantum states. At about the same time in Cambridge, another young physicist, Paul Dirac, was working toward a similar theory. Heisenberg and his allies were quite self-consciously jettisoning the trappings of classical physics and trying to base their procedures on a wholly new observational foundation.

A different approach to the anomalies of quantum theory was also being developed based on the suggestion of the young French aristocrat Louis de Broglie. Inspired by Einstein's suggestion in his 1905 paper that light occasionally behaved like a particle, de Broglie suggested in 1923 that under certain circumstances it might be possible to treat particles (electrons specifically) as if they were waves. He suggested that the electrons orbiting a nucleus could be described as existing in a stationary wave with the different possible orbitals then being defined as the range of possible frequencies at which that stationary wave could oscillate. The suggestion was taken up and expanded a few years later by the Viennese physicist Erwin Schrödinger. Schrödinger's particular achievement in his formulation of wave mechanics (as he called his theory) in 1926 was to derive a wave equation for the hydrogen atom, showing that it was possible to calculate stationary wave states that corresponded to each of Bohr's orbital levels. Where Heisenberg saw himself as quite self-consciously doing away with classical physics, Schrödiger regarded his wave mechanics as a continuation of the classical tradition. It was clear, however, that, as the physicist Wolfgang Pauli argued and Schrödinger acknowledged, wave mechanics and quantum mechanics were, formally at least, different but equivalent mathematical expressions of the same state of affairs. What remained unclear was just what that state of affairs was.

Schrödinger himself offered one early response to the question of how to interpret this new physics. He suggested that the wave packets described by his theory held together over time and that particles should be visualized as simply being tightly held together wave packets. In that case, there was no discontinuity between classical and wave mechanics. A more radical interpretation was offered by Max Born. In his view, the best way to understand quantum mechanics was by invoking statistics. In a paper published in 1926 on the quantum mechanics of a beam of particles being scattered by a center of force, Born suggested that the best way to interpret the equations was as expressions of probabilities. In other words, what his equations showed in terms of the effect on individual particles colliding with the center of force was not what happened but what probably happened. Where Schrödinger wanted to preserve the link with classical approaches by ditching particles, Born wanted to preserve the usefulness of particle-based physical explanations while defining a concrete meaning for wave equations. His conclusion was that the wave equations were expressions of probability distribution. Increasingly, battle lines were being drawn up around this issue: What did quantum mechanics mean? What kind of picture of the world did it project?

The protagonists gathered in Copenhagen in 1926 and 1927. Bohr, Schrödinger, and Heisenberg met in October 1926 when Schrödinger gave a lecture there at Bohr's invitation on the foundations of wave mechanics. Heisenberg had already heard him give a similar talk in Munich and was horrified by his attempts to produce a classical interpretation of quantum mechanics. Schrödinger was likewise unimpressed by Bohr's and Heisenberg's leaps between quantum states and Born's probability interpretation. Heisenberg was back in Copenhagen in early 1927, still working on a satisfactory physical interpretation of the new physics. The result was an abandonment of the laws of classical causality and the establishment of the uncertainty principles. According to Heisenberg, it was not possible in the quantum world to state definitively that a particular state of affairs would definitively cause another state of affairs. Before the event, all that could be known were probabilities. This was because there were limits as to what could, in principle, be known about any state of affairs. It was impossible to know both the location and the momentum of a particle with equal precision. Similarly, it was impossible to know the energy state of an object and the time at which it was in that state with equal precision. The focus was on the observable phenomena. Bohr's way of putting it was that the question of whether an electron was a particle or a wave was no longer relevant. What mattered was whether and under what circumstances it behaved like a particle or a wave.

The Copenhagen interpretation was and remains controversial. Schrödinger never accepted it—hence the famous paradox of Schrödinger's cat. In this paradox, Schrödinger described a hypothetical experiment in which a cat, confined in a box, was subjected to a process that either would or would not kill it, depending on the outcome of a particular event at quantum level, such as a vial of poison being released only if it was triggered by the emission of a single electron from an atom. According to the Copenhagen interpretation, the decisive quantum event could not meaningfully be said to have taken place until its outcome was actually observed. Until then, all that could be said was that there was a superposition of quantum states. But that would mean that until somebody opened the box and looked inside, the cat could not meaningfully be said to be either dead or alive. It would exist in a superposition of states, both dead and alive. Schrödinger regarded this as a reductio ad absurdum argument, revealing the inherent absurdity of the Copenhagen position (Wheaton 1983; Darrigol 1992).

Another famous dissident was Albert Einstein, who never accepted that quantum mechanics was really "the secret of the Old One . . . that *He* does not play dice." Some historians have argued that the wholesale rejection of classical notions of causality that underpinned the Copenhagen interpretation can be traced to the cultural pessimism of postwar Germany's Weimar Republic. According to this view, quantum mechanics should be seen in the same light as the philosophical, literary, and artistic rejection of classical forms of rationality that followed Germany's defeat in the Great War (Forman 1971). There is clearly some truth in the suggestion, though it does little to explain quantum mechanics' success elsewhere or its continued hold on contemporary theoretical physics. The explanation for that is more likely to lie—as we have argued it does for relativity theory—in the appeal of new, powerful, and esoteric mathematical technologies to a new (almost the first so trained) generation of theoretical physicists and in the power of the institutional traditions that generation established. It is also worth noting the relatively small size and the mobility of the group involved in founding quantum mechanics. They knew each other; they traveled constantly among each others' research institutions and met frequently at recently inaugurated international events, such as the Solvay Conferences. In that respect, quantum mechanics was successful precisely because it was a team effort.

Big Physics

By the 1920s, Ernest Rutherford, by then J. J. Thomson's successor as director of the Cavendish Laboratory in Cambridge, was well-established as one of the world's foremost investigators of the atom's interior. The apparatus he and his coexperimenters used was—by the modern standards for such experiments with which we are more familiar today—deceptively modest and simple. Rutherford and his team bombarded sheets of metal foil with radiation from a radioactive source such as radium. Their goal was to find out how the path of the radiation changed as it passed through the foil so they used phosphorescent screens to capture the individual scintillations as the particles of radiation struck. The problem with studying the trajectories and properties of these subatomic particles was simple—how to detect them? Rutherford's Manchester colleague Hans Geiger had developed a number of different techniques to record the incidence of radiation. Working at the Physikalisch-Technische Reichsansalt after 1912, he developed what became known as the Geiger counter for counting alpha particles. The Cambridge graduate C. T. R. Wilson developed another important device. In the process of trying to produce artificial clouds in the laboratory, he found that tiny drops of water would collect around individual ions, leaving a visible trace. Using Wilson's cloud chambers, as they were called, it was possible actually to trace the movements of individual particles of radiation.

Maybe the greatest triumph of the Cambridge school of nuclear physicists that built up around Rutherford was James Chadwick's identification of a new subatomic particle—the neutron. In 1928, the German physicists Walter Bothe and Herbert Becker had found that when a sample of the metallic element beryllium was bombarded with alpha particles, it gave off an electrically neutral radiation, which they took to be gamma rays. A few years later in 1932, Irene Joliot-Curie (Marie Curie's daughter) and her husband Frederic found that this radiation caused protons (positive subatomic particles, taken at the time to be one of the constituents of the nucleus along with equal numbers of electrons) to be emitted from a paraffin target. Chadwick repeated the Joliot-Curies' experiments using other elements as well as targets. By comparing the energies of the charged particles emitted by the different targets he concluded that the electrically neutral radiation was not gamma rays but a stream of neutral particles of approximately the same mass as the proton. This was the neutron. Not only did the discovery—for which Chadwick won the Nobel Prize in 1935—provide more information about the structure of atoms, it also provided a powerful new tool for further research. Being electrically neutral, streams of neutrons were highly penetrative and could be used to delve even further into the atom's heart.

In 1928 the Soviet physicist George Gamow published an explanation of alpha particle radiation in terms of quantum mechanics. It was one of the first efforts to apply the new tools of theoretical physics to understanding the subatomic particles and processes that the radioactivists had been investigating for the past decade. Gamow showed that alpha particle emission was not the result of some random and arbitrary instability in the atomic nucleus but a straightforward consequence of the laws of quantum mechanics (an effect now known as quantum tunneling). During the 1930s, theoretical physicists were increasingly interested in finding ways of interpreting the new information provided by nuclear physicists—particularly the new information about the interior of the nucleus that could be gleaned using the newly discovered neutron. Heisenberg suggested that the contents of the nucleus were held together by a new kind of force, that these nuclear forces must act only at very short ranges, and that they were about a million times stronger than the electrostatic forces holding the atom together. From the mid-1930s onward, Niels Bohr elaborated his theory of the nucleus in which it was regarded as similar in many ways to a drop of liquid. Bohr and his co-worker Fritz Kalchar argued that just as drops of liquids vibrate when force is applied to them, so does the atomic nucleus, and that those different states of vibration could be regarded as quantum states.

With the outbreak of war, many nuclear and theoretical physicists found themselves working for their respective sides' war efforts. Heisenberg played a key role in Nazi efforts to produce nuclear weaponry. Einstein was one of the instigators of a letter to Franklin Roosevelt, the U.S. president, which was instrumental in bringing about the Manhattan Project. By the end of the Second World War, a great deal more was known about nuclear physics than at its outset. The bombings of Hiroshima and Nagasaki had made the consequences of splitting the atom terrifyingly explicit. On both sides, too, the war efforts had resulted in unprecedented resources in manpower and money being directed toward nuclear physics. For the first time, physics was starting to become a matter of large-scale collective effort (see chap. 20, "Science and War"). When nuclear physicists met at the Cavendish Laboratory in Cambridge in 1946 for their first conference since the beginning of the war, their field appeared to be booming. The number of elementary subatomic particles had certainly proliferated. The list now consisted of electrons, mesons, neutrons, neutrinos, photons, positrons, and protons. Mesons had been predicted by the Japanese physicist Hideki Ukawa in 1935 as a means of explaining the transmission of nuclear forces. They were identified in cosmic ray studies a few years later. Positrons (positively charged electrons) had been theoretically predicted by Paul Dirac at Cambridge and found at CalTech in the early 1930s. Neutrinos were hypothetical particles, invoked to preserve the conservation of energy in certain interactions involving beta particles. They were not universally accepted at first. Bohr initially preferred to abandon the principle of the conservation of energy rather than accept the existence of particles of whose existence there was no other evidence. By about 1936, however, he was leaning toward acceptance of the physical reality of neutrinos.

By the 1940s, experiments in nuclear physics were rapidly leaving behind the tabletops on which they were first carried out. Experimental apparatus during the 1920s and early 1930s was relatively small scale. The main piece of apparatus that Chadwick used in identifying the neutron was only six inches long. His was the last discovery of a subatomic particle to take place using apparatus like this. By the 1950s and 1960s, chasing subatomic particles needed massive equipment and equally massive investments of labor and money. The trend was well underway by the beginning of the Second World War. When the Italian physicist Enrico Fermi carried out the first controlled nuclear chain reaction of 1942, he needed a laboratory the size of a squash court (it was, in fact, a squash court beneath the University of Chicago's football stadium). After the war, Fermi became head of the Institute of Nuclear Physics at Chicago, where in 1951 he played a key role in developing their synchrocyclotron, a massive piece of equipment in which subatomic particles were accelerated to high velocities before hitting a target so that their properties and constitution could be studied. It was one of the first of a new generation of increasingly powerful experimental apparatuses. By the later 1950s, instruments like this were already several meters in diameter. It was this kind of experiment that was starting to make terms such as "elementary" or "fundamental" increasingly dangerous words in particle physics.

By the early 1960s, two kinds of elementary particle were generally recognized. There were hadrons—particles like protons and neutrons that made up the nucleus—and leptons—like electrons. By 1964, however, this picture was starting to fall apart. Experiments with ever more powerful particle accelerators seemed to suggest that hadrons were not elementary particles after all. They were made of other particles, eventually dubbed quarks. The suggestion was first made on theoretical grounds by an American physicist, Murray Gell-Mann, working at the California Institute of Technology, along with the Russian-born George Zweig, then working at the Conseil Européen pour la Recherche Nucléaire Laboratory in Switzerland. There were three kinds of quarks: "up," "down," and "strange" quarks. Different combinations of quarks came together to produce the range of hadrons. Quarks rapidly became very useful theoretical entities. They could be used to explain a great deal about the different quantum states of nuclear particles. The question of whether quarks really existed was nevertheless a matter of considerable debate. Many physicists argued that quarks were simply useful ways of organizing information rather than being real physical objects. Part of the problem was that quarks were difficult to find, despite the fact that given their properties—particularly the fact that they were supposed to have fractional electric charges—they should be relatively conspicuous. It was well into the 1970s before their physical reality was generally accepted (Pickering 1986).

The kind of physics that produced quarks was increasingly esoteric and technical. It also needed massive resources. The European contribution to particle physics by the 1950s needed international cooperation. The CERN particle accelerators built in Switzerland near the border to France were (and still are) literally huge enterprises, with instruments several kilometers in diameter. These massive enterprises have also required massive manpower. By the early 1960s, it is estimated that there were about 685 practicing particle physicists in Europe and an additional 850 in the United States. By the 1970s, the European figures had more than quadrupled and the American figures had doubled. Such projects were matters of national prestige as well. Successive American and European governments throughout the 1960s and 1970s poured increasingly large amounts of money into high energy particle physics (fig. 11.5). This was a very far cry from Rutherford's or Chadwick's tabletop experiments at the Cavendish Laboratory a half-century or so previously. High-energy particle physics was collaborative science par excellence. It also came to manifest a high degree of separation between experimenters and theoreticians. Where J. J. Thomson or the Curies at the beginning of the twentieth century combined theory and experi-

FIGURE 11.5 The site of a late twentieth-century particle accelerator (courtesy of Fermilab, Batavia, IL). Comparing this picture with the apparatus illustrated in fig.11.2 gives a graphic example of the change in scale in experimental physics over the intervening century.

ment in their activity, that combination became increasingly rare during the second half of the century. Doing theory or doing experiments came to require wholly different kinds of expertise.

Conclusions

The founders of relativity theory and quantum mechanics at the beginning of the last century certainly regarded themselves as engaged in a revolutionary process. They were overturning classical physics and replacing it with a wholly new intellectual edifice. In many ways, though, it was precisely through this dismantling that the idea of classical physics as a coherent and self-contained body of thought was established in the first place. It was defined as being what the new physics was not. This rift with the past was not, however, as inevitable or clear-cut as some of its proponents, at least, argued. We have seen that there were clear continuities between developments in relativity and quantum theories and previous approaches. Some of the new physics' own founders had mixed feelings about the abandonment of the old certainties. As we have seen, both Einstein and Schrödinger, for example, never reconciled themselves entirely to the withdrawal of physics from causality. Even Niels Bohr was considerably more ambivalent about the prospect than was Heisenberg—the real enthusiast for uncertainty. Throughout the century, physics also became an increasingly esoteric practice (or more accurately, set of practices). Becoming a physicist required years of extended and dedicated training. This only seems unsurprising to us because this is the scientific culture we live in too. It is easy to forget that nothing like it had existed before. Physics became an increasingly fragmented business as well, with experimenters and theorists inhabiting different institutes and different worldviews. New specialisms, such as solid state physics, developed, which crossed over old boundaries between academic and industrial science.

It is clear, moreover, that it is impossible to separate out the intellectual and institutional stories of twentieth-century physics. The institutions where physics was practiced had a massive impact on what physics was. The kind of highly skilled, intensive, and mathematically abstruse practice that theoretical physics became during the course of the twentieth century depended entirely on the existence of intensive, specialized research, and training institutes where it largely took place. It was an activity that could not take place without the cadres of thoroughly trained, specialized, and dedicated professional physicists that such places produced. Experimentation, too, was no longer the province of an individual scientist with a small team of helpers and technicians. An experiment at CERN or Fermilab required the mobilization of hundreds, if not thousands, of scientific workers. Physics became big business during the course of the twentieth century, requiring resources on a hitherto unprecedented scale. The number of people calling themselves professional physicists swelled by orders of magnitude during the course of the century. This was not an incidental feature in the development of modern physics. Without those resources and institutions, physics as it was practiced simply would not have been possible. The institutional shape of modern physics was an indispensable prerequisite of its intellectual content.

References and Further Reading

Cassidy, David. 1992. *Uncertainty: The Life and Science of Werner Heisenberg*. New York: Freeman.

Darrigol, Oliver. 1992. *From C-Numbers to Q-Numbers: The Classical Analogy in the History of Quantum Theory*. London and Berkeley: University of California Press.

Earman, John, and Clark Glymour. 1980. "Relativity and Eclipse: The British Expeditions of 1919 and Their Predecessors." *Historical Studies in the Physical Sciences* 11:49–85.
Forman, Paul. 1971. "Weimar Culture, Causality, and Quantum Theory, 1918–1927: Adaptation by German Physicists and Mathematicians to a Hostile Intellectual Environment." *Historical Studies in the Physical Sciences* 3:1–115.
Galison, Peter. 1987. *How Experiments End.* Chicago: University of Chicago Press.
Galison, Peter, and Bruce Hevly, eds. 1992. *Big Science: The Growth of Large-Scale Research.* Stanford, CA: Stanford University Press.
Heilbron, John, and Thomas Kuhn. 1969. "The Genesis of the Bohr Atom." *Historical Studies in the Physical Sciences* 1:211–90.
Jungnickel, Christa, and Russell McCormmach. 1986. *The Intellectual Mastery of Nature.* Vol. 2. Chicago: University of Chicago Press.
Keller, Alex. 1983. *The Infancy of Atomic Physics.* Oxford: Clarendon Press.
Kragh, Helge. 1990. *Dirac: A Scientific Biography.* Cambridge: Cambridge University Press.
Kuhn, Thomas S. 1978. *Black Body Theory and the Quantum Discontinuity, 1894–1912.* Oxford: Clarendon Press.
Nye, Mary Jo. 1996. *Before Big Science.* Cambridge, MA: Harvard University Press.
Pais, Abraham. 1982. *Subtle Is the Lord: The Science and the Life of Albert Einstein.* Oxford: Clarendon Press.
———. 1991. *Niels Bohr's Times in Physics, Philosophy and Polity.* Oxford: Clarendon Press.
Pickering, Andrew. 1986. *Constructing Quarks.* Chicago: University of Chicago Press.
Segré, Emilio. 1980. *From X Rays to Quarks: Modern Physicists and Their Discoveries.* San Francisco: W. H. Freeman.
Wheaton, Bruce. 1983. *The Tiger and the Shark: The Empirical Roots of Wave-Particle Dualism.* Cambridge: Cambridge University Press.
Whitaker, Edmund. 1993. *History of the Theories of Aether and Electricity.* Vol. 2. London: Nelson.

CHAPTER 12

REVOLUTIONIZING COSMOLOGY

We tend to take the modern view of the cosmos and our place in it very much for granted. Modern astronomers regard the planet Earth as being an undistinguished planet, orbiting a fairly unremarkable star on the outer fringes of an unexceptional galaxy—one of an indefinitely large number of galaxies in an indefinitely large universe. In the words of Monty Python in *The Meaning of Life:*

> Our galaxy itself contains a hundred billion stars.
> It's a hundred thousand light years side to side.
> It bulges in the middle, sixteen thousand light years thick,
> But out by us, it's just three thousand light years wide.
> We're thirty thousand light years from galactic central point.
> We go 'round every two hundred million years,
> And our galaxy is only one of millions of billions
> In this amazing and expanding universe.

This view of the universe and of humans' place in it is, however, of very recent origin. Until the 1930s, there was no consensus among astronomers concerning the size and shape of the Milky Way (our own galaxy) or the planet Earth's position within it. There was no consensus over the question of either whether the Milky Way was a unique structure in the universe or whether other galaxies even existed. According to one astronomer at least, "the realization . . . that our galaxy is not unique and central in the universe ranks with the acceptance of the Copernican system as one of the great advances in cosmological thought" (Berendzen, Hart, and Seeley 1976).

From this perspective, then, the emergence of the modern view of the cosmos ranks as a scientific revolution comparable to one of the defining events of the Scientific Revolution itself. There are certainly parallels between the change in perspective entailed by the development of modern cosmology and the Copernican revolution—as it has traditionally been portrayed, at least. Copernicus challenged late medieval presumptions about humanity's place in the cosmos by removing the earth from the center of the universe. Modern cosmology completed the task and removed the last vestiges of human uniqueness by relegating even the galaxy that we inhabit to the backwaters of the universe. There are certainly senses in which this twentieth-century cosmological revolution might be taken as a classic case study of a Kuhnian scientific revolution. In particular, as we shall see, it illustrates Kuhn's point concerning the subjectivity of observational evidence. Astronomers engaged in debates about the size and shape of the universe interpreted data differently depending on their various views of what the cosmos was really like, just as Kuhn suggests that different observers with different views as to what is "really there" might see either a duck or a rabbit in the same picture (Kuhn 1962). It is also a good example of more recent sociological points concerning the importance of issues such as training, institutional affiliation, and personal relationship in the resolution of scientific controversies (Barnes 1974; Collins 1985).

As we have seen earlier, the predominant ancient Greek view of the universe was that it was finite, with the earth as its center and bounded by the sphere of fixed stars. By the late Middle Ages and Renaissance, this picture was coming under increasing attack with the advent of Copernicus's heliocentric system. As far as Newton was concerned, space—and therefore the universe—was infinite. During the eighteenth and nineteenth centuries, a range of opposing views concerning the structure of the universe was developed. Some, like Immanuel Kant, argued that nebulae represented other galaxies like the one in which the earth was located. Others argued that nebulae were clouds of gases from which other solar systems like our own would eventually develop. During the second half of the nineteenth century, new tools such as photography and spectroscopy were used to look deeper into space and to identify the elements that made up celestial objects. By the first decades of the twentieth century, arguments concerning the size and shape of the universe hinged on different views concerning the nature and distance of nebulae. The consolidation of Einstein's new theory of general relativity during the 1910s and 1920s also had important ramifications for arguments concerning the size of the cosmos. Einstein reckoned that he could use his relativistic field equations to understand the geometrical structure of space and time. Einstein's universe was static. Others disagreed, suggesting that evidence showed the universe to be expanding.

By the middle of the twentieth century, two opposing models of an expanding universe had been developed. According to one view, it was possible to use observations of the universe's rate of expansion to extrapolate back through time to the universe's beginnings. This was what came to be known as the "big bang" theory of the universe. Big bang proponents argued that all the matter currently in the universe was originally concentrated at one point. It was the explosion and subsequent expansion of that point—the original big bang—that had brought the modern cosmos into being. Opponents of the big bang, such as the British astronomer Fred Hoyle, argued that the universe had no discrete beginning. It had always existed and would continue to exist indefinitely. New matter was continually being produced throughout the universe to fuel its steady expansion. This was the "steady state" model of the universe. By the closing decades of the twentieth century, however, the big bang model of the universe was increasingly dominant. Increasingly in accounts of the universe, the modern cosmos was described as being populated by bizarre entities like black holes, pulsars, and wormholes. By the end of the twentieth century, new technologies had been developed that allowed astronomers to claim to be able literally to see back to the very beginnings of the cosmos.

The Shape of the Universe

Is the universe something that can meaningfully be said to have a shape or a size? According to ancient Greek views of the cosmos, the answer would presumably have been yes. The universe was spherical, with the earth at its center and the orb of fixed stars as its outer boundary. As this basic Aristotelian model of the universe was adopted and adapted in medieval Europe, outside the sphere of the fixed stars was Heaven. By the end of the Scientific Revolution and the gradual adoption of first Copernicus's heliocentric universe and then Kepler and Newton's views of the mechanism of the heavens, the crystalline celestial spheres had long been abandoned as real physical entities (Kuhn 1966). By the middle of the eighteenth century, astronomers were largely agreed that Sir Isaac Newton's theory of gravitation provided the best explanation for the movements of celestial objects. Newton's universe was infinite, absolute, and unchanging. It had come into being at the moment of creation. It had no boundaries of any kind, it simply stretched to infinity. Beyond the confines of the earth's own system, where the earth along with the other planets orbited around the central sun, there was nothing except the stars, distributed more or less uniformly and in infinite numbers. From this perspective, it was certainly not at all clear that questions about the shape and size of the universe made any kind of sense at all.

In 1750, however, the Englishman Thomas Wright published *An Original Theory or New Hypothesis of the Universe,* in which he proposed a specific structure for the universe. In Wright's model, the universe consisted of two concentric spheres with the stars sandwiched between them. At the center of the universe was the throne of God. Wright had observational evidence to support his model. The highly luminous band of stars visible in the night sky—the Milky Way—was the result of looking along the tangent of the spheres. The German philosopher Immanuel Kant, better known as the author of the *Critique of Pure Reason,* published his *Universal Natural History and Theory of the Heavens* in 1755, in which he argued that the Milky Way was only one of a number of similar "island universes" scattered throughout the cosmos. Having read a slightly ambiguous account of Wright's theory, he understood him as suggesting that the Milky Way was a disk of stars seen lengthways and adopted the suggestion. When the Anglo-German astronomer William Herschel—famous for his discovery of the planet Uranus—started mapping the heavens with his powerful new telescopes and identifying a number of glowing stellar clouds, or nebulae, in the skies, these were often identified as being island universes. Herschel himself originally agreed that nebulae were extragalactic systems of stars but later observations led him to doubt the claim (Hoskin 1964).

William Herschel's compendious observations of nebulae provided important evidence for a theory of the solar system's origins that became increasingly popular in some astronomical circles during the first half of the nineteenth century. The so-called nebular hypothesis put forward by the French physicist Pierre-Simon Laplace argued that nebulae were massive clouds of gaseous matter that formed the birthplaces of stars and planets. The swirling clouds of gases gradually coalesced over time, forming clumps of matter orbiting around a central mass. These eventually evolved into planets orbiting a star. The nebular hypothesis was particularly popular in Britain, where it was advocated by such radical popularizers as John Pringle Nichol and Robert Chambers. In his notorious *Vestiges of the Natural History of Creation* published in 1844, Chambers used the nebular hypothesis to argue that the universe was in a state of continuous evolution and progress, suggesting that the same applied to human beings and their societies. The nebular hypothesis depended on the claim that nebulae were clouds of stellar gas rather than collections of stars. The Anglo-Irish astronomer Lord Rosse, during the 1840s, famously used the enormous seventy-two-inch reflecting telescope built at Birr Castle, his Irish family seat, to resolve the

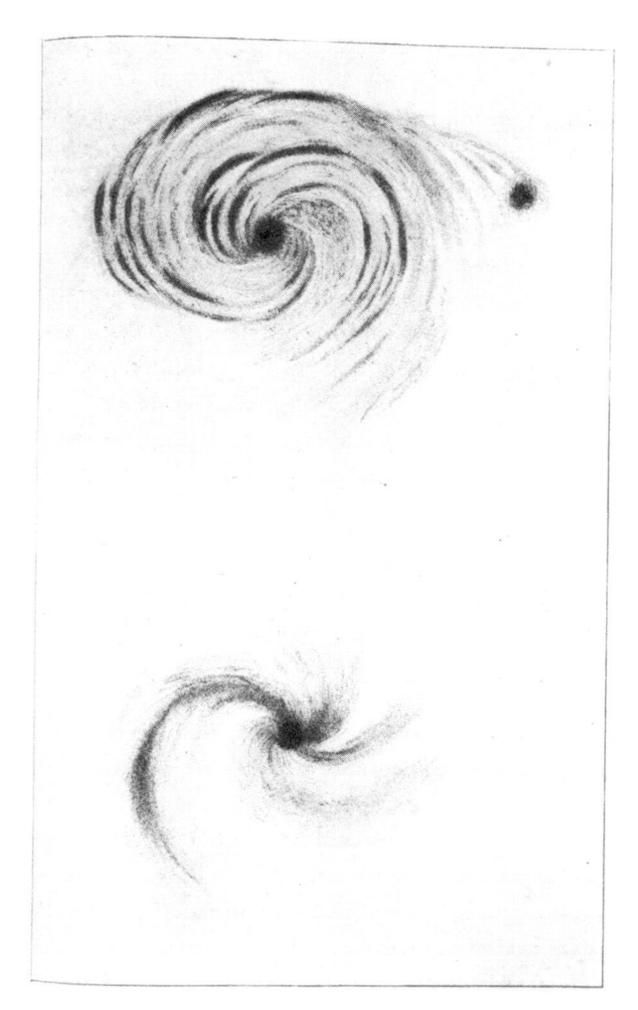

FIGURE 12.1 Lord Rosse's depiction of a spiral nebula as seen through the Leviathan of Parsonstown.

Orion Nebula into its constituent stars in an effort to disprove the nebular hypothesis (fig. 12.1). Despite Rosse's efforts, however, doubts remained as to whether all nebulae could be resolved into collections of stars or whether some were "true" nebulae made up of clouds of gases (Jaki 1978).

Developments in photography and spectroscopy during the second half of the nineteenth century also provided new ammunition for ongoing debates about the true constitution of nebulae and other celestial objects. Some astronomers hoped that photography might be able to capture features of distant objects in the night skies that fallible human eyes might miss or misinterpret. Chemicals reacting to light might prove more sensitive than mere eyesight and provide a permanent and objective record of what was really there. They might be able to distinguish between clusters of stars and clouds of gases in a way that human senses could not. Spectroscopy, the other addition to astronomers' armory during this period, had its origins in the observation that different substances burned with different colors or gave off differently colored electric sparks when used as electrodes. When such light was viewed through a prism it gave a spectrum unique to each particular element. The German instrument-maker Josef von Fraunhofer had also noted that light from the sun exhibited characteristic lines in its spectrum when viewed through a prism (Jackson 2000). By turning their spectroscopes on celestial objects and comparing the spectra they produced to those produced by terrestrial elements, astronomers could try to identify the elements that made up stars and nebulae. As we shall see below, by examining the shift in these spectral lines toward the red end of the spectrum (so-called redshift), thought to be caused by light sources moving away from the earth, astronomers could even come up with estimates of the speeds at which distant stars and other celestial objects were moving through the heavens. By the beginning of the twentieth century, photography and spectroscopy were standard tools of observational astronomy, vital to the task of distinguishing different kinds of objects in the night skies.

During the first few decades of the twentieth century, there were two dominant and competing theories concerning the nature of nebulae, both of which had important implications for astronomers' views concerning the size and shape of the universe. According to one view, at least some nebulae—particularly spiral nebulae—were galaxies similar to our own Milky Way. According to the other view, nebulae were dense clusters of stars or gaseous clouds within the confines of the Milky Way. In deciding between these two opposing views, much rested on astronomers' differing views concerning the size of the Milky Way, the solar system's position within it, and the distances between the solar system and the various nebulae. Matters came to a head in a famous encounter in Washington, DC, in 1920—the so-called great debate—between Harlow Shapley of the Mount Wilson Observatory and Heber D. Curtis of the Lick Observatory. Shapley argued that our own galaxy was of a massive size, about 300,000 light years in diameter, with the galactic center about 65,000 light years from Earth. Globular star clusters and spiral nebulae were part of the galaxy and did not constitute separate systems of stars. Curtis, in contrast, argued for a considerably smaller local galaxy (about 30,000 light years in diameter) and suggested that spiral nebulae were best understood as distant galaxies. The "great debate" did little to resolve the issue. Debate concerning the size and structure of the universe continued throughout the 1920s and beyond (Smith 1982).

Both sides in the debate could point to quantities of observational evidence supporting their respective positions. Much depended on various estimates of the distances of the different celestial features from Earth. There was, of course, no direct way of measuring such distances, so astronomers typically made use of a range of approximations based on features such as the apparent magnitude (brightness) of stars of different types and the appearance of their spectra. In the early 1920s it did look, however, as if the key piece of evidence was in the hands of those who opposed the theory that nebulae (or at least some nebulae) were separate galaxies. The Dutch astronomer Adriaan van Maanen claimed that he could identify "proper motion" on the part of components of spiral nebulae. Van Maanen was a highly respected observational astronomer working at the prestigious Mount Wilson Observatory (fig. 12.2) and had come to the conclusion that proper motion was detectable in the arms of spiral nebulae on the basis of careful comparison of nebular photographs taken over long periods. Opponents of the separate galaxies theory argued that if proper motions of this magnitude were detectable in objects as far away as the spiral nebulae were supposed to be by proponents of the theory, then the spirals' arms must be moving at speeds in excess of the speed of light. Such a proposition was clearly ridiculous, and the nebulae must therefore, in fact, be considerably closer, as required by those who argued that they were within the Milky Way itself.

Despite van Maanen's apparently conclusive evidence, proponents of the separate galaxies theory by and large stuck to their guns. In 1923, a new observation by the young American astronomer Edwin Hubble seemed to provide decisive evidence in their favor. Working at Mount Wilson Observatory (like van Maanen) and using what was then the world's most powerful telescope, Hubble identified a Cepheid variable star in the Andromeda Nebula. Previous studies of Cepheid variables by the Harvard astronomer Henrietta Swan Leavitt in 1908 had identified a constant relationship between the period of a Cepheid variable (the time between its moments of highest luminosity) and its luminosity. That meant that measurements of a Cepheid variable star's period could be used to gauge its absolute luminosity. Its absolute luminosity compared with its apparent luminosity (how bright it appeared in the night sky) could then be used to approximate its distance since different objects with the same absolute levels of brightness

FIGURE 12.2 Mount Wilson Observatory as it appeared in the early twentieth century. This is where many of the astronomical observations used to decide the size of the universe were made.

appear relatively less bright the further away they are. Hubble could therefore use his discovery of a Cepheid variable in the Andromeda Nebula to calculate its approximate distance. He calculated this distance as about 300,000 parsecs (a parsec being 3.26 light years)—much farther than van Maanen or Shapley argued. At distances like these it seemed inconceivable that nebulae like the Andromeda Nebula could be part of the Milky Way galaxy (fig. 12.3).

Astronomers were left with two apparently highly trustworthy sets of observations that were nevertheless contradictory. If van Maanen was to be believed, his measurements of the internal proper motions of spiral nebulae indicated that they must be relatively nearby (fig. 12.4). If Hubble, on the contrary, were to be believed, spiral nebulae like the Andromeda Nebula were well outside the plausible boundaries of the Milky Way. By the end of the 1920s, most astronomers agreed that the separate galaxy theory—the "island universe" hypothesis, as it was known—had won the day. They found Hubble's Cepheid variables more convincing than van Maanen's photographic evidence of proper motions. In the end, it was a matter of de-

FIGURE 12.3 An early twentieth-century photograph of a distant nebula.

ciding what kind of observational evidence—and which individual astronomers—was to be considered most trustworthy.

The island universe model was also used as the foundation for yet another transformation of the traditional worldview. In studying the light coming from distant galaxies, astronomers noticed that the spectral lines (described above) were shifted toward the red end of the spectrum. The most obvious explanation of this was in terms of the Doppler effect, in which the frequency of a wave motion is affected by the velocity of the

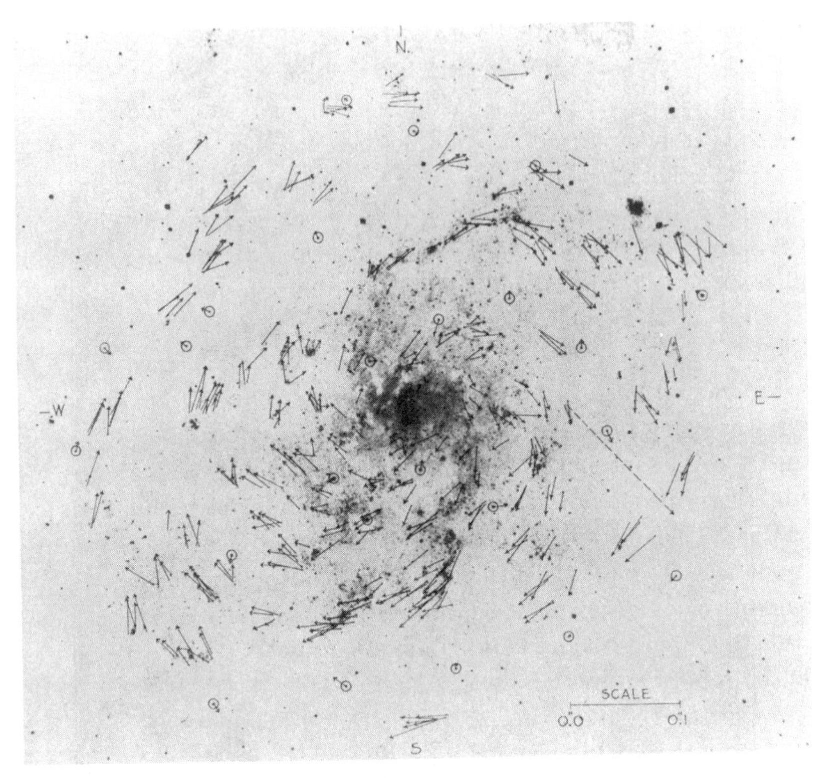

FIGURE 12.4 Van Maanen's observations of internal nebular motions.

body emitting the wave (in the case of sound, this produces the familiar drop in tone when a whistling train passes by an observer standing by the track). This explanation of the "redshift" implied that galaxies are receding from us. In 1929, Hubble went further by proposing a law governing the relationship between the distance of a galaxy from the earth and its velocity of recession. Not only did we live in an expanding universe, but the more distant are the galaxies we see, the faster they are moving away from us.

By the 1930s, astronomers were therefore largely agreed about the size and shape of the universe and had begun to see it as a dynamic, not a static, system. The Milky Way was recognized as being only one among a huge number of similar galaxies, with the earth and its solar system located near the outer rims of one of its spiral arms. No longer was the galaxy that human beings inhabit to be considered as the center of the universe. From that perspective, the transformation might certainly be regarded as truly revolutionary in the same sense that the Copernican revolution was.

Whether the participants in the debate saw matters in the same apocalyptic terms is another question.

Einstein's Universe

New observational technologies and techniques were not the only source of insight into the shape of the universe. New theoretical developments in physics at the beginning of the twentieth century also had a major impact on the way in which astronomers understood the cosmos. As we have already seen, many historians of physics have characterized the changes that took place in physics at the beginning of the twentieth century as being revolutionary. The traditional worldview associated with Newton was swept away and replaced with a new, relativistic physics (see chap. 11, "Twentieth-Century Physics"). The view of space and time as being absolute regardless of the observer's position and velocity was abandoned and replaced with the standpoint that time and space were relative to the position and velocity of the observer. The key figure in this transformation was the German physicist Albert Einstein. Einstein's special theory of relativity, published in 1905, and the general theory of relativity (dealing with accelerating systems) published a decade later had a profound impact on the new discipline of theoretical physics. The implications of Einstein and his followers' views for understanding the structure of the universe were quickly recognized by astronomers (Pais 1982). After all, two of the key pieces of evidence for the theory of general relativity—the anomalous shift in the perihelion (nearest point to the sun) of the planet Mercury and the observed bending of light during an eclipse found by the astronomer Arthur Eddington—were themselves astronomical in nature.

Einstein himself quickly recognized that his theories had important implications for the ways in which astronomers understood the universe. In the years following his announcement of the general theory of relativity, he worked at finding solutions to his relativistic field equations that would provide a stable description of the universe's structure. The universe as described in Einstein's field equations had a non-Euclidean geometry. In other words, it did not follow the classical geometric laws whereby, for example, a straight line is always the shortest distance between two points. Einstein's space was curved. The solution to Einstein's field equations was a finite, unbounded four-dimensional space. This can be understood by analogy with a three-dimensional sphere. An entity living on the surface of such a sphere would, if it traveled for long enough in the same direction, arrive back at its point of origin. It would also be possible, in principle, to traverse every point on the sphere's surface. That surface must therefore be finite. At the same time, at no stage would the entity encounter a boundary, so the surface is also unbounded. According to Einstein, this was the way the universe was in four dimensions. Einstein was also firmly convinced that the universe must be static—unchanging in its structure. He therefore introduced an extra component—the cosmological constant—into his field equations to ensure this feature. Einstein famously later described the cosmological constant as the greatest mistake he had made in physics.

Not everyone was satisfied with Einstein's solution to his field equations. In 1917, the Dutch astronomer Willem de Sitter proposed an alternative geometric model of the universe but one that also obeyed Einstein's relativistic field equations. After studying at the University of Groningen, de Sitter had spent several years working at the Royal Observatory at the Cape of Good Hope in South Africa before returning to the Netherlands and eventually becoming professor of astronomy at Leiden University in 1908. His main research interests lay in celestial mechanics, but from 1911 on, he became increasingly interested in the astronomical implications of the theory of relativity. Unlike Einstein's universe, the model that de Sitter proposed was infinite. Its equivalent in three dimensions would be a saddle shape stretching to infinity in each direction. Like Einstein, de Sitter was convinced that any model of the universe must be static. In order to maintain this characteristic in his model he had to assume that the universe contained no matter. Clearly, the real universe did not obey this assumption, but de Sitter argued that the overall density of matter in the universe was sufficiently low for his model to provide a reasonable approximation. Einstein was particularly worried by this feature of de Sitter's solution to his equations. The suggestion that a massless universe was possible seemed to him to imply that space itself had absolute properties—a view at odds with his own interpretation of relativity theory.

De Sitter's model of the universe had one feature in particular that caught the interest of some astronomers, particularly the British astronomer Arthur Eddington. If atoms were introduced into this mathematical model at large distances from each other, it seemed that, as a result of time dilation, any light emitted by them would appear to an observer as being of lower frequency than it actually was. Translating this into the real universe, the suggestion was that light from distant sources would appear to be shifted toward the red end of the spectrum. Similarly, it seemed that point masses inserted into this hypothetical mathematical universe would spontaneously begin to accelerate away from each other as a result of the cos-

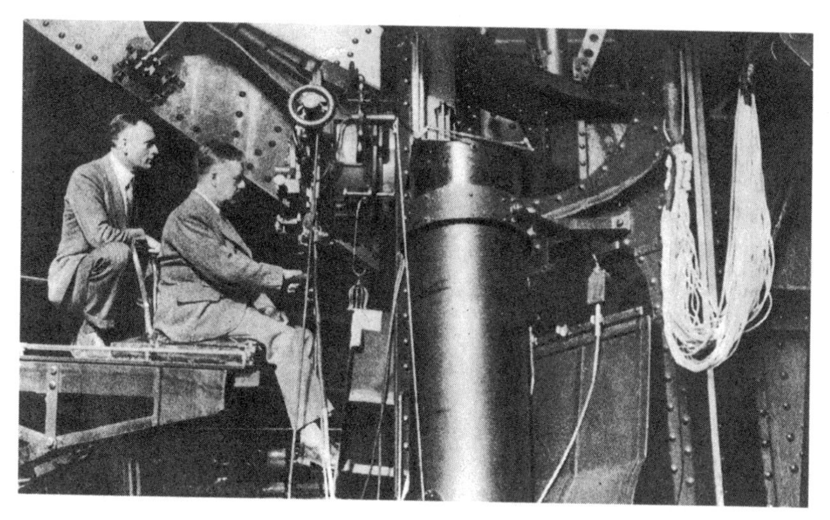

FIGURE 12.5 Edwin Hubble and James Jeans making astronomical observations, *Fortune* (July 1932).

mological constant that de Sitter, like Einstein, had inserted into his equations. In his *Mathematical Theory of Relativity* published in1923, Eddington suggested that these features of de Sitter's model might be useful in solving the problem of the large radial velocity (apparent velocity away from the earth) of many spiral nebulae. In the first place, de Sitter's model would explain the apparent movement as the result of the general tendency of matter in his model to move away from each other. In the second place, estimates of radial velocity were usually based on measurements of the shift toward the red end of the spectrum (redshift) of distant objects as the result of velocity. If de Sitter was right, then some, at least, of that observed redshift was the result of distance and time dilation, rather than velocity, so the spiral nebulae were not really moving away at such large velocities after all (Smith 1982).

Eddington also made another observation on de Sitter's model: "It is sometimes urged against de Sitter's world that it becomes non-statical as soon as any matter is inserted in it. But this property is perhaps rather in favor of de Sitter's theory than against it." Eddington was starting to move toward the position that the universe might be expanding rather than static. In 1929, the American astronomer Edwin Hubble (fig. 12.5) presented a paper before the National Academy of Science demonstrating, on the basis of observations, a straightforward linear relationship between the radial velocity and the distance of spiral nebulae, the relationship now known as Hubble's Law. According to Hubble, at least, he had embarked on the research leading to the new generalization at least partly as an attempt to test de Sitter's model of the universe. Most astronomers interpreted Hubble's Law as strong evidence in favor of an expanding rather than a static universe (Crowe 1994). Einstein was sufficiently concerned to visit Hubble at the Mount Wilson Observatory before announcing in 1930 that he had given up on the static universe and the cosmological constant that went with it. According to one story, when Einstein and his wife visited the observatory they were shown the telescopes and it was explained to Einstein's wife that they were used to discover the structure of the universe. Elsa Einstein responded, "Well, well, my husband does that on the back of an old envelope" (Berendzen et al. 1976). The story may well be apocryphal, but it demonstrates nonetheless the growing intellectual and professional differences between theoreticians and observational astronomers and the different techniques they adopted to approach the same questions.

Big Bang or Steady State?

By the 1930s, astronomers and physicists increasingly agreed that the universe appeared to be expanding. This was what Einstein's relativistic field equations, shorn of the cosmological constant, appeared to suggest. It was also the conclusion that many drew from Hubble's observations of the relationship between the velocity and distance of spiral nebulae. Some theorists started suggesting that if the universe was expanding, then it must have had a discrete beginning. They argued that by extrapolating backward through time the universe's current rate of expansion it would be possible to arrive at a point when all the matter at the universe was concentrated at one point (Kragh 1996). The explosion of this point represented the origins of the universe. A mathematical model of a universe expanding from a single point had been put forward by the Soviet physicist Alexander Friedmann in the early 1920s. Neither its author nor anybody else suggested, however, that the model was anything other than a mathematical curiosity. In 1927, the Belgian astronomer Georges Lemaître, a student of the British astronomer Arthur Eddington at Cambridge, before studying for a PhD at the Massachusetts Institute of Technology, did come up with a physical model of an expanding universe. It was not until the 1930s, however, that Lemaître's model was taken seriously. Lemaître suggested that the universe had started as a massive single atom. This single atom would have been highly unstable and would have broken up "by a kind of super-radioactive process" producing an expanding universe (Kragh 1996).

During the 1940s, another Soviet scientist, the nuclear physicist George

Gamow, started working on his own version of the big bang theory of the universe and its origins. Gamow's interest in cosmology was prompted by his researches in quantum mechanics and nuclear physics. Gamow had made a name for himself in 1928 with his theory of quantum tunneling, explaining the emission of alpha particles from radioactive matter. Along with colleagues such as Fritz Houtermans and Robert Atkinson, Gamow soon concluded that his quantum tunneling theory could also be used to help understand nuclear processes taking place inside stars. Particularly after the discovery of new subatomic particles during the early 1930s, the stars increasingly came to be regarded as testing grounds for new theories in nuclear physics (see chap. 11, "Twentieth-Century Physics"). During the 1940s, Gamow was particularly concerned to produce a theory that would account for the origins of the heavy elements, and since it seemed increasingly unlikely that they could have been produced inside stars, he turned to the big bang for an alternative scenario. Gamow first suggested that the universe had originally consisted of a cold (comparatively speaking) and thick soup of neutrons that expanded, forming more complex configurations that eventually produced the known chemical elements through the emission of beta radiation. In 1948, along with Ralph Alpher and Hans Bethe, Gamow submitted a revised version of his big bang theory to the *Physical Review* (the so-called αβγ paper). Bethe had in fact not made a significant contribution to the paper—his name was included in order to preserve the αβγ "joke." In this new version, the universe had started life as a hot and highly compressed neutron gas that had started decaying into protons and electrons, eventually producing the modern universe.

As far as many of its early promoters were concerned, one good reason for supporting the big bang theory of the universe's origins was its theological significance. While some, like Gamow himself, explicitly avoided theological arguments, others embraced them. Edward Arthur Milne, professor of mathematics at Manchester University and inveterate opponent of Einstein's theory of relativity, argued in 1947 that anything other than a universe created from a single point was a logical contradiction. Similar claims were made by the mathematician and historian of physics Edmund Whittaker, who argued that knowing that the universe had a distinct beginning in time proved the existence of God as the first cause of the universe. It is worth noting that Georges Lemaître, one of the first astronomers to produce a physical big bang theory, was himself a Catholic priest. In 1951, Pope Pius XII delivered an address to the Pontifical Academy of Sciences in which he explicitly appealed to the big bang theory of the universe as a scientific endorsement of the Catholic Church's position. According to the pope, there was nothing new for Christians in the latest cosmological theories. They were simply a restatement of the opening sentence of Genesis: “In the beginning God created heaven and earth” (quoted in Kragh 1996).

This kind of explicit linking of cosmological theory with religion provided at least one reason for the discomfort with the big bang theory felt by the advocates of an increasingly powerful alternative—the so-called steady state theory of the universe. Steady state theory was first put forward by three Cambridge graduates, Hermann Bondi, Thomas Gold, and Fred Hoyle, during the late 1940s, just as Gamow’s theories concerning the big bang were taking shape. Hoyle, in particular, was an outright atheist who felt that religious views had no place in scientific discussions and argued that big bang theory only made sense in a religious context. According to Bondi, Gold, and Hoyle’s new theory, the universe had always existed and always would. As the universe expanded, new matter was continually created to fuel the expansion. In two papers in the *Monthly Notices of the Royal Astronomical Society* in 1948, one by Hoyle and the other jointly authored by Bondi and Gold, they set out the principles of their new theory. In particular, they introduced what Hoyle called the “wide cosmological principle” and Bondi and Gold called the “perfect cosmological principle,” stating that the universe was homogenous and unchanging on the large scale through both space and time. In 1949, Hoyle gave a series of radio broadcasts to the BBC in which he expounded his steady state theory. In 1950, the talks were published as a book, *The Nature of the Universe,* which sparked widespread controversy. Many astronomers felt that Hoyle’s representation of the state of cosmology had been far too partial and favorable to his own steady state theory.

The controversial new steady state theory gathered only a few new supporters throughout the 1950s, particularly outside its promoters’ own close-knit Cambridge circle. At the same time, supporters of the big bang theory found few new theoretical arguments that they could use to argue for their own theory’s superiority. Many astronomers took little interest in these kinds of grand cosmological theories, taking the view that they had little relevance to the everyday astronomical business of observing and cataloging. Observationally, it seemed that there was little evidence available to help choose between the two theories. In the early 1960s, however, new measurements of the universe’s background radiation seemed to big bang theorists to give their views of the universe’s origins a decided edge. In 1961, the Cambridge radio astronomer Martin Ryle presented the results of the latest survey of extragalactic radio sources, suggesting that their range of energies supported a big bang rather than steady state view of the universe. Many supporters of the big bang (including Ryle himself) regarded this as the decisive nail in the coffin of the steady state theory. The steady state theory's advocates disagreed, suggesting that further refinement of Ryle's result would bring them back into line with steady state theory predictions. The discovery of quasars during the first half of the 1960s also seemed to pose a problem for the steady state theory. These stellar objects only seemed to exist at huge distances away in time and space—an observation at variance with steady state theory's assumption of the universe's homogeneity in time and space.

Textbook histories of astronomy and cosmology often present these observations from the 1960s as decisive refutations of steady state theory and triumphant vindications of the big bang. Historical reality is, of course, rather more complex. Most of the steady state theory's adherents—certainly its founding fathers—remained convinced that these were no more than local difficulties that would eventually be solved by way of further observational and theoretical refinements. Hoyle, for example, put forward an alternative theory concerning the physical nature of quasars that would allow them to be understood as local rather than distant objects. By the second half of the 1960s, however, steady state theory was an increasingly marginalized area with its proponents seeming increasingly at odds with the mainstream of their profession. The controversy has still not disappeared entirely. Hoyle and his advocates continued and still continue to argue in favor of the steady state. This episode is an instructive example of the historical (and philosophical) difficulties involved in identifying decisive events that exclusively determine the outcome of scientific debate. What big bang theorists regarded as increasingly desperate ad hoc measures to defend a bankrupt theory were regarded by their steady state opponents as simply further refinements of a highly productive and powerful theoretical framework and suggestions for further elaboration.

Black Holes and the Modern Cosmos

During the last quarter of the twentieth century, cosmologists succeeded in transforming their discipline into a popular science, though of course a strong tradition of popular cosmology had existed since at least the beginning of the century as well (see chap. 16, "Popular Science"). The process culminated in many ways with the publication of the theoretical physicist Stephen Hawking's *Brief History of Time* in 1988. For much of the century, even, most astronomers regarded cosmology—and theoretical cosmology in particular—as a highly esoteric subject, far divorced from the concerns of mainstream astronomy. According to one eminent astronomer during the early 1960s, "there are only 2½ facts in cosmology" (quoted in Kragh 1996). The two he had in mind were the observation that the night sky is dark and Hubble's observation of the recession of the galaxies. The half fact was that the universe was evolving. The joke was symptomatic of a widespread view among astronomers that the theoretical models hypothesized by cosmologists were based on very little hard astronomical evidence and were therefore of little use in understanding known astronomical phenomena. From the early 1960s, there were an increasing number of new astronomical phenomena to understand, too, as astronomers turned to new technologies to examine the night skies. New techniques such as radio astronomy, itself based on surveillance and early warning systems developed during the Second World War, produced large amounts of novel information in need of theoretical interpretation (see chap. 20, "Science and War"). By the 1980s, the new view of a cosmos composed of a variety of bizarre and hitherto unknown objects and areas where the known laws of physics came apart at the seams was catching the public imagination. It was helped on by a renewed vogue for popular TV science fiction such as the *Star Trek* series.

During the late 1950s and early 1960s, a number of astronomers reported observations of unusual starlike objects that appeared to have peculiar properties. In 1963, the Dutch astronomer Maarten Schmidt studied the spectrum of one of these objects and concluded that its light was redshifted to a high degree, indicating that it was at an immense distance. This also meant that the object must be putting out an enormous amount of energy. Further observations suggested the same to be the case with others of these "radio stars" that were soon renamed "quasi stellar sources," or "quasars" for short. The fact they all seemed to be at immense distances was itself, as we have seen, of theoretical significance, casting doubt on the viability of the steady state theory of the universe. Cosmologists also set about trying to understand what could the source of the massive amounts of energy given off by these quasars. During the late 1960s another mysterious set of energetic objects was added to the cosmic population. In 1967 the Cambridge graduate Jocelyn Bell, working at Cambridge's radio astronomy observatory, noticed a series of regular but intermittent signals coming from an unknown source. She described them as flashing like a "Belisha Beacon" (the popular term for the flashing orange light at a British pedestrian crossing). After excluding all possible terrestrial sources of contamination (and some extraterrestrial ones, including the possibility of little green men), she and her PhD supervisor Anthony Hewish concluded that a hitherto unknown kind of stellar object, which they dubbed a "pulsar," emitted the signals. In 1974 Hewish and Martin Ryle, the head of Cambridge's radio astronomy observatory, received a Nobel Prize for Bell's discovery.

In 1968 the steady state theorist Thomas Gold suggested that pulsars could be explained as being rapidly spinning neutron stars. Theoretical cosmologists had predicted that entities such as neutron stars might exist as the result of stars of a certain size collapsing in on themselves under the influence of gravity as the outward push of their radiation slowed down over time. It was starting to look as if some of the stranger objects postulated by cosmologists might have observational equivalents in the real astronomical universe. In 1916 the German mathematician Karl Schwartzchild had proposed a solution to Einstein's relativistic field equations in which there were points where the curvature of space-time became infinite. At such points the force of gravity would also become infinite and no light could escape. Schwartzchild's speculations were regarded as interesting mathematical curiosities for the next several decades until, during the 1960s, the American physicist John Wheeler set out to investigate the circumstances in which they might exist in the real universe. In 1968, Wheeler coined the phrase "black hole" to describe a massive star that had hypothetically collapsed under the force of its own gravity and been compressed to such a degree that it formed a singularity of the kind described by Schwartzchild. The properties of such black holes became an increasingly important topic of theoretical research for a new generation of theoretical cosmologists such as Stephen Hawking, who first postulated the hypothesis that black holes might emit radiation in 1973 (Hawking 1988).

By the end of the 1980s, not only professional astronomers but large sections of the public as well were increasingly familiar with the cosmological menagerie of black holes, neutron stars, white dwarfs, and wormholes. Stephen Hawking's bestselling *Brief History of Time* was a major factor in this rise in public interest in cosmological theorizing. Hawking's bestseller was only the crest of a wave of similar titles, such as John Gribbin's *In Search of the Edge of Time* and P. C. W. Davies's *God and the New Physics*. Another factor was the (eventual) success of the Hubble space telescope, named after the pioneering astronomer Edwin Hubble and designed to transmit images of the distant universe of hitherto unparalleled clarity back to planet Earth. When the space telescope was first launched in 1990 by the American space agency NASA, astronomers soon realized that major design flaws in the reflecting mirror (it was the wrong shape) rendered it largely useless for the purposes for which it had been originally designed. Once those faults had been corrected, however, TV viewers in the Western world were bombarded by spectacular images of the faraway cosmos comparable to the fictional space vistas seen through the bridge viewer of *Star Trek*'s Starship Enterprise (Smith 1993). The result was to make large parts of the once esoteric lexicon of theoretical cosmology part of the everyday vocabulary of significant sections of at least the European and North American public.

Conclusions

The universe was transformed beyond recognition during the course of the twentieth century. At the end of the nineteenth century, space and time were generally understood to be absolute categories, unchangeable and unvarying in their properties regardless of the position and speed of the observer. Few if any astronomers seriously considered the possibility of an universe that—in terms of its observable contents, at any rate—extended very far beyond those visible using the then existing technology. For all intents and purposes, the universe was synonymous with the Milky Way galaxy. This understanding changed radically during the twentieth century's opening decades. New techniques and technologies—as well as new theoretical worldviews—made it possible for astronomers to produce convincing estimates of stellar distances. The end result was a view of the Milky Way as only one relatively undistinguished galaxy among an innumerable host of others. Einstein's theory of general relativity brought new meaning to the question of the shape of the universe. Considerations drawn from Einstein's theories led theoretical cosmologists to think about the age and duration of the universe in novel ways. At about the same time, new observational evidence led astronomers to rethink their view of the universe as an unchanging, largely static entity. The universe as it entered the twenty-first century was a very different place, inhabited by very different beasts, from the one that began the twentieth century.

So to pose our familiar question, was this a revolution? In many ways it seems difficult to avoid the conclusion that, yes, it was. There can certainly be no doubt that a thoroughgoing overhaul of astronomers' understanding of the nature of the universe and humanity's physical place in it took place during the twentieth century. At the same time, however, the complexities of the history summarized here are an indication of the difficulties involved in imposing such a category on the past. While it may seem relatively obvious that a significant change did take place over the century or so covered in this chapter, it would be far harder to pinpoint any particular episode or point in time as the decisive moment. It would be just as difficult to identify any particular new theoretical insight or observational discovery or technique as being the decisive trigger for such a transformation in worldview as well. A full account of the transformation in cosmological understanding we have outlined here would need to look at developments in the institutions and professional structures of astronomy and physics as well as at changes in ideas and practices. We would need to look at the kind of training new generations of astronomers received and the material and cultural resources available to them. In short, if there was a cosmological revolution at all, we would need to understand it as a revolution in the culture of cosmology as much as a revolution in its content.

References and Further Reading

Barnes, Barry. 1974. *Scientific Knowledge and Sociological Theory.* London: Routledge.

Berendzen, R., R. Hart, and D. Seeley. 1976. *Man Discovers the Galaxies.* New York: Science History Publications.

Collins, Harry. 1985. *Changing Order.* London: Sage.

Crowe, Michael J. 1994. *Modern Theories of the Universe: From Herschel to Hubble.* New York: Dover.

Hawking, Stephen. 1988. *A Brief History of Time.* London: Bantam.

Hoskin, Michael. 1964. *William Herschel and the Construction of the Heavens.* New York: Norton.

Jackson, M., 2000. *Spectrum of Belief: Joseph von Fraunhofer and the Craft of Precision Optics.* Cambridge, MA: Harvard University Press.

Jaki, Stanley. 1978. *Planets and Planetarians: A History of Theories of the Origins of Planetary Systems.* Edinburgh: Scottish Academic Press.

Kragh, H. 1996. *Cosmology and Controversy: The Historical Development of Two Theories of the Universe.* Princeton, NJ: Princeton University Press.

Kuhn, Thomas S. 1962. *The Structure of Scientific Revolutions.* Chicago: University of Chicago Press.

———. 1966. *The Copernican Revolution.* Chicago: University of Chicago Press.

Pais, Abraham. *Subtle Is the Hand: The Science and the Life of Albert Einstein.* Oxford: Clarendon Press.

Smith, R. 1982. *The Expanding Universe: Astronomy's "Great Debate," 1900–1931.* Cambridge: Cambridge University Press.

Smith, R. 1993. *The Space Telescope.* Cambridge: Cambridge University Press.

CHAPTER 13

THE EMERGENCE OF THE HUMAN SCIENCES

COULD HUMAN NATURE AND SOCIETY be studied by the methods of science? In the seventeenth century hardly anyone would have accepted this possibility. The Christian religion taught that the human spirit was of supernatural origin, its mental and moral faculties beyond the reach of natural law and, hence, outside the realm of science. Descartes based his commitment to the mechanical philosophy of nature on the assumption that the human mind was quite separate from the mechanism of the body. This dualistic position left the study of the mind and of social interactions to philosophers and moralists, not scientists.

There are other ways of explaining why the human or behavioral sciences could not have emerged in the period following the Scientific Revolution. The historian Michel Foucault (1970) argued that it was only after the modern state emerged in the nineteenth century that it became possible to recognize human behavior as something that had to be understood and controlled. Social deviants (as defined by the state) had to be identified and incarcerated in jails and mental hospitals. The masses of ordinary people had to be monitored and educated to fit the new industrial society. There can be little doubt that the emergence of psychology, anthropology, and sociology as independent scientific disciplines owed a great deal to their potential utility for the governors of industry and the modern state. But the process by which they were created was slow. As areas of interest, they were becoming well-defined by the mid-nineteenth century, although still very much associated with their origins in philosophy and moral theory. But the foundation of academic disciplines with scientific pretensions was delayed until the early decades of the twentieth century.

The problem was that there were rival methods of seeking to understand human behavior in scientific terms. The most obvious assault on the Cartesian separation of mind and body came from those who favored materialism and what we would now call a methodology of reductionism. Encouraged by the hope that the workings of the body were yielding to the methods of scientific investigation, they predicted that the nervous system and the brain would be understood in similar terms. To the materialists, the mind was no more than a by-product of the functioning of the body—and society would be understood by a simple extension of the same approach to include the interactions between individual minds. Evolutionary theories bolstered the same hope: if humans had emerged from animals, they could be understood in the same terms as animals or, at least, by extending the categories of nature to include new levels of awareness that emerged as progress generated more complex structures. Such ideas were deeply troubling to traditional ways of thought and were taken up by radicals seeking to overthrow or reconfigure the foundations of the social order (see chap. 18, "Biology and Ideology").

The reductionist approach certainly played a role in the early stages of the emergence of the modern human or behavioral sciences. In the nineteenth century, the philosopher Herbert Spencer exploited the evolutionary perspective to make important contributions to the fields of psychology and sociology. Spencer was also aware of new developments in neurophysiology. But the problem with the reductionist perspective was that it could all too easily be exploited to deny any autonomy to the study of human nature. If we are just machines, then there would be no need to institute separate sciences devoted to the understanding of human mental and social activities. The final stages in the consolidation of the human sciences arose not from reductionism but from a deliberate reaction against it. In the late nineteenth century, experimental techniques were developed to study mental processes without reference to the corresponding physiological processes going on in the brain. Shortly after 1900, psychologists began to reject the model offered by evolutionism and insist that the study of behavior should become an autonomous scientific discipline. This rejection of biology played a key role in the institutionalization of psychology in the academic system. At the same time, anthropologists and sociologists also staged a revolt against biology, insisting that evolutionary models offered no relevant insights into the functioning of human cultures and societies. The emergence of the social sciences as independent disciplines occurred through a deliberate rejection of the model that would have made those studies scientific only by subordinating them to biology and, ultimately, to physics and chemistry (for modern surveys, see Smith [1997]; Porter and Ross [2003]).

Psychology Becomes a Science

The study of mental processes was originally a part of philosophy. Efforts to understand the functioning of the human mind depended on introspection, the philosopher's self-conscious attempt to analyze his or her own thoughts and sensations. It was assumed that many of these functions lay outside the realm of natural law—the moral faculty or conscience, for instance, depended on the freedom of the will, apparently the very antithesis of the deterministic functioning of the physical world. This did not mean that the philosopher could not draw conclusions about the nature of the mind, although it did leave the field open to dramatic differences of opinion as new philosophical schools emerged. In the late seventeenth century, the "sensationalist" philosophy of John Locke elaborated the empiricism of the Scientific Revolution into an influential theory of how the mind worked. For Locke and his followers, the mind of an infant was a blank slate, a tabula rasa, on which experience wrote to generate an understanding of the laws of nature and the habits needed to function in the natural and social worlds. Sensations that normally appeared together were linked by the "association of ideas" to give habitual thought and behavior patterns in the individual who experienced them. This empiricist philosophy saw the mind as a learning machine but without specifying the mechanisms in the brain responsible for the mental processes it postulated.

By the late eighteenth century, political philosophers such as Jeremy Bentham were building a reformist social system known as utilitarianism on the basis of associationist psychology (Halévy 1955), according to which, individuals would be adapted to their social environment by exploiting their desire for pleasure and their dislike of pain to condition their habits according to the ends their rulers desired. If the rulers were enlightened, they would adjust the laws to encourage a society in which individual behavior worked for the "greatest good of the greatest number." Associationist psychology thus became linked to the laissez-faire free-enterprise system of government favored by the rising middle classes. The "gloomy science" of political economy sought to define the limits to social improvement imposed by the natural world. Thomas Malthus's principle of population, which so impressed Darwin, arose from this attempt to synthesize what we today would call psychology, sociology, and economics.

The sensationalist/associationist tradition did not go unopposed. Some philosophers, Descartes included, thought that the individual mind was created already containing innate ideas that did not need to be generated by experience. This made a curious parallel with the studies of many naturalists, who were convinced that animals were created with instinctive behavior patterns designed to adapt them to their environment (just as they were designed with the appropriate physical adaptations). The mind was, thus, far more than a passive learning machine, a viewpoint taken much further by the philosopher Immanuel Kant, who argued that the mind actually imposes the categories of space and time on the flow of sensations it receives. This position generated the idealist philosophy that became popular in nineteenth-century Germany, according to which the mind plays an active role in creating the external world it experiences. Sensationalism and idealism defined two radically different perceptions of the mind that would be debated through the following centuries: the image of the passive learning machine versus the more active model in which the mind has a structure predetermining how it will perceive and interact with the external world.

New developments in the biological sciences offered ways of resolving this conflict, but only at a price that conservative thinkers were reluctant to pay. The reductionist methodology was invoked by radical materialists who argued that the mind was no more than a by-product of physical activities going on within the brain and nervous system. If this were so, the individual brain could have some predetermined patterns laid down in the structure inherited from its parents but might also have the capacity to learn from experience by relating nervous impulses that routinely occurred together. In the early nineteenth century, the movement known as phrenology exploited these implications of the claim that the brain is the organ of the mind to offer a radical social message. As described in the chapter "Biology and Ideology," this attempt to bring psychology into the world of natural law was marginalized within the elite scientific community but retained considerable popular appeal. One thinker who was inspired by phrenology was Herbert Spencer, who was determined to create a new social philosophy for the age of aggressive free-enterprise capitalism. Spencer soon realized that the Lamarckian theory of evolution offered an even better hope of breaking the deadlock in philosophical psychology. In his *Principles of Psychology* of 1855 (four years before Darwin's *Origin of Species*) he proposed an evolutionary theory of mind that linked individual self-improvement to the general idea of biological and social progress (Richards 1987; Young 1970). Spencer realized that if the Lamarckian theory of the inheritance of acquired characteristics could be applied to the mind, then acquired mental characters—habits learned by the individuals of one generation—might be translated into instincts inherited automatically by their descendants. New habits and even new mental abilities developed by individual activity and initiative would become permanent characters of the species. Mental and ultimately social progress thus became inevitable as the consequence of millions of individual acts of self-improvement.

Although Darwin attempted to switch attention to his theory that natural selection could act to transform instincts as well as physical characters, Spencer's Lamarckian view of mental evolution dominated the late nineteenth century (see chap. 6, "The Darwinian Revolution"; see also Boakes 1984). Darwin's own disciple in the area of mental evolution, George John Romanes, adopted the Lamarckian explanation of instincts. Romanes also endorsed another theory widely associated with Lamarckism: the recapitulationist vision of individual development repeating the evolutionary history of the race (Gould 1977). On this model, the psychological development of the child passed through stages equivalent to the ascending series of mental improvements acquired by animal species in the course of evolution's long progress toward humankind.

Evolutionism thus offered the opportunity for psychology to be seen as a branch of science rather than philosophy, but only if the new discipline followed the lead offered by biology. The American psychologist G. Stanley Hall explained the psychological traumas of the teenager as the equivalent of a key stage in mental evolution in his book *Adolescence* (1911). Animals became the models for human mental processes, although the mental characters of animals were often exaggerated by reliance on anecdotal evidence from untrained observers. The behavior of both humans and animals was explained in terms of instincts, including social instincts, shaped by the process of evolution. By understanding these instincts, the psychologist could offer the managers of industry and the state a new tool to control their workforce. In the early twentieth century, American psychologists exploited the newly developed intelligence tests to provide what seemed like convincing evidence of extensive mental deficiency among the lower classes (Gould 1981). This evidence was widely quoted in support of the eugenics movement's claims that the breeding of the genetically "unfit" should be restricted by the state (see chap. 18, "Biology and Ideology").

Conventional histories of psychology (e.g., Boring 1950) are often rather evasive about the whole evolutionary episode, preferring to concentrate on a parallel development in Germany that is seen as laying the foundations of an experimental approach to the study of behavior. German physiologists had begun to study the functioning of the sensory nervous system in the middle decades of the nineteenth century, and in 1879 Wilhelm Wundt created a laboratory at Leipzig devoted to a "physiological psychology" in which the mental processes of human subjects were studied using mechanical apparatus to control the presentation of sensory stimuli and the recording of responses. Wundt himself proclaimed that psychology had at last become a science in its own right, and his disciples transferred the experimental approach to other countries, especially America. But the tradition that sees his laboratory as the foundation stone of modern experimental psychology ignores the fact that Wundt himself still encouraged introspection as a valid method of studying the higher human faculties, while many other psychologists openly retained the more philosophical approach still visible in the writings of Spencer and the evolutionists. A noted example is William James, whose *Principles of Psychology* in 1890 became a classic text for those who welcomed the new techniques but did not want to see them sweep away the old.

The result was a protracted struggle between those who retained psychology's traditional links with philosophy and moral theory and the young turks who strove to build a new discipline that would have all the attributes of a true science. By 1900, psychology was beginning to acquire an identity that would separate it from philosophy, but there was little agreement among its practitioners as to how far the separation should be pushed at the intellectual and methodological level. Journals, societies, and university departments were being founded, but the competing interest groups struggled for control of the nascent apparatus of academic power. Partly as a result of this conflict, the institutionalization of psychology as a science took some time to achieve. The barriers were overcome more rapidly in the United States, where the expansion of the university system around 1900 made it easier to establish new departments. There were soon more psychology laboratories in America than in Germany. In Britain, too, the creation of an academic framework for psychology was slow, only half a dozen professorships being established by the 1920s. The founding of the American Psychological Association in 1892 preceded the establishment of a Society for Experimental Psychology in Germany by twelve years and the British Psychological Association by nine (Cravens 1978; Degler 1991).

In the end, though, the creation of psychology as an independent discipline came to rest on the claim that it was an experimental science, not a branch of philosophy or evolutionary biology. From around 1910 onward, the rhetoric of experimental rigor began to dominate psychology text-

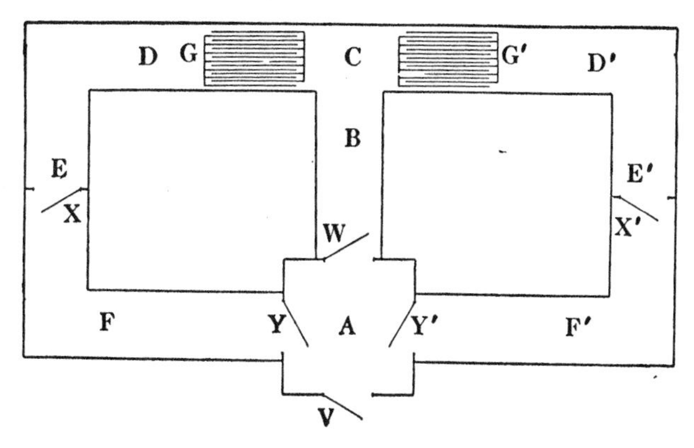

FIGURE 13.1 Diagram of a maze used for experiments on animal behavior, from John B. Watson's *Behavior: An Introduction to Comparative Psychology* (1914), 87. The animal starts at *C* and has to make its way to the home box at *A*, choosing its route according to the stimulus being given to it, in this case alternative sounds from tuning forks. If it chooses the correct route, it obtains food at *F* before reaching the home box, but if it chooses the wrong route it gets an electric shock from the punishment grills at *G*. Not surprisingly, the animals soon learn which route the experimenter links to each tone.

books. One of the most visible and controversial expressions of this movement was the behaviorist psychology pioneered in America by John B. Watson in his 1913 article "Psychology as the Behaviorist Views It." To distance himself from the subject's origin in introspection, Watson insisted that the whole concept of consciousness should be excluded from psychology. Only the observed aspects of behavior were to be taken into account. Watson also repudiated the evolutionary model. Although he and his followers favored animals as models for human behavior, they made no use of the idea of an evolutionary sequence. The rats that were conditioned to acquire new habits in the classic maze-running experiments were just learning machines, simpler perhaps than humans but working according to the same principles (fig. 13.1). The biological drives (for food, etc.) used to provide rewards and punishments were common to all organisms, humans included, and were of little interest to the psychologist. The impact of behaviorism, even in America, has been exaggerated, but by the 1930s a generally experimentalist approach had swept away the remnants of the subjects origins in philosophy and moral theory. The new approach dominated scientific psychology because it reinforced the claim that the discipline offered a means both to explain and to control human behavior. When Watson left academic life following a scandal in 1920 (he had an affair with a student), he turned to the advertising industry on Madison Avenue. His view of human nature offered the prospect of influencing consumers through the scientific design of the stimuli presented to them in advertisements. This vision was caricatured in Aldous Huxley's novel *Brave New World* (1932), in which a future human race is conditioned to accept a rigid social hierarchy by the application of behaviorist-style psychological manipulation.

Just to confuse the situation, though, there was a rival form of the "new" psychology, equally anxious to distance itself from the old tradition but deriving its evidence from psychiatry and the study of mental illness rather than from experiment. This was the analytical psychology pioneered by Sigmund Freud and taken up by his disciples (and eventual rivals) Alfred Adler and Carl Jung. Freud had begun his career studying the nervous system of animals but soon switched to medical psychology and became interested in the possibility that deviant behavior is produced by conflicts between the conscious mind and a buried stratum of unconscious drives and desires. He became convinced that the unconscious was a reservoir of the darker side of the personality, driven by sexual impulses that the conscious mind strove to suppress. Neuroses were purely psychological illnesses with no basis in disorders of the nervous system: they were due solely to conflicts arising between the conscious and unconscious levels of the mind. After experimenting with drugs and with hypnotism as means of resolving these conflicts, Freud eventually developed his technique of analysis through free association, allowing the patient to reveal his or her buried feelings through the controlled relaxation of the normal boundaries of polite conversation on the analyst's couch.

Like the behaviorists, Freud and his followers were adamant that they had thrown off the yoke of biology to create an autonomous science of psychology. But the analytical approach was not "scientific" because it used experiment, and it certainly did not reject the notion of consciousness. Its claim to be a science rested on its willingness to challenge conventional opinions about human nature and to face up to the unpleasant truths that an honest study of the subconscious revealed. In fact, as explained in "Biology and Ideology" (chap. 18), it was yet another product of the evolutionary model, with the addition of a pessimistic assumption that the more highly evolved conscious levels of the mind might not be able to control the older animal instincts still preserved in the unconscious (Sulloway 1979). Whether it was truly a science was open to question then and remains a subject of debate today (see Cioffi 1998; Webster 1995). Freud's vision of the unconscious exerted a profound effect on twentieth-century thought and was widely used as a basis for psychotherapy, but it failed to gain a place in the academic psychology departments that had become wedded to the alternative, experimentalist vision of the discipline's future. To the critics, Freud's views were a product of his own imagination projected onto the disturbed behavior of his patients. There was no objective method for verifying his particular vision of the unconscious, and the movement he founded soon broke up as Adler and Jung developed rival interpretations of our hidden drives and desires. For many critics today, the analytical movement represents a profoundly unscientific approach to human nature that gained credence only because its pessimistic vision of the human situation resonated with the cultural anxieties of the early twentieth century.

Anthropology and the Study of Non-Western Cultures

The evolutionary paradigm had an especially strong influence on Western thinkers' perception of other cultures and societies. In the early decades of the nineteenth century, it became widely accepted that modern European society had evolved from primitive origins: early tribes of hunter-gatherers had invented agriculture and eventually founded great empires, now replaced by modern industrial civilization (Bowler 1989). For liberal thinkers such as Spencer, social and cultural progress was a consequence of individual initiative and enterprise over many generations. The evolutionary view of culture was massively reinforced in the 1860s by discoveries in prehistoric archaeology, which confirmed that there had been an extensive Stone Age preceding the emergence of more highly structured societies. But in the absence of any evidence beyond stone tools, how would European archaeologists reconstruct the cultures of their own distant ancestors?

To European (and soon American) traders and colonists spreading their influence around the world, the answer to this question seemed obvious. They encountered great empires in places like India that were reminiscent of the empires of Egypt and Rome, familiar from their own history. In more remote regions such as Australia they found "savages" still living in the Stone Age. Anthropology, the study of other cultures, could thus be called in to bolster the evolutionary sequence: savages were cultural relics of the past, frozen at a stage of development passed through by Europeans thousands of years earlier (Burrow 1966; Stocking 1968, 1987). For Edward B. Tylor, the first academic anthropologist to teach in Britain (at Oxford) and Lewis Henry Morgan, who had studied American Indian languages, it seemed obvious that there was a linear hierarchy of cultural evolution, which the white race had ascended more rapidly, leaving the "lower" races

FIGURE 13.2 Australian aboriginal marriage ceremony, the frontispiece to John Lubbock's *The Origin of Civilization and the Primitive Condition of Man* (1870). The ceremony involves a symbolic "capture" of the woman from another tribe, which for Lubbock was a clear sign that the aborigines still retained the instincts of our primitive ancestors.

stuck at earlier levels of the sequence. For the archaeologist John Lubbock, a strong supporter of Darwinism, the culturally primitive races were biologically more primitive, too—little better than the ape-men whose remains would eventually be found in the fossil record to fill in the "missing link" in human origins (fig. 13.2). Few of these anthropologists ventured to foreign countries for information about the peoples whose cultures they downgraded so casually. They relied on reports by traders, soldiers, and missionaries, whose views naturally reflected the prejudices of the white colonists who were conquering and, in some cases, virtually exterminating the savages they encountered.

Anthropology thus gained some visibility as a subject, but it remained a subsection of a wider project defined by the old philosophical tradition as modernized by evolutionism. James George Frazer's classic *The Golden Bough* (popular after the publication of a three-volume condensation in 1900) brought the evolutionary model to a wider public by explaining the classical myths of Greece and Rome as survivals of an earlier era. As yet there were few professional anthropologists—most studies of the native peoples of North America, for instance, were done by scientists working for the Bureau of Ethnology, created by John Wesley Powell of the U.S. Geological Survey.

The transition to an independent academic discipline came, as with psychology, as the result of a deliberate attempt to throw off the shackles of evolutionism. This occurred most abruptly in America, where the German immigrant Franz Boas created a powerful postgraduate department at Columbia University to train specialist anthropologists (Cravens 1978). Boas insisted on fieldwork in the modern tradition: the anthropologist had to live with his or her chosen culture to absorb its complexities through direct experience. He repudiated the notion of an evolutionary hierarchy, insisting the all cultures should be treated as equally complex responses to a basic human need. He also rejected the claim that culture was predetermined by biological instincts. Like the behaviorists in psychology, he insisted that learning was dominant over biological inheritance. Culture was a distinct layer of activity that could not be explained in biological terms. In the words of Boas's disciple Hans Kroeber, culture was the "superorganic." One of Boas's most influential students was Margaret Mead, whose *Coming of Age in Samoa* (1928) explicitly challenged G. Stanley Hall's view that the adolescent trauma was biologically generated. Mead claimed to find no such trauma in Samoan teenagers, suggesting that the effects studied by Hall were a product of sexual repression in Western culture. Boas and his students also came into conflict with the biologists who were still defending the claim that the nonwhite races were genetically inferior to Europeans.

In Britain there was a similar rejection of the linear evolutionary model in the early twentieth century (Stocking 1996). The end product of this reaction was a school of anthropology known as functionalism, led by the Polish immigrant Bronislaw Malinowski, who taught at the London School of Economics in the 1920s. The functionalists studied societies through intensive fieldwork, trying to work out how cultures help people cope with their physical and economic environment. They brought anthropology much closer to the emerging field of sociology (discussed below). Their form of anthropology was particularly restricted because it rejected not only the input of biology but also any interest in the history of the cultures studied. British social anthropology thus severed the link with archaeology, preserved in the United States because Boas encouraged his students to be aware of how each culture was shaped by its own local history. (Curiously, this is a very Darwinian insight—the anthropologists were rejecting the linear evolutionary hierarchy, not the more environmentalist focus of Darwin's own vision of evolution.) Another difference between British and

American anthropology was that while Boas and his students stressed the need to record information about cultures under threat from the expansion of modern industrialism, the British school offered itself as a training ground for colonial administrators who needed to understand how the "native" culture worked in order to govern more effectively (Kuklick 1991).

Sociology: The Science of Society

It was perhaps easier for anthropologists to insist that culture could not be explained solely in terms of psychology or biology than it was to make the same case for the laws governing social interactions. In the end, sociology became a distinct field of study in much the same way as anthropology, but the influence of the evolutionary paradigm was even harder to shake off. In the early nineteenth century, it was by no means clear to those who studied the functioning of human beings en masse that there might be laws of social activity that could not be reduced to the laws governing individual behavior. The political economists of the utilitarian school, including such influential thinkers as Bentham and Malthus, worked within an individualist, laissez-faire ideology that encouraged the view that economic and social activity could be understood and even regulated through an appreciation of how the individual acted within the community. Individual behavior reflected social and economic pressures but did so in a way that could be explained by associationist psychology coupled with the iron laws of economics (including Malthus's principle of population). Even today, this way of thinking discourages the idea that society operates according to laws that transcend those of individual psychology. Margaret Thatcher, the Conservative prime minister of Britain in the 1980s, famously declared that there was no such thing as "society"—there were only masses of individuals seeking their own self-interest.

It was the French philosopher Auguste Comte who coined the term "sociology" and insisted that it denoted an area of science that would have laws of its own. Comte's *Cours de philosophie positive* (*Course of Positive Philosophy* [1830–42]) defined a new approach to science that abandoned the search for causes and insisted that the only goal should be the establishment of laws relating observable phenomena. He accepted that although living things were clearly driven by the laws of physics and chemistry, biology would nevertheless have laws of its own that could not be reduced to the lower level. Sociology, too, would have to seek out the laws governing human interactions without simply assuming that those laws could be explained in terms of the physiology of the body (Comte rejected the notion of an intervening level corresponding to individual psychology).

Techniques were becoming available in the mid-nineteenth century that showed how the new science envisioned by Comte might gain its information. The Belgian statistician Lambert Quetelet began to collate information gathered from the whole population, including figures for the rates of crime, suicide, and so forth that showed that such activities occurred with remarkable regularity in all societies. Darwin was impressed by Quetelet's efforts to show how any character within a population could be understood in terms of variation from a mean value (Quetelet had pioneered the concept of the "average man"). Later in the century, Darwin's cousin, Francis Galton, gathered masses of information on human mental and physical variation and began the development of statistical tools for analyzing the data. His work played a role in the creation of modern Darwinism but, more immediately, provided the basis on which the eugenics movement argued for a hereditarian basis for character differences.

The emergence of a true sociology rested ultimately on the repudiation of Galton's biological emphasis and the realization of Comte's goal of an autonomous science of society. The first steps in this transformation arose from the nineteenth century's obsession with the idea of historical development or evolution. In their very different ways, Karl Marx and Herbert Spencer elaborated a science of society in which the emergence of higher levels of organization in the course of time was seen as an inevitable consequence of social dynamics. Marx's revolutionary vision saw the proletariat (the workforce of the new industrialized economy) as a product of social history and as a force that would ultimately transform society into a socialist paradise by expropriating the capitalists who had initiated the Industrial Revolution. By the end of the century, his "scientific socialism" had become a source of enthusiasm for left-wing thinkers, and in the mid-twentieth century the Soviet regimes in Russia and elsewhere promoted the claim that Marxism was the only explanation of social dynamics consistent with the methodology of science. Outside the Soviet bloc, however, Marxism gained no institutional foundation, although it remained an important source of critical arguments for those who opposed the prevailing ideology of capitalism.

It was Herbert Spencer who put Comte's program into practice in a way that provided capitalism with its own scientific framework, but he approached the task through his commitment to the theory of evolution. Spencer accepted that there were laws of social activity beyond mere individual psychology, but he saw the higher level of behavior as emerging from the lower under the influence of more general laws of universal development. Spencer's vision of social dynamics took for granted the individualism of the old utilitarian tradition but made individual effort and initiative the driving force of change. Because he saw competition as the stimulus that generated individual self-improvement and hence social progress, Spencer and his followers became know by their opponents as "social Darwinists" (see chap. 18, "Biology and Ideology"). His sociology certainly inspired the evolutionary movement, but it had more diverse roots in biology. In particular he stressed the "organic metaphor" in which society was modeled on an individual organism: just as the many specialized organs of the body unconsciously cooperate so that the body can perform its higher-level activities, so the individuals within society perform their specialized jobs for the good of the whole (although their immediate motive is self-interest). Spencer's popular *The Study of Sociology* of 1873 made the case for an independent science of social action, reinforced later by his more substantial *Principles of Sociology*.

Despite these initiatives, Spencer's sociology was seen as an integral part of his evolutionary philosophy. This inevitably blunted the force of his message that sociology should be a scientific discipline independent of psychology and biology. As in anthropology, the emergence of a specialist discipline of sociology with the academic apparatus of university departments, journals, and societies depended on a deliberate repudiation of the evolutionary model. In Europe, this transition occurred quite abruptly in the 1890s, when scholars such as Émile Durkheim in France began to insist that the laws governing society could not be explained by reducing them to a lower level of activity. Like Comte, Durkheim had little interest in psychology. He set out to discover the laws by which social conditions shaped behavior. In a study of suicide published in 1897, he ignored the psychology of the individual, using statistics to show how different social situations affected the suicide rate. Durkheim was a passionate secularist who was deeply concerned with the question of how people develop a sense of purpose within their society—a higher suicide rate was an indication that the society did not encourage a sense of solidarity. By the turn of the century, Durkheim had begun to create an influential school of sociology in France. A journal, the *Année sociologique,* was founded in 1898, and in 1902 Durkheim's move from Bordeaux to Paris brought him to the center of French intellectual and academic life. His own professorial chair, originally in education, was renamed to include sociology in 1913. Even so, the Durkheim school was influential more through its wider effect on European intellectuals than through the creation of a formal research program. Few academic departments of sociology were founded, and the subject remained marginal in many European countries. In the 1930s it was actively discouraged by the Nazi and Fascist regimes that came to power in Germany, Italy, and Spain.

As with psychology and anthropology, it was in the rapidly expanding university system of the United States that sociology found its most secure position (Cravens 1978; Degler 1991). Here, within a society being rapidly transformed by immigration and industrialization, the proclamation of a scientific study of social action offered new hope of controlling the complex forces at work. Universities could appeal both to big business and to government for support in developing a science that offered the hope of heading off discontent or even revolution. Sociology would retain its traditional role as a tool of political action, but it would sharpen that tool by the application of a rigorous scientific method in the gathering and analysis of its information. It was the need for specialization that made it necessary to stress the subject's distinct character and lack of dependence on the old evolutionary model. The role of the sociologist as an expert to whom the governing classes should turn for advice was stressed by Franklin H. Giddings, who became the first professor of sociology in an American university, at Columbia in 1894. William Harper at the University of Chicago got the tycoon John D. Rockefeller interested in supporting the social sciences, while Daniel Coit Gilman at Johns Hopkins University also used the appeal of a new science to raise funds. The *American Journal of Sociology* was founded in 1895 and the American Sociological Society in 1905.

By the early decades of the new century, sociology joined anthropology and psychology within the American academic system as firmly established partners within the new areas of the human or behavioral sciences. Tensions remained between the old humanistic traditions retained by thinkers such as William James and the need to stress the subjects' scientific character by emphasizing experiment (in the case of psychology) and the statistical analysis of objectively gathered information (in sociology). To some extent, the provision of funding by big business imposed constraints on the new sciences and raised questions of academic freedom. Those who provided the funds wanted science, not moral philosophy, and they wanted it in a form that could be used as a tool for social manipulation. In a very real sense, the ideological niche that the human sciences exploited to gain a hold on early twentieth-century American academic life would continue to define both the disciplines involved and the challenges they would face.

Conclusions

The human or behavioral sciences were by no means obvious products of the emergence of modern science. Indeed, they were a very late response to the social and professional opportunities that Foucault perceived as being offered by the more highly organized modern state. It was hard for many to believe that human behavior could be governed by law and hence understood by the methods of science. And even if that were possible, the lure of reductionism and evolutionism offered ways of explaining human nature without creating independent disciplines for studying the mind and social activity. Pioneering work by Spencer made such sciences as psychology and sociology a possibility, though without realizing the need for research programs that would eschew the wider synthesis offered by evolutionism and new developments in neurophysiology. In the end, it was the drive for professional autonomy within a rapidly expanding academic system that led American psychologists, anthropologists, and sociologists to lead the way in severing links that had been integral to those subjects' early development. The methods of scientific biology might be borrowed but not that area's theoretical paradigms. The claim that genuine sciences of human behavior could be formulated tapped into sources of funding and influence that allowed these sciences to be recognized as important players in the game of academic politics. The more it became possible to stress the scientific credentials of the new subjects, and their independence from the old tradition of moral philosophy, the more support could be obtained. By contrast, the human sciences in Europe were slower to attain the professional identity enjoyed by their American colleagues and slower to lose the links back to moral and philosophical concerns.

The tensions generated within the American system came to a head in the Cold War, when money from the military-industrial complex poured into the social as well as the physical sciences. The result was to drive areas such as psychology and sociology ever more firmly into the camp of those who stressed their scientific credentials and their usefulness in the areas of social control (Simpson 1998). There was a predictable backlash from radical groups from the 1960s onward, even more visible in Europe, where the human sciences had at last begun to follow the American model of professionalization.

In the end, perhaps the most interesting question is: Did the drive to create a scientific approach to the study of human nature achieve its goal? For all the money and effort poured into creating a body of practical information on the topic, many scientists in better established areas remain suspicious, pointing to a lack of theoretical coherence that undermines the analogy with the "hard" sciences. Psychology, at least, has built on its experimentalist credentials and has more recently fused with the expanding field of neurophysiology to create what is now called cognitive science. This might be seen as a new phrenology, a genuine science linking the mind and the brain, although it is also informed by developments in evolution theory. The so-called evolutionary psychology of Stephen Pinker (1997) and others seeks to identify modules in the brain that have been shaped for particular tasks in perception or cognition by natural selection. The results are highly controversial because they reopen the debate over the biological determination of human behavior. At the other extreme, anthropology prides itself on its objectivity in the study of non-Western cultures but seldom makes use of an explicitly scientific methodology. In between, sociology regards itself as the preeminent social *science* but is seldom acknowledged as such by the scientific community as a whole. The American Association for the Advancement of Science has a category of membership for the social, economic, and political sciences, but it is smaller than those for anthropology or psychology and is dwarfed by those representing the physical and biological sciences. The association's journal *Science* routinely publishes research articles on cognitive science and commentaries (but seldom research articles) on anthropology and even archaeology, but sociology gets little coverage. To some extent, the human sciences have sold their birthright as products of an earlier moral and philosophical discourse for a scientific outlook that has been immensely profitable but is still viewed with suspicion by those who see themselves as the guardians of what it takes to be "scientific."

References and Further Reading

Boakes, R. 1984. *From Darwin to Behaviourism.* Cambridge: Cambridge University Press.

Boring, Edwin G. 1950. *A History of Experimental Psychology.* 2d ed. New York: Appleton-Century-Crofts.

Bowler, Peter J. 1989. *The Invention of Progress: The Victorians and the Past.* Oxford: Basil Blackwell.

Burrow, J. W. 1966. *Evolution and Society: A Study in Victorian Social Theory.* Cambridge: Cambridge University Press.

Cioffi, Frank, 1998. *Freud and the Question of Pseudoscience.* Chicago: Open Court.

Cravens, Hamilton. 1978. *The Triumph of Evolution: American Scientists and the Heredity-Environment Controversy, 1900–1941.* Philadelphia: University of Pennsylvania Press.

Degler, Carl. 1991. *In Search of Human Nature: The Decline and Revival of Darwinism in American Social Thought.* New York: Oxford University Press.

Gould, Stephen Jay. 1977. *Ontogeny and Phylogeny.* Cambridge, MA: Harvard University Press.

———. 1981. *The Mismeasure of Man.* Cambridge, MA: Harvard University Press.

Halévy, Elie. 1955. *The Growth of Philosophic Radicalism.* Boston: Beacon Press.

Foucault, Michel. 1970. *The Order of Things: The Archaeology of the Human Sciences.* New York: Pantheon Books.

Kuklick, Helena. 1991. *The Savage Within: The Social History of British Anthropology.* Cambridge: Cambridge University Press.

Pinker, Stephen. 1997. *How the Mind Works.* New York: Norton.

Porter, Theodore, and Dorothy Ross, eds., 2003. *The Cambridge History of Science.* Vol. 7, *The Modern Social Sciences.* Cambridge: Cambridge University Press.

Richards, Robert J. 1987. *Darwin and the Emergence of Evolutionary Theories of Mind and Behavior.* Chicago: University of Chicago Press.

Smith, Roger. 1997. *The Fontana/Norton History of the Human Sciences.* London: Fontana; New York: Norton.

Simpson, Christopher, ed. 1998. *Universities and Empire: Money and Politics in the Social Sciences during the Cold War.* New York: New Press.

Stocking, George W., Jr. 1968. *Race, Culture and Evolution.* New York: Free Press.

———. 1987. *Victorian Anthropology.* New York: Free Press.

———. 1996. *After Tylor: British Social Anthropology, 1888–1951.* London: Athlone.

Sulloway, Frank. 1979. *Freud, Biologist of the Mind: Beyond the Psychoanalytic Legend.* London: Burnett Books.

Webster, Richard. 1995. *Why Freud Was Wrong: Sin, Science and Psychoanalysis.* London: Harper Collins.

Young, Robert M. 1970. *Mind, Brain and Adaptation in the Nineteenth Century.* Oxford: Clarendon Press.

PART II

THEMES IN THE HISTORY OF SCIENCE

GANIZATION OF SCIENCE

At one level, science is a very personal activity: credit is obtained by getting everyone to acknowledge your priority as a discoverer. But this process necessarily entails social interaction because the discovery must be communicated to others who must be persuaded to accept it and the theoretical conclusions associated with it. Thus the scientist needs to be part of an organization for disseminating and adjudicating ideas and information. From the time of the Scientific Revolution onward, the system of communication has increasingly been formalized through the creation of scientific societies that meet regularly and disseminate results through published journals. These societies have other functions besides mere communication, though. Very often they have served a gatekeeping role by defining who is accepted into the scientific community as a whole or into a particular research school. Membership of a formal society can easily be regulated so as to keep out those whose views do not fit in, while the publication of results has almost always been subject to a refereeing process that guarantees that only research conducted within an approved format is accepted. At various points in time, this selectivity has been used to isolate potential candidates whose views have been deemed unacceptable, as when the scientific journals of the 1980s refused to publish James Lovelock's "Gaia" hypothesis (see chap. 9, "Ecology and the Environmental Sciences"). Here the scientific community closed ranks against someone whose theory was thought to verge on mysticism, leading to charges that it was enforcing a rigid dogma and suppressing alternative viewpoints. The highly specialized scientific societies of the modern research community also reinforce professional identities in a way that has often reduced scientists' ability to communicate on wider issues. Whatever its benefits, the emergence of an organized scientific community cannot be seen merely as a practical move to improve the dissemination of knowledge.

Scientific societies have also been used by scientists in their attempts to communicate with interested or potentially interested parties in the outside world. Here there has often been an explicitly self-serving motive: as science has become ever more expensive, those outside have to be convinced of its value so they will provide the necessary resources. The source of the patronage has changed, from royalty and the nobility in the seventeenth century to governments, industry, and the general public in the modern world, but the need to "sell" science remains the same. Scientists have also sought influence within the academic community, gradually expanding the proportion of the university system devoted to scientific activities and shaping the community through the creations of departments of increasing specialization. This serves the double purpose of providing salaries and research opportunities for existing scientists and controlling the education of the students who are entering the profession. The modern history of science has focused much of its attention on the creation of scientists' professional identities though the founding of university department and government-funded institutes specializing in particular disciplines. Indeed, the very existence of an identifiable scientific discipline is now held to rest on the successful creation of such an institutional framework, and wider theories that do not lend themselves to this activity have been to some extent marginalized. At this level of analysis, for instance, there was no such thing as "evolutionary biology" until the 1940s because Darwin's late nineteenth-century followers did not found departments specializing in that topic. Such a move certainly helps to focus our attention on what scientists were actually doing, as opposed to their rhetoric aimed at the general public, but it risks loosing sight of broader initiatives that gain influence by transforming a wide range of existing activities.

The historians' focus on the emergence of the apparatus characteristic of the modern scientific community also creates problems when we seek to understand earlier phases in the development of science. The specialization and professionalization of science are key components of its success, but they took a long time to become consolidated at the level we take for granted today. In any period up to the late nineteenth century, we have to accept that a significant proportion of scientific research was being done by people who were not professionals in the modern sense. They were "gentlemanly specialists," to use the term adopted by Rudwick (1985) to denote the geologists of the early nineteenth century—men who were leading figures in their field but who did not gain their income from science and would have been suspicious of anyone who did. Nathan Reingold (1976) calls them "cultivators" in order to sidestep the modern definitions of "professional" and "amateur" that imply that the amateur is somehow inferior. Darwin is a classic example of a scientist who had no need to earn his living, and we should remember that his leading champion, T. H. Huxley, had to struggle hard to find a paying position in the London of the 1850s. Huxley's generation engineered a takeover by the professionals who did need a salary to support themselves and were all the more keen to attract government and industrial support to science for this reason. At first they made common cause with the remaining gentlemanly specialists, but their aim was to join and eventually control the elite social circle that managed science behind the scenes.

To a great extent the developments outlined above were a necessary consequence of science's success as a social activity. As Derek De Solla Price pointed out in the 1960s, by almost any measurable indicator, science has been growing at an exponential rate since the seventeenth century, and this means that "80 to 90 percent of the scientists who have ever lived are alive now" (Price 1963, 1). This expansion has taken place largely because science has become useful to governments and industry, and the organization of science has been shaped by the need to encourage and influence this support. As a result, the character of science itself has changed, as suggested by Price's distinction between the "little science" of the early centuries and the "big science" of today. Little science was done by individuals, often for their own amusement and at their own expense. Big science is done by research teams using equipment so expensive it can only be financed by governments or major industrial concerns in the expectation of practical results (or in some cases sheer prestige). But the changing structure of scientific organizations reflects more than the desire to engage with the public demand for new technologies; it also responds to the different needs of specialized professionals needing to communicate with one another and define their own disciplinary territories.

This chapter begins with a survey of how science first became organized during the Scientific Revolution of the seventeenth century, illustrating how some aspects of the modern community emerged in conditions very different from those of later centuries. In the eighteenth century, these developments were consolidated as a recognizable scientific community began to emerge. It was in the early nineteenth century, though, that many of the institutions we take for granted today were forged. The educational reforms of the revolutionary and Napoleonic governments in France placed greater emphasis on science, followed soon afterward by the creation of the modern form of research university in Germany. Scientists began to unite on a national scale to demand greater recognition for what they did and greater resources from governments. By the end of that century, educational reforms had greatly increased the size of the scientific community and its degree of professionalization, while at the same time government and industry at last began to concede that support for scientific research was of national importance.

The Scientific Revolution

Late medieval scholars routinely traveled between the universities that had sprung up in towns across Europe. The universities were centers for the scholastic philosophy based on the writings of Aristotle, and as such we do not usually think of them as major centers from which the New Science was disseminated. Yet most of the key figures in the Scientific Revolution (see chap. 2) were educated at universities, and some spent significant parts of their careers in university posts (Pyenson and Sheets-Pyenson 1999). Copernicus studied medicine and canon law at several Italian universities, Galileo taught mathematics at Pisa and Padua, and Newton spent much of his career at Cambridge. The anatomist Andreas Vesalius studied at Louvain and taught at Padua. The existing curriculum placed severe restrictions on how science could be studied, but the recognized subjects of medicine, mathematics, and philosophy were interpreted broadly and gave some leeway within which the New Science could be practiced. Botany was taught in medical faculties because most medicines were still derived from plants. The universities of the seventeenth century, thus, should not be dismissed as irrelevant to the rise of the New Science (Feingold 1984). Equally important was the foundation of new educational institutions to provide education of a more practical nature. A leading example is that of Gresham College founded in 1597 through the will of the London merchant Sir Thomas Gresham, which had professorships in astronomy, geometry, and medicine. Within the Catholic Church, the Jesuits were active in promoting astronomical work, although they distanced themselves from the more radical theoretical notions of the Copernicans.

Although most of the leading figures associated with the Scientific Revolution were educated at universities, the subjects they studied were sometimes unconnected to their later interests in natural philosophy, and many had no subsequent base within the university system. Some were independently wealthy, as in the case of the chemist Robert Boyle. Others sought patronage from wealthy figures, who would employ them either because they were genuinely interested in the new learning or because it boosted their prestige to have famous scholars associated with their court or household. In 1610, Galileo left Padua to become philosopher and mathematician to the grand duke of Tuscany. He also sought the patronage of leading figures in the Church, a move that went disastrously wrong when he lost favor with the pope and was tried by the Inquisition (Biagioli 1993). The astronomer Tycho Brahe built his observatory at Hveen under the patronage of the Danish king Frederick II and moved to Prague to work under the emperor Rudolph II when Frederick's son withdrew support after his father's death. Johannes Kepler, who began as apprentice to Tycho, also served under Rudolph II. Court patronage was an uncertain business, but it was the recognized system in the Renaissance—and it was a long time before more democratic governments would provide equivalent levels of support. Even in the late seventeenth and eighteenth centuries, the patronage of the wealthy was important, especially for naturalists who could describe and catalog collections of animals and plants. John Ray left his Cambridge fellowship when he was offered support by the wealthy Francis Willoughby.

The New Science depended on interactions between the major figures and a host of interested parties who had to be persuaded to accept both experimental discoveries and theoretical innovations. Communication was vital so that consensus could be reached (sometimes only after acrimonious controversy), and it was necessary to define a community of reputable individuals whose judgment was to be trusted in such matters. At a time when many were suspicious of the New Science, its supporters also needed to band together for mutual support. From the start, there were local societies where a sufficient number of figures lived in a single location to support regular meeting or other forms of interaction. Galileo was proud of his membership in the Accademia dei Lincei (the lynx-eyed), and his followers helped form the Accademia del Cimento in Florence, to which leading figures such as G. A. Borelli and Francesco Redi belonged (Middleton 1971; for a general account of seventeenth-century societies, see Ornstein [1928]). But these were not permanent institutions, the first truly lasting scientific organization being the Royal Society of London, founded in 1660 and given its royal charter two years later (Boas Hall 1991; Hunter 1989). It had been preceded by informal meetings at Oxford attended by figures such as Robert Boyle, Christopher Wren, and Robert Hooke, but incorporation as a recognized public body gave the Royal Society a significant level of status (although Charles II provided no funds and was himself suspicious

FIGURE 14.1 The frontispiece to Thomas Sprat's *History of the Royal Society* (London, 1667). Francis Bacon, whose empiricist philosophy was idealized as the basis for the New Science, sits to the right of the bust of the society's patron, King Charles II. Various scientific instruments, including Robert Boyle's air pump, appear in the background.

of the new learning). As proclaimed in Thomas Sprat's *History of the Royal Society* (1667), the group highlighted the empiricist philosophy of Francis Bacon as an alternative to scholasticism and insisted that philosophical and political divisions should not intrude into their discussions (fig. 14.1). The society employed curators, one of whom was Hooke, to perform demonstration experiments. But more important was the society's function in reporting observations and discoveries. Their secretary, Henry Oldenburg, maintained an international correspondence with scientists, and the *Philosophical Transactions* was begun as the first scientific journal.

All this sounds very positive, but members of the society were anxious to define their status as the arbiters of what counted as the New Science. Since everyone could not perform experiments for themselves, the trustworthiness of reports was vital, and only gentlemen were thought reliable enough—the artisans who actually did the work on Boyle's air pump, for instance, are never mentioned in his reports. Those who challenged the philosophical basis of experimental philosophy or expressed suspicion of how the inner circle imposed the values of that philosophy on the New Science were rigorously excluded (Shapin and Schaffer 1985). The new society thus acted very much as a gatekeeper to exclude anyone not deemed socially or philosophically acceptable from participation in the scientific enterprise. Far from being religiously or ideologically neutral, the members of the Royal Society had a very definite social agenda. They may not all have been Puritans, as some scholars allege, but they professed a liberal Anglicanism and support for the restoration monarchy and the new mercantile basis for creating wealth (see chap. 15, "Science and Religion"). The one thing the society did not have was access to government funding, and this left it to some extent at the mercy of its wealthier members whose interest in science was often superficial. Only in one area did Charles II take the New Science seriously: astronomy offered the hope of better navigational techniques that were crucial to British trade abroad. On the advice of a commission including Wren and Hooke, the Royal Observatory at Greenwich was built in 1675–76 and its first Astronomer Royal, John Flamsteed, installed. Even so, Flamsteed had to spend a considerable amount of his own money on equipment.

In France the highly centralized government of Louis XIV made the situation very different. After the failure of several local societies, scientists petitioned Louis's minister J. B. Colbert for state support, and the Académie Royale des Sciences met for the first time in the royal library in 1666 (fig. 14.2; see Hahn 1971). Salaried positions were established in mathematics (including astronomy) and natural philosophy (where the emphasis was on the physical sciences). Eminent figures such as Christian Huygens came to Paris to take up these positions. Here was state support on a notable scale, although the academy was expected to produce useful results, especially in areas such as navigation. The academy was reorganized in 1699 when an observatory was constructed. Although well-funded at first, the

FIGURE 14.2 Louis XIV visiting the Académie des Sciences, the frontispiece to Denis Dodart's *Mémoire pour servir a l'histoire des plantes* (Paris, 1676). The Paris academy was dependent on the patronage of the French king and had to convince him that its activities were useful to the state. Various scientific instruments are being demonstrated to the king.

academicians' activities were highly regulated in a way that did not always allow them the freedom to pursue original research, and the funding itself became more limited as Louis's wars generated economic hardship in France. Even so, the Académie Royale offered a model that would be copied by rulers elsewhere in Europe during the following century, while the Royal Society pioneered the looser kind of organization whose structure and interests were defined more by the scientists themselves.

The Eighteenth Century

The eighteenth century saw some developments in science education, but they were very unevenly distributed. Dutch and German universities became active centers of research and teaching, especially in the physical sciences. Leiden was especially important for the study of electricity, and it was here that Petrus van Musschenbroek invented the condenser or Leyden jar in 1746 (Heilbron 1979). Scottish universities were also active in the teaching of medicine, while a natural philosophy chair was established at Edinburgh in 1776. Linnaeus promoted his new system for classifying species from his botanical garden at Uppsala but was still based in the medical faculty because there was no framework for teaching natural history. Elsewhere there was little effort to bring the New Science into the curriculum, Oxford and Cambridge being notable examples of universities that offered little science teaching until well into the nineteenth century. The practical approach to science education that had been pioneered at Gresham College did, however, begin to spread more widely. Many of the independent German states gained much of their income from mining and began to set up mining academies where both geology and engineering were taught. A. G. Werner promulgated his Neptunist theory of the earth from his base at the mining academy in Freiburg and attracted students from all over Europe.

In France there was little university teaching of science, although the government did set up a technical school, the École des Ponts et Chaussés, for the training of military engineers. The Académie des Sciences continued as a state-supported center for research, and the Jardin du Roi, the royal botanical and zoological garden, was established to house the king's collections (fig. 14.3). As its superintendent, the comte de Buffon was well-positioned to promote his encyclopedic *Histoire Naturelle* (and well-protected from the Church when his speculations challenged orthodoxy a bit too openly; see chaps. 5 and 6). The Royal Society of London declined, to some extent, as its enthusiastic founders died and were replaced by gentlemen with only a passing interest in science. It took on a new lease of life later in the century, however, under the presidency of Sir Joseph Banks, who used his links with the admiralty to coordinate a worldwide program of scientific exploration modeled on his own expedition with Captain Cook (Makay 1985). Banks was so anxious to protect the society's position at the center of British science that he actively blocked the foundation of the more specialist societies that were being called for as the scientific community expanded. The one exception was the creation of the Linnaean Society in 1788, which was established around Linnaeus's own

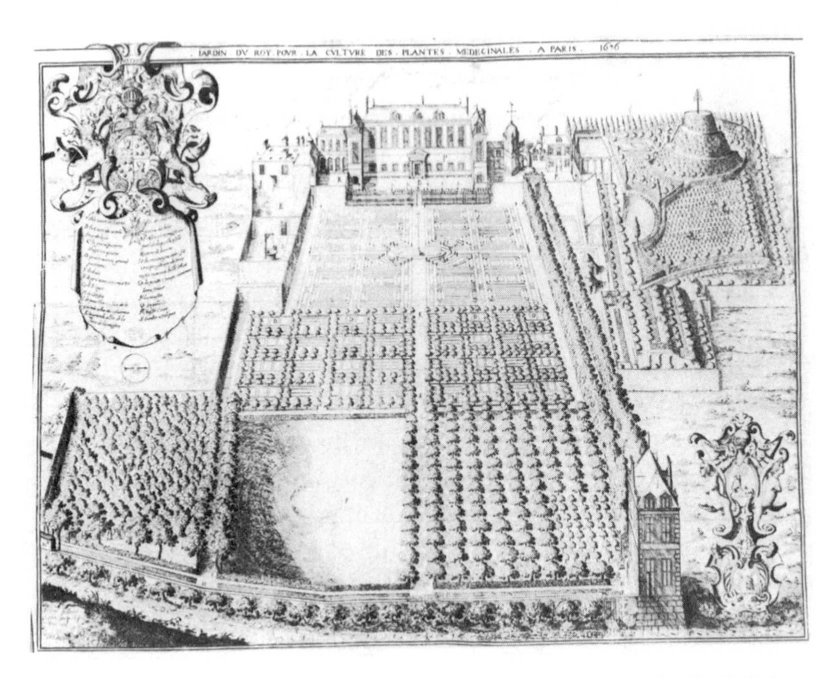

FIGURE 14.3 The Jardin des Plantes, from Frederic Scalberge, *Le Jardin du Roi* (Paris, 1636). Botanical gardens had originated in the medieval universities when they were important for training doctors to recognize the plants from which many medicines were derived. They were still important centers of science in the seventeenth and eighteenth centuries, when new plants from all over the world were being brought to Europe. The Jardin des Plantes can still be visited today—look for the statues of Buffon and Lamarck and the buildings where Cuvier worked.

collection of plants and animals, acquired after his death by one of Banks's wealthy students, James Smith. The society became the country's leading focus for study and publication in natural history, although its origins in an act of patronage by a wealthy collector illustrates the fact that such institutions were still very much clubs for gentlemanly specialists. Over the next few decades, the expanding popular interest in natural history led to the foundation of local societies in towns throughout the country, often based on local elites that used their association with science to highlight their authority as the arbiters of culture.

Historians of British science have focused much attention on an entirely new kind of society that emerged in the new industrial town of Birmingham in the 1760s. This was the Lunar Society (it met on the night of the full moon so its members could find their way home) in which a number of eminent figures associated with the application of science to industrial technology met to share ideas. Members included James Watt and Matthew

Boulton, now collaborating in the manufacture of steam engines, Josiah Wedgewood, whose pottery business was one of the success stories of the early Industrial Revolution, and Erasmus Darwin, who had a strong interest in mechanical improvement on top of his work in medicine and the life sciences. The chemist Joseph Priestly also joined in after he moved to Birmingham in 1780. These were wealthy but practical men who met to promote their common interest in exploiting science as the basis for useful knowledge, in effect reviving the principles of the old Royal Society now that the original had become a club for the London social elite. The Lunar Society was relatively short-lived, but it highlighted tensions that would become apparent in the following century as the influence of those who actually used the New Science grew at the expense of those who only studied it as a hobby.

The Nineteenth Century

It was in the nineteenth century that the growing involvement of science with industry and government, coupled with the resulting expansion and specialization of the scientific community, produced the institutions we take for granted today. This did not happen overnight because there were powerful social forces at work that restricted the developments required for these interactions to work efficiently. The tendency for science to be regarded as the preserve of a social elite held back the process of professionalization and the search for public funding. Universities long used to training the elite in classics resisted the move to incorporate scientific teaching and research into their curricula. In Britain and America, the ideology of laissez-faire made even the new breed of successful industrialists suspicious of government support for science. In a free enterprise society, it was held, those who benefited from the research should pay for it—but industrialists only wanted to finance research that was immediately useful and had no interest in pure research that might not yield profits until a generation later. For this reason, the centralized governments of France and Germany led the way in public funding of science, with Britain and America struggling to catch up in the later part of the century. Eventually, however, the role of science in promoting national wealth and prestige was recognized and both the educational system and the internal working of the scientific community began to adjust to the new reality.

France experienced a rapid burst of change in its educational and scientific institutions following the outbreak of the revolution in 1789. By 1793 the new government had replaced the old Jardin du Roi with the

Muséum d'Histoire Naturelle, which was intended for display, teaching, and research. With influential professors such as J. B. Lamarck, Georges Cuvier, and É. Geoffroy Saint-Hilaire, it became the model for research and education in natural history for the whole of Europe. The Académie des Sciences was also totally reorganized to give a new but equally centralized system for recognizing scientific distinction (Crossland 1992). New institutions, the École Polytechnique and the École Normale Supériore, were established for technical research and education, with a host of famous figures as professors. Napoléon built on these foundations, confirming the image of science as a practical activity that should be at the service of the nation. In the early decades in of the nineteenth century, these institutions made Paris the mecca of the scientific world, although the highly centralized French system was extremely rigid and was not well placed to respond as other nations' industrial developments began to reduce the country's preeminence as a world power.

Several German universities had become active in the eighteenth century, and although some were closed down during the Napoleonic invasions, there was a wave of foundations and refoundations in the early nineteenth century. The division of the German-speaking region into a number of states created a situation in which each sought to rival its neighbors in the acquisition of scientific and scholarly talent (Ben-David 1971). It was here that the modern research university was created, with professors being expected to both do research and train their graduate students to become researchers in their own right. The doctoral degree (PhD) became the symbol that proclaimed that a student was capable of independent research. Justus von Liebig's department of chemistry at Giessen established this system in the 1820s, and it was soon being copied in other subjects and in other universities. By the middle decades of the century, Germany was replacing France in the leadership of European science, and—significantly—the German industrial system was beginning to expand as scientific research opened up new technological possibilities in fields such as the production of chemical dyestuffs.

In Britain, Scottish universities remained active in science, but the ancient universities of Oxford and Cambridge still made no effort to introduce science into the curriculum. Cambridge did offer a rigorous training in mathematics, and there were eminent professors such as the geologists Adam Sedgwick and William Buckland—although they did no undergraduate teaching. It was not until the 1850s that these universities were forced by government commissions to accept science teaching as part of their degree programs. Even then progress was slow until further reforms in the 1870s, after which Cambridge's Cavendish Laboratory soon became a leading center of research in physics. Meanwhile University College, London, was founded for Nonconformists (only members of the Church of England could enter Oxbridge), and it, too, eventually gained some eminence in science education. Inspired by Liebig's students, the government founded the Royal School of Chemistry in 1845. Partly as a spin-off from the creation of the Geological Survey (discussed below) it also established the Royal School of Mines in 1851, and it was here that the young T. H. Huxley eventually gained his first job (Desmond 1994, 1997). In the 1870s, Huxley introduced a famous course in biology for schoolteachers, using his young disciples as demonstrators. Eventually these institutions coalesced into the Imperial College of Science and Technology.

By the later part of the century, America began very rapidly to catch up with the latest developments in science education. The Johns Hopkins University in Baltimore was founded as a research university on German lines, and soon there were a number of private universities following the same model. Meanwhile, the land-grant colleges in the Midwestern states offered publicly funded education that included the sciences and also sponsored research into areas of biology relevant to farming interests. By the end of the century there had thus been a consolidation of scientific research and teaching in the universities and senior technical colleges around the developed world. The opportunities for jobs in science had expanded enormously, much to the benefit of a younger generation, such as Huxley, who wanted to enter science from the lower ranks of society and needed a paying position. During periods of rapid expansion, the universities also provided opportunities for the establishment of new research programs, and by allowing professorships and departments to be founded they encouraged the recognition of more specialized scientific disciplines. Significantly, however, women at first found it difficult to gain access to a scientific education—even Huxley opposed allowing women into medical schools. Gradually, however, such barriers were broken down, often at first through the creation of specialist colleges for women students (Rossiter 1982).

The highly centralized governments of France and Germany provided a mechanism by which state funds could be channeled into scientific research and education if the ruling elite approved. In Britain and America, however, the popularity of the free-enterprise system among the rising class of industrialists made it far more difficult for scientists to tap the state for resources (Rupke 1988). On this model of government, the state had no role to play in such activities: if anyone wanted to do pure research for amusement, they would have to be wealthy enough to support themselves, and if the research had practical implications, then the industrial firms who would benefit should fund it. Such a philosophy was immensely short-sighted because it ignored the fact that much research turns out to be useful only after its implications have been explored for some time. Scientists began to argue that there was a level of pure research that the state should fund because its potential benefits were enormous but too uncertain for private firms to commit their funds to it. Charles Babbage's *Reflections on the Decline of Science in England* (1830) lamented the indifference of the British government toward science and insisted that if the subject were to be developed properly a scientific profession needed to be established, composed of paid and properly funded researchers. An example of the difficulty faced by those seeking to move in the direction marked out by Babbage can be seen in the creation of the Geological Survey of Great Britain. The advantages of such a survey for the mining industry were pointed out by Henry De la Beche in the 1830s, but the government thought the mining companies should pay for it, while the companies were not interested in anything that did not lead to the immediate discovery of workable mineral deposits. Thanks to persistent lobbying, De la Beche did get temporary state support, and gradually the Geological Survey was established as a permanent institution. But throughout the rest of the century it was only with the greatest reluctance that the British government was persuaded to fund science (Alter 1987). To some extent, the process was accelerated in the aftermath of the Great Exhibition of 1851, profits from which were used to create a number of scientific institutions in South Kensington (which at the time was still considered a suburb of London).

There were similar problems in the United States (Dupree 1957). Several states set up their own geological surveys and a few did important geological research, but most were bedeviled with penny-pinching legislatures that demanded instant practical advantages for local industry. The U.S. Geological Survey was established within the military in 1879 to survey the potential resources of the west and achieved notable results, especially under its second director, John Wesley Powell, who explored the Grand Canyon (Manning 1967). But at the federal level, too, there was constant pressure to save money and a consequent reluctance to fund what was perceived as "pure" science, especially the study of fossils. In 1886, Congress established the Allison Commission, which criticized activities such as the Geological Survey and recommended savage cuts in funding. Over the next couple of decades, however, a number of government scientific departments did come into existence, although there was still no coordinated science policy. Most of these departments were concerned with surveying and environmental work, or with medicine, but the Bureau of Standards also set up a physics laboratory.

The scientists themselves were aware of the potential for their field to expand and were anxious to tap the resources offered by governments, industries, and educational establishments. They were aware of the practical value of many aspects of science, but they also saw themselves as the guiding force of modern society—where people had once sought advice on social issues from the churches, now science would provide the appropriate source of expertise. Science was expanding anyway, as the proliferation of more specialist societies and journals indicated. Significantly, the term "scientist" was coined in 1833 by William Whewell, although it would be some time before it came into general use. Many scientists felt that the field would expand far more rapidly if society could be persuaded to take it more seriously and provide appropriate funding. So national bodies were needed to lobby government and industry for funding and to ensure that the scientists themselves were given control of how the money would be spent. At the same time, the expansion was changing the character of the scientific community. The need for coordination was making science less individualistic, and eventually the expense of doing "big science" would begin to move beyond the resources of an individual, however wealthy. In the early decades of the nineteenth century science was still dominated by gentlemen amateurs. Much scientific work was still privately funded, and even where government support was sought, the elite wanted to keep control in its own hands. But the nature of the scientific elite that sought to control this process was changing. Clever people from outside the social elite were struggling to gain an education and sought jobs where they could both do research and support themselves and their families. The elite was becoming a profession in the modern sense, no longer gentlemen amateurs but paid servants of the state and society.

Early nineteenth-century developments have been extensively studied. In Britain, the expansion of the scientific community could be seen in the emergence of specialist societies and journals that were dedicated to the interests of those with common research interests (Cannon 1978; Cardwell 1972; MacLeod 2000). After the death of Banks, the Royal Society of London went into decline and a number of specialist societies took over the leadership of British science. Most active was the Geological Society of London, founded in 1807, which for years was said to have the most exciting debates in the capital. In the late 1830s, Charles Darwin served as its secretary and coordinated a surprisingly modern system of sending papers out to be refereed in order to determine which should be published. This was very much a group led by Rudwick's gentlemanly specialists—like Darwin, they were all independently wealthy and belonged to the social elite (although De la Beche lost his family fortune, which is why he was so anxious to create a state-funded geological survey). The Zoological Society was founded in 1826, again as a gentlemen's club—there would be bitter debates before the general public would be allowed into what became the London Zoo. The Chemical Society followed in 1840, and here the interests of those who hailed the science's practical applications would play a greater role.

This practical dimension also shaped the scientific community's concern for public relations. In Britain, the Royal Institution was founded in 1799 by wealthy patrons interested in science as an agent for technological progress (Berman 1978). Humphry Davy made his reputation there as an experimenter in chemistry and electricity and as a public lecturer. A generation later the institution came under the control of elite radicals seeking to reshape society along utilitarian lines. It still supported a research laboratory, where Michael Faraday took over from Davy and built his own reputation in the study of electromagnetism, and its public lectures continued to play a major role in popularizing scientific discoveries and convincing the upper classes that social problems could have a technological solution (fig. 14.4). But scientists around the country were seeking to promote these interests on a national scale, not just to the London elite. Babbage's complaint about the decline of science in England was made after he attended a meeting of the Society of German Scientists and Physicians, founded in 1822 to unite the scientists of the many separate German states. Stimulated by his complaints, the gentlemanly specialists decided that an equivalent British society was needed to raise the profile of science, and in 1831 the first meeting of the British Association for the Advancement of Science met in York (Morell and Thackray 1981). The association met annually in a different provincial city—initially it deliberately avoided London—in order to provide a forum for scientists to interact and to plan efforts to lobby the government for funding and other means of support. It also promoted local interest in science and provided a forum in which local cultivators could mix with national figures. Eventually it did gain some support from the government to finance individual research projects led by its senior members.

Much of the association's work went on behind the scenes, however, as the gentlemen who served as its governing elite sought to impress their influence on the way things developed. There was an informal group, the Red Lions, where the inner circle met, and in later years the development of British science was shaped by an equally informal X Club, in which Huxley and his fellow specialists sought to manipulate government and academic

FIGURE 14.4 Opening of the Davy-Faraday Laboratory, Royal Institution, 1897, from the *Illustrated London News* (2 January 1897). This new laboratory, named after the institution's two most famous early scientists, was funded by the chemist and industrialist Professor Ludwig Mond. In a high-profile ceremony, it was formally opened by the prince of Wales, and a demonstration was given by Professor James Dewar, who worked at the institution on the liquefaction of gases and had invented the vacuum flask in the process.

FIGURE 14.5 By the early twentieth century, American science had expanded enormously. This is a formal photograph of the meeting of the American Association of Anatomists, held in Toronto in 1937 (Canadian scientists usually found it convenient to join American societies). There are well over a hundred men in this picture, but only a handful of women.

appointments to ensure that loyal scientists got the positions of influence (Barton, 1990, 1998; MacLeod 2000). This group was also responsible for the founding of the journal *Nature* in the 1870s. By that time, the elite was gradually being taken over by the new breed of professionals, determined to see science become the new source of knowledge and expertise in the running of society and anxious to have that expertise recognized in the form of state financial support. Figures such as Huxley routinely worked themselves into nervous breakdowns through their overloaded schedules of research, teaching, public speaking, and service on government commissions (Desmond 1994, 1997).

Similar developments took place in America, where the greater geographical separation of the cities had encouraged the emergence of local societies for the promotion of science. To provide national coordination, the American Association for the Advancement of Science was founded in 1848 (Oleson and Brown 1976). Among the country's most eminent scientists, there were serious divisions; a group known as the "Lazzaroni" operated as an equivalent to the X Club, but its influence was resented by a number of major figures who were not part of the network. During the Civil War, members of the Lazzaroni, including Joseph Henry of the Smithsonian institution and A. D. Bache of the Coast Survey, encouraged the government to set up a National Academy of Sciences through which leading

figures could offer advice. After the war, government funding dropped off, but the academy survived by turning itself into a more broadly based elite group. Throughout the rest of the century it had little impact on government policy, which was more often directed through individual research programs and surveys, but it would survive to play a significant role in the reconfiguration of American science in the twentieth century (fig. 14.5).

Conclusions: Science and the Modern World

By the early decades of the twentieth century, an embryonic form of the scientific community as we know it today had emerged. By then, the vast majority of scientists were paid professionals working in universities, government research facilities, or industry. The educational system had expanded both to house pure and applied research and to supply the trained graduates who would continue the expansion of science. From the early twentieth century onward, an increasing proportion of these scientists would be talented individuals from outside the "developed" world, attracted to Europe or America by the superior facilities for education and research. Funding for this expansion was provided by government and industry, often focused on research that was deemed to be of practical benefit to society. Research foundations, privately funded by wealthy individuals, began to make some impact and could actually shape the course of research, as when the Carnegie Foundation, set up by the industrialist Andrew Carnegie in 1902, began to support the new science of genetics. But the future would increasingly lie with government and applied industrial research.

Much of the expansion in research has been at a fairly low level of creativity, perhaps inevitably in a situation where an increasing proportion of it is meant to solve the immediate problems of industry. It is estimated that the number of mediocre scientists rises as the square of the number of truly creative ones (Price 1963, chap. 2). At the same time, there has been an enormous expansion of cooperative research, leading to an increasing proportion of multiauthored papers (it is not unusual now to see more than a dozen names listed on a paper). New specializations are created at an ever-increasing rate, often with their own small societies and journals but also as informal networks of researchers who value this collaboration more than they do their home institution.

There can be little doubt that one of the main forces that has shaped the development of the modern scientific community is the expansion of the link with the military and associated industries (see chap. 20, "Science and War"). In the years prior to World War I, government support for science was still limited and uncoordinated in most countries, leading to outcries at the start of the war that the state was wasting its scientific resources. Although the military applications of science were still very limited in the course of that war, new institutions were established to ensure some level of cooperation. In World War II, the whole situation was transformed, especially in America. Big science finally became unmistakable as government and industry co-opted scientists into major projects such as the development of radar and the atomic bomb. Following the war, the scientist-turned–science adviser Vannevar Bush promoted the need for the American government to continue its support for science, leading to the creation of the National Science Foundation in 1950. The continued tensions of the Cold War ensured that the involvement of scientists with the military-industrial complex remained high through into the later twentieth century, a very large proportion of physicists being employed directly or indirectly on projects funded through this source. The senior scientists who end their careers as directors of major research groups have to become administrators as much as scientists and need the political skills necessary to interact successfully with the government that supplies the funds.

The expectations of the early nineteenth-century scientists who collaborated to promote the utility of science to industry and government have thus been fulfilled but perhaps not in the way they hoped. The scientific community has expanded and has developed the structures needed to function in a world where science can often be done only by having access to the level of resources provided by government and big business. Those earlier scientists would probably be dismayed to find that it was only under the pressure of war that the cooperative structures were finally worked out and that now a significant fraction of science is devoted to improvements in military technology. In one respect, however, their concerns have reemerged in the modern world as the public has become suspicious of the extent to which science has fallen into the pockets of the military-industrial complex. One consequence of the professionalization of science was a rejection of the nineteenth-century ideal in which scientists were part of the intellectual elite and sought to influence public opinion through nonspecialist writing and public speaking. By the early twentieth century, many research scientists regarded such involvement with public debate as incompatible with their scientific objectivity. That situation has now begun to change as movements challenging the authority of science have grown in influence. National bodies such as the American and British Association for the Advancement of Science once again see themselves as having a major role to play in maintaining public interest and confidence in science. Having for a while retreated into a world of professional isolation, scientists now recognize that getting their message across to ordinary people is vital for the future health of their profession. In this respect, at least, the lessons learned by those earlier generations need to be relearned by the modern profession.

References and Further Reading

Alter, Peter. 1987. *The Reluctant Patron: Science and the State in Britain, 1850–1920.* Oxford and Hamburg: Berg; New York: St. Martin's Press.

Barton, Ruth. 1990. "'An Influential Set of Chaps': The X Club and Royal Society Politics, 1864–85." *British Journal for the History of Science* 23:53–81.

———. 1998. "'Huxley, Lubbock, and Half a Dozen Others': Professionals and Gentlemen in the Formation of the X Club, 1851–1864." *Isis* 89:410–44.

Ben-David, Joseph. 1971. *The Scientists' Role in Society.* Englewood Cliffs, NJ: Prentice-Hall.

Berman, Morris. 1978. *Social Change and Scientific Organization: The Royal Institution, 1799–1844.* Ithaca, NY: Cornell University Press.

Biagioli, Mario. 1993. *Galileo Courtier: The Practice of Science in the Culture of Absolutism.* Chicago: University of Chicago Press.

Boas Hall, Marie. 1991. *Promoting Experimental Learning: Experiment and the Royal Society, 1660–1727.* Cambridge: Cambridge University Press.

Bruce, Robert V. 1988. *The Launching of Modern American Science, 1846–1876.* Ithaca, NY: Cornell University Press.

Cannon, Susan F. 1978. *Science in Culture: The Early Victorian Period.* New York: Science History Publications.

Cardwell, D. S. L. 1972. *The Organization of Science in England.* New ed. London: Heinemann.

Crossland, Maurice. 1992. *Science under Control: The French Academy of Sciences, 1795–1914.* Cambridge: Cambridge University Press.
Desmond, Adrian. 1994. *Huxley: The Devil's Disciple.* London: Michael Joseph.
———. 1997. *Huxley: Evolution's High Priest.* London: Michael Joseph.
Dupree, A. Hunter. 1957. *Science in the Federal Government: A History of Policies and Activities to 1940.* Cambridge, MA: Harvard University Press.
Feingold, Mordechai. 1984. *The Mathematicians' Apprenticeship: Science, Universities and Society in England, 1560–1640.* Cambridge: Cambridge University Press.
Hahn, Roger. 1971. *Anatomy of a Scientific Institution: The Paris Academy of Sciences, 1666–1803.* Berkeley: University of California Press.
Heilbron, John. 1979. *Electricity in the Seventeenth and Eighteenth Centuries: A Study of Early Modern Physics.* Berkeley: University of California Press.
Hunter, Michael. 1989. *Establishing the New Science: The Experience of the Early Royal Society.* Woodbridge, Suffolk: Boydell Press.
MacLeod, Roy. 2000. *The "Creed of Science" in Victorian England.* Aldershot: Variorum.
Makay, David. 1985. *In the Wake of Cook: Exploration, Science and Empire, 1780–1801.* London: Croom Helm.
Manning, Thomas G. 1967. *Government in Science: The United States Geological Survey, 1867–1894.* Lexington: University of Kentucky Press.
McClellan, James E., III. 1985. *Science Reorganized: Scientific Societies in the Eighteenth Century.* New York: Columbia University Press.
Middleton, W. E. Knowles. 1971. *The Experimenters: A Study of the Accademia del Cimento.* Baltimore: Johns Hopkins University Press.
Morell, Jack B., and Arnold Thackray. 1981. *Gentlemen of Science: The Early Years of the British Association for the Advancement of Science.* Oxford: Oxford University Press.
Oleson, Alexandra, and Sanborn C. Brown, eds. 1976. *The Pursuit of Knowledge in the Early American Republic: American Scientific and Learned Societies from Colonial Times to the Civil War.* Baltimore: Johns Hopkins University Press.
Ornstein, Martha. 1928. *The Role of Scientific Societies in the Seventeenth Century.* Chicago: University of Chicago Press.
Price, Derek J. De Solla. 1963. *Little Science, Big Science.* New York: Columbia University Press.
Pyenson, Lewis, and Susan Sheets-Pyenson. 1999. *Servants of Nature: A History of Scientific Institutions, Enterprises and Sensibilities.* London: Fontana; New York: Norton.
Reingold, Nathan. 1976. "Definitions and Speculations: The Professionalization of Science in America in the Nineteenth Century." In *The Pursuit of Knowledge in the Early American Republic: American Scientific and Learned Societies from Colonial Times to the Civil War,* edited by Alexandra Oleson and Sanborn C. Brown. Baltimore: Johns Hopkins University Press, 33–69.
Rossiter, Margaret W. 1982. *Women Scientists in America: Struggles and Strategies to 1940.* Baltimore: Johns Hopkins University Press.
Rudwick, M. J. S. 1985. *The Great Devonian Controversy: The Shaping of Scientific Knowledge among Gentlemanly Specialists.* Chicago: University of Chicago Press.
Rupke, Nicolaas, ed. 1988. *Science, Politics and the Public Good.* London: Macmillan.
Shapin, Steven, and Simon Schaffer. 1985. *Leviathan and the Air-Pump: Hobbes, Boyle, and the Experimental Life.* Princeton, NJ: Princeton University Press.

CHAPTER 15

SCIENCE AND RELIGION

Mentioning science and religion together conjures up an immediate image of conflict and confrontation. Everyone remembers the trial of Galileo by the Inquisition and the furor that surrounded (and still surrounds) Darwin's theory of evolution. Yet a moment's reflection tells us that this image of conflict cannot be the whole story. Many of the great scientists of the past were sincerely religious, and there have always been theologians willing to argue that religious faith must be flexible enough to accommodate the new discoveries made by science. As James R. Moore (1979) points out in his analysis of the Darwinian debates, the image of a perpetual "war" between science and religion was deliberately constructed by late nineteenth-century rationalists who wanted to enlist science as an ally in their campaign to have all religious belief swept away as outdated superstition. J. W. Draper's *History of the Conflict between Religion and Science* (1875) was a founding document in this tradition. T. H. Huxley also portrayed science as a force that would steadily undermine the tenets of organized religion—although as an agnostic (indeed he coined the term) he accepted that science could not disprove the existence of a creator. As Moore's study of the Darwinian debate shows, however, there were many scientists whose faith required them to think carefully about the new theory, and many liberal theologians willing to see evolution as the unfolding of God's plan of creation. That more liberal tradition is still at work today, energetically promoted by organizations such as the John Templeton Foundation.

There are many different areas of science, some of which are more likely to generate problems for religious believers than others. But there are also many different forms of religious belief, some of which more easily accommodate new theories about the nature of the universe. Eastern religions such as Buddhism and Hinduism are linked to cosmologies that do not require the kind of recent creation implied by the Christian Bible, nor do those religions postulate a spiritual division between humanity and the rest of nature. Judaism, Christianity, and Islam are theistic religions that postulate a God who is deeply involved with his creation, and their sacred records define cosmologies that have proved difficult to reconcile with some scientific developments. They suggest that not only did God design and create the world but, also, that he works supernaturally within it to achieve his ends. But some scientists have been deists, believing in a remote God who designed the universe but does not concern himself with the details of what happens in it. Even within the Christian tradition, there is enormous variation between the Roman Catholic Church's reliance on a carefully monitored tradition of faith, the fundamentalism of some Protestant churches that focuses on the text of the Bible, and the liberal tradition mentioned above. Conflicts may arise between particular theories and particular religious traditions, but outside these flashpoints a far more constructive dialogue goes on. In some cases the two areas may simply coexist without interacting in any significant way. But to treat this "coexistence" model as the norm ignores the fact that Christianity does make specific claims about God's relationship to his creation and about human nature. Even without a literal reading of biblical texts, these claims ensure that some areas of science will almost inevitably generate tensions that, at the very least, will require extensive dialogue for them to be resolved. The important point that emerges from these issues is the need for the historian to take a contextualized view of the relationship, exploring the different modes of interaction that have come into play at different times and places (for comprehensive studies, see Brooke 1991; Lindberg and Numbers 1986, 2003).

Theological issues obviously play a role in determining how scientific theories are received both by other scientists and by the general public. But the historian also has to bear in mind the possibility that scientists' religious beliefs actually shape the kind of science they do. Stanley Jaki (1978) has argued that the Christian concept of God as a lawgiver played a vital role in establishing the concept of natural laws that could be understood by rational analysis. Kepler's search for mathematical harmonies within the motions of the planets was clearly influenced by his faith in a God who imposed a rational order on the world. This example shows that such influences can be positive as well as negative, forcing us to beware of oversimplifications such as the assumption that if a theory is used to defend a religious belief, it must necessarily represent bad science. There are numerous examples of theories, especially in the earth sciences, that historians used to dismiss as harmful to the progress of science, explaining their popularity entirely in terms of their adherents' desperate desire to defend a particular religious belief. Later research has often shown that these "theologically distorted" theories actually played a positive role in the development of positions still accepted today.

In this chapter, we will focus on several themes of concern to historians of science. The question of biblical literalism is certainly important, although it should not be approached in a way that suggests a simpleminded confrontation between text and theory. We must also explore the suggestion that certain kinds of religious belief, often linked to particular social values, are more supportive of science than others. A major concern is the possibility that science may contribute to a "natural theology"—a way of understanding God by studying his creation—and the threats that some theories seem to offer to that prospect. Here Darwin's theory is crucial because even if evolution were seen as God's method of creation, the natural selection of random variation seems far more like unplanned trial and error. And Darwin's theory is just one among many that have created problems by affecting our view of human nature. If the human mind is just the by-product of the mechanical operations of the brain, the whole notion of moral responsibility, and with it the concept of sin, seems threatened. Although developments in physics have often been portrayed as supporting the mechanistic view of things, the twentieth century saw the emergence of new theories that directly challenged that view and were welcomed by religious thinkers as a sign that scientific materialism was only a passing phase.

The Problem of Literalism: Cosmology

Like many other religions, Christianity has its sacred text, the Bible, which is taken to have been written under divine inspiration. But unlike some other texts, the Bible tells a historical story of deep spiritual significance. Its purpose is to guide the believer toward the true faith and right conduct, but it also makes references—sometimes explicitly, sometimes incidentally—to matters that concern science. The events it refers to often involve miracles, apparent violations of the laws of nature by supernatural agency. These miracles could be regarded as exceptions to the normal rule, allowing the scientist to study the uninterrupted laws of nature at all other times. But as science built up confidence in the uniformity of natural laws, there was a potential for conflict as the more militant scientists became skeptical about the plausibility of the exceptions mentioned in the Bible. To be fair, some liberal religious thinkers also feel uncomfortable with the assumption that the creator has for some reason reduced the scale of his willingness to interfere with the world in modern times. For this reason, direct conflict over the plausibility of miracles has been peripheral to many of the debates with science.

More serious are the direct references to the structure and origin of the universe in the sacred text. The question of how these references are to be understood is crucial in determining the potential for conflict with the relevant areas of science. It would be easy for a modern person to assume that all early Christian scholars must have insisted on taking the word of scripture literally, thereby locking themselves into a particular model of the world and its origin. In fact, however, the Catholic Church has always approached the sacred text through a body of scholarly interpretation that has accumulated through the centuries. Many of the early Church Fathers were not literalists—they realized that the words written down centuries or millennia before were intended to be read by ordinary people and might have to be interpreted more flexibly by scholars. This did not mean that it was easy to change the interpretation of a text if science suggested the literal reading was wrong, as Galileo found out to his cost. But the possibility of reinterpretation was always there if the Church could be persuaded it was necessary. It was the theologians of the Protestant Reformation who rejected this tradition of exegesis and focused attention more narrowly on the Word of God that, because it must be read by everyone for him- or herself, had to be taken far more seriously and, hence, literally.

The first area of potential confrontation was the transition from the geocentric worldview of the medieval period to Copernicus's heliocentric theory (see chap. 2, "The Scientific Revolution"). The trial of Galileo has come to symbolize the painful nature of this transition and, for many, stands as a symbol of the Church's determination to resist the advance of science in order to defend traditional orthodoxy (fig. 15.1; see De Santillana 1958). There was certainly an issue of biblical literalism here, since conservative theologians were only too anxious to point to occasional passages in scripture that seemed to imply that the earth is stationary, especially Josh. 10:13, in which Joshua tells the sun to stand still. In his *Letter to the Grand Duchess Christina* (1615) Galileo had tried to respond to this argument by insisting that the Bible was not an astronomy text, having been written in commonsense language for ordinary people. In fact he implied that science ought to play a major role in shaping the interpretation of the sacred text, which certainly did not endear him to his conservative opponents. But the hostility to Galileo was a product of far more than a narrow-minded literal-

FIGURE 15.1 Galileo before the Vatican Council; oil painting by Robert Fleury (Réunion des Musées Nationaux, Louvre, Paris/Art Resource, New York). This image of Galileo being brought to heel by an overpowerful Church reflects the mythology that grew up around his trial as it became a symbol of science's links with free thought.

ism. Over the centuries, the Church had come to accept the Aristotelian worldview in which the earth is at the center of a hierarchical universe, with the heavens surrounding us in perfect order. To treat the earth as just another planet going round the sun threatened a comforting image of humankind as the center of God's creation. It also raised the disturbing prospect that if the other planets were like the earth, they might be inhabited by rational beings, whose spiritual status and relationship to the savior would be highly problematic. A controversial study by Pietro Redondi (1988) claims that the trial was really a cover for a deeper level of opposition to Galileo aimed at his espousal of a mechanical worldview. When Galileo and the other Copernicans tried to persuade theologians that they should accept the new theory of the cosmos, there was far more at stake than a simple reinterpretation of a few passages of scripture.

Modern commentaries all agree that the trial cannot be understood as a simple conflict between scientific objectivity and religious obscurantism. There were many different factions in the Church, some supportive of Galileo, some hostile. Galileo was told that he could teach the Copernican theory "as a hypothesis," that is, as a mathematical device for predicting planetary motions, but must not present it as physically true. In his *Dia-*

logue on the Two Chief World Systems (1632), he not only disobeyed this injunction but included passages that seemed to ridicule the pope. In these circumstances, the authorities had little choice but to take action and force a recantation. Galileo was not tortured (although he was warned about the possibility of this) and his subsequent imprisonment consisted of house arrest in his own villa, so the more lurid tales about his punishment can be discounted. There are many historians who believe that if Galileo had only been a little more diplomatic, he might have persuaded the Church to soften its opposition and could have paved the way for a far more positive relationship with the New Science.

There were Protestants, too, who objected to the Copernican system. Both Luther and Calvin made disparaging remarks about the theory, but only in passing—they did not institute a systematic line of opposition. Protestants were free to make up their own minds, and increasingly this meant that they appreciated the reasons for moving to the new cosmology. It was Kepler, a Protestant, whose vision of God as the designer of a rational cosmos helped to make Copernicanism more convincing (see below). At the same time, however, we should not forget that the Catholic Church, too, encouraged science, especially in areas that did not raise controversial implications. The Jesuits were active in astronomy and many other areas of science, although they preferred the old cosmology. There remains, however, a more pervasive sense that during the course of a century or more, the focus of science shifted from southern Europe to the north and, hence, to the areas dominated by Protestantism. Even in France, it has been claimed, the scientific community drew more from the Protestant minority than the Catholic majority. This sense that Protestantism offered a more congenial culture within which science could develop has been articulated most fully in the case of seventeenth-century Britain.

Puritanism and Science

Britain provides a clear illustration of the social changes that accompanied the Protestant Reformation. The seventeenth century saw the rise of a prosperous middle class making its living from trade and increasingly anxious to challenge the authority of the king and aristocracy. This polarization came to a head with the English civil war, which placed the liberals temporarily in control under Cromwell and lost King Charles his head. Religion was involved because the conservative political forces were also conservative in religion, either openly or implicitly favorable to Catholicism, while the middle classes were Protestant, often belonging to the evangelical wing of Protestant thinking identified, at the time, as Puritanism. There has been a longstanding assumption that Protestantism favored the rise of capitalism, linked in part to the so-called Protestant work ethic. But this line of thinking was also extended to science by Robert K. Merton, who argued that the English Puritans were strongly inclined to support the New Science and formed the nucleus of a group called the invisible college that eventually achieved respectability through the formation of the Royal Society of London (Merton 1938; see also Cohen 1990; Webster 1975; Westfall 1958). Merton's analysis can be linked to a broader view of Christianity's involvement with science and technology that points to a widespread assumption that humanity can use science to regain the power over nature lost by Adam and Eve when they first sinned (Noble 1997; see chap. 16, "Science and Technology").

The "Merton thesis" has been much debated by historians of science and is at best accepted now only in a modified form. The logic of the argument rested on the assumption that Puritans were inclined to support the study of nature because they saw it as a way of understanding the creator's handiwork but also because it offered the hope of improved technology, vital to their hopes of industrial and social progress. That these motivations were an important component in support for the New Science is beyond doubt. But historians have raised questions about the detailed application of Merton's thesis to the situation in seventeenth-century England, pointing out that many of the early members of the Royal Society were not, in fact, Puritans (although it is a matter of some judgment as to what exactly counts as Puritanism, especially in a period when it was sometimes safer not to be too open in expressing one's opinions). In a more general sense, however, some historians are prepared to give qualified support to the claim that Protestant values did help to create a culture within which science, especially practical science, could thrive. Following the restoration of Charles II, it was moderate Anglicans who did most to promote Newtonianism as the basis for a vision of a world order in which the social hierarchy was flexible enough to allow room for individual initiative.

Literalism Again: Genesis and Geology

Protestant scholarship was, however, responsible for another major effort to limit the range of scientific theorizing: the emphasis on the literal truth of the Genesis creation story. This would have a significant impact on the rise of geology (see chap. 5, "The Age of the Earth") and ultimately on reactions to evolution theory. It was in the mid-seventeenth century that Archbishop James Ussher published his now notorious calculation that the earth must have been created in 4004 b.c. At one level, this did rest on a literal reading of Genesis, since it assumed that only seven days separated the creation of the universe from the creation of Adam. But Ussher's work on ancient chronology was a respected contribution to a substantial scholarly debate, so it is hardly surprising that the notion of a recent creation was taken seriously at the time. We have seen how most of the theories of the earth proposed around 1700 were designed to keep within this restricted time span, although this barrier was steadily eroded in the course of the following century (Greene 1959).

Another aspect of this literalist approach was the assumption that Noah's flood must have been a real event that might be appealed to in theories designed to explain the evident changes that have taken place on the earth's surface. Thomas Burnet, William Whiston, and John Woodward all used the flood to explain the origin of mountains and of fossil-bearing rocks (fig. 5.1, p. 107). But there were important differences between them as far as religious thinkers were concerned. Woodward followed the traditional assumption that the flood was a divine punishment brought about by supernatural means. But both Burnet and Whiston adopted the new materialist approach by explaining the flood as a natural consequence of physical changes taking place within the cosmos. This meant that their theories did not match the Genesis story exactly, and Burnet cautioned against trying to construct too close a relationship between the Bible and a particular theory, which might turn out to be wrong: "'Tis a dangerous thing to ingage the authority of Scripture in disputes about the Natural World, in opposition to Reason; lest Time, which brings all things to light, should discover that to be evidently false which we had made Scripture to assert" (Burnet [1691] 1965, 16). Burnet was criticized by theologians on the grounds that a catastrophe brought about by natural causes would have been inevitable—so how could it be a punishment for human sinfulness? He had to respond by arguing that an omnipotent God could foresee the history of the human race and design the physical world so that the laws of nature would bring about a convulsion in the earth at just the right time. But this technique did little to inspire confidence in these theories, and over the next century the flood came to play a much less significant role in geological thinking. Endorsing an event that appeared in Genesis was certainly not part of the program favored by Buffon and other naturalists of the eighteenth-century Enlightenment.

The flood story did, however, reemerge around 1800 in the context of conservative reaction to the threat of Enlightenment radicalism, especially as that ideology was seen to have precipitated the French Revolution. In Britain it once again became fashionable in conservative circles to appeal to science as a means of salvaging something from the biblical view of creation. James Hutton's uniformitarian geology, with its denial of a creation and of any catastrophic events in earth history, was a popular target. Two geologists in particular, Jean-André Deluc and Richard Kirwan, responded to Hutton by modifying the rival Neptunist position (Gillispie 1951). Both saw the retreating-ocean theory as compatible with the belief that the earth had a beginning that might be identified with the creation. And both sought to show that the theory could explain a global deluge in the comparatively recent past. Deluc, in particular, thought that the collapse of the land covering deep caverns into which the ancient ocean had retreated had caused not only a flood but a complete restructuring of the earth's crust as well. It would be easy to dismiss his claims as a frantic effort to resist the march of scientific geology, yet he identified some phenomena for which Hutton had no explanation (later geologists would invoke the ice age). It must also be stressed that Deluc's position was not typical of the prevailing version of Neptunism promoted by A. G. Werner and his followers—they made no provision for a reemergence of the ocean once it had disappeared.

The same warning is necessary when evaluating the last serious effort by geologists to support the idea of a global deluge: William Buckland's 1823 *Reliquiae diluvianae.* Buckland was reader in geology at Oxford, a notoriously conservative university, and he had to show that his science was no threat to religion. Like Deluc, Buckland had studied phenomena that were inexplicable in uniformitarian terms—there was no obvious way that observable causes could have filled a cave high in the hills with mud (see fig. 5.6, p. 117). His mistake was to assume that the effect was universal, like the biblical flood, and over the next ten years even Buckland himself had to concede that he had overstepped the mark in this respect. Far from being poor science, Buckland's work on the remains of the hyenas buried in the cave at Kirkdale was a model of the new comparative anatomy. And his model of earth history had the flood appearing at the very end of a vast succession of geological transformations not mentioned in the Bible—indeed he was openly criticized for this by more conservative thinkers. By the 1830s it was becoming widely accepted that the earth did have an extensive history. Those who sought accommodation with Genesis often followed the suggestion offered by Buffon in the previous century, in which the "days" of creation correspond to geological epochs. By the time Darwin published his theory of evolution in 1859, the revolution in geology had ensured that there was little support for opposition based on a literal reading of Genesis. Only in the 1920s did the "young earth" form of creationism reemerge as the basis for continued resistance to Darwinism.

Natural Theology

It would be easy to portray biblical literalism as a factor that has always created problems for science. But as Peter Harrison (1998) has pointed out, there is another side to the story. The Protestants wanted everyone to read the Bible for themselves, and to make this possible, they deliberately stripped the sacred text of the layers of interpretive commentary provided by the Catholic Church. One effect of this was to remove a host of symbolic and allegorical meanings read into the stories and images provided by the Bible: the words now had to mean exactly what they said. One consequence of this was the literal reading of the Genesis creation story, but Harrison argues that another effect was a parallel tendency to strip nature itself of the symbols once associated with it. It was no longer fashionable, as it had been in the medieval period, to describe each animal species along with its heraldic and astrological significance, its appearance in myth and folktale, and other human inventions. Biblical literalism may thus have played a role in focusing naturalists' attention onto the need to describe each species just as it appeared in nature, with immense consequences for the emergence of a scientific natural history.

But this did not mean that describing nature was without religious significance, since the world was assumed to be a divine contrivance, designed and created by a rational and benevolent God. Attention began to focus on natural theology, the study of God though the investigation of his handiwork. All aspects of science could be brought into this project, from cosmology to the study of microscopic forms of life. In cosmology, Kepler's efforts to see a rational pattern in the planetary orbits provides the best illustration of the movement's significance, although Newton, too, saw the cosmos as a divine construct. It was in natural history, however, that the search for design came into its own. New studies in anatomy and with the microscope were revealing the complex structures of living bodies, and the mechanical philosophy encouraged naturalists to visualize these structures as machines. Since there was no sense of deep geological time, it was impossible to think in what we would call evolutionary terms, and in any case, the assumption that Genesis should be taken literally encouraged the belief that species were divinely created just as we see them today. In these circumstances, to describe the complexity and utility of organic structures was to illustrate the wisdom and benevolence of their creator.

The astronomers of the seventeenth century inherited the belief that the cosmos was an orderly system governed by mathematical regularities. Copernicus sought to portray his heliocentric system as a better representation of that divine order, and for those who accepted the system's physical reality there was an immediate need to show that it would lead to a better understanding of the pattern of creation. Galileo sought physical arguments to support heliocentrism, but to Johannes Kepler, the astronomer's most important duty was to refine the mathematical study of the planetary orbits to reveal their underlying laws. As a Protestant, Kepler took the idea of God as the designer of the cosmos very seriously, and as a Platonist he took it for granted that the divine order could be expressed in mathematical terms. The significance of this belief in motivating his decades-long search for the laws of planetary motion cannot be underestimated. But the most revealing aspect of his search is his "discovery" of a pattern that modern astronomers would dismiss as an illusion. In his *Mysterium cosmographicum* of 1596, he showed that in the Copernican system, the spacing of the six planetary orbits can be explained by showing that the spheres defined by the orbits are separated by the six regular Platonic solids (tetrahedron, cube, etc.—these are the five solids that can be constructed with all faces of identical shape; see fig. 15.2). There was no physical reason for this pattern to exist—although Kepler was not averse to proposing physical powers to move the planets around their orbits. The pattern made sense only as a design intended by the creator for us to discover and marvel at his rational plan for the cosmos. Kepler never lost interest in this model, and it thus serves as a valuable illustration of the belief system that motivated his search for the laws of planetary motion.

Kepler's geometric solar system made no sense at all in the cosmology proposed by Descartes, where the planets simply drifted into the sun's vortex at random. But Newton's search for the forces governing the planetary orbits was also conducted within the assumption that the whole system was a divine construct. He did not think that it would be possible to describe a physical process by which the planets could have been brought into their present orbits. Hence the structure must have been planned this way by God—although Newton was prepared to admit the need for occasional miracles to correct cumulative deviations in the orbits. By the mid-eighteenth century, however, the Cartesian program of searching for a physical cosmogony (a physical process that would generate the cosmos as we see it today) had already generated two possible explanations. One was Buffon's 1749 theory in which the planets were struck from the sun by a colliding comet. The other was the "nebular hypothesis" proposed by Immanuel

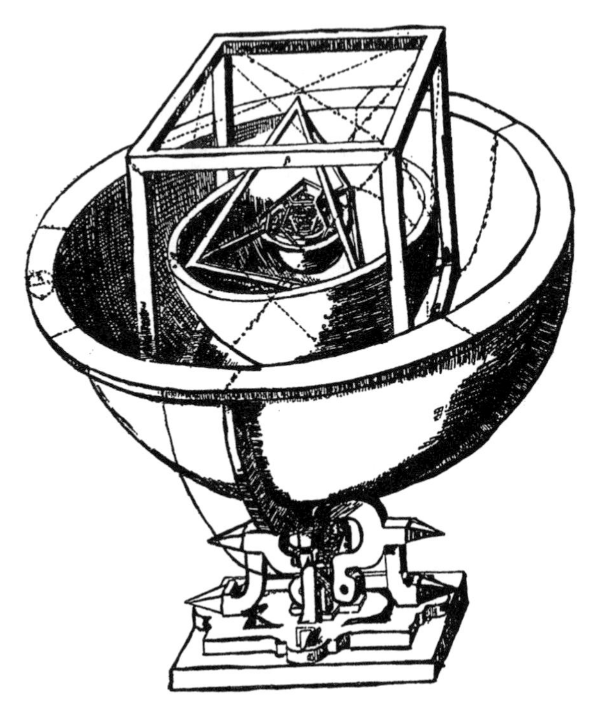

FIGURE 15.2 Kepler's geometrical model of the solar system, from his *Mysterium cosmographicum* (1596). Kepler knew only the six planets visible to the naked eye (including the earth) and linked this to the fact that there are only five "perfect" solids, that is, solids that can be constructed with all faces of the same shape. He argued that a rational God used the solids to determine the spacing of the planetary orbits, and here shows how the spheres defined by those orbits could be separated by the solids, thus: Saturn (*cube*), Jupiter (*tetrahedron*), Mars (*dodecahedron*), Earth (*icosahedron*), Venus (*octahedron*), and Mercury—the innermost orbits and solids are too small to be visible in this illustration.

Kant and refined by Pierre-Simon Laplace, which derived the sun and the planets from the collapse of a vast rotating dust cloud under its own gravity. In each case, the laws governing the process were mathematical, but the pattern of planetary orbits could not be predetermined by any geometrical abstraction because it would depend on the size and consistency of the original cloud. At best, the natural theologian could only marvel at the exact distancing of the earth from the sun, which ensured a climate fit for life. But more radical thinkers were already speculating about different forms of life that might inhabit the other planets, a prospect that might delight those whose theology was liberal enough to include multiple creations but that many Christians found profoundly disturbing.

It is no surprise, then, that the attention of natural theologians began to focus ever more closely on the earth rather than the cosmos. For seventeenth-century thinkers such as Robert Boyle and John Ray, the New Science provided ample opportunity for repudiating the claims of materialists who were already claiming that the world was merely an unplanned assembly of particles moving around at random. Boyle was one of the leading "virtuosi" promoting the New Science in Britain and made substantial contributions to physics and chemistry. He was a fervent advocate of the mechanical philosophy, which he used to discredit the traditional worldview in which magical powers were attributed to natural objects. To Boyle, these alleged powers were a denial of God's true creativity: if matter was inert, the particles driven only by the laws of motion, then matter by itself could create nothing and all meaningful structures in the world must have been designed and created by God. Boyle reluctantly admitted that the Deity occasionally interfered with the world through miracles—after all, Christianity rests on the miraculous events recorded in the Bible—but he insisted that with these rare exceptions the laws of nature held absolute sway over the world. The laws could only preserve structures imposed by an initial supernatural creation; they could create nothing by themselves. Although he did little work in natural history, Boyle conceded that it was in the study of living things that we see the clearest evidence of the creator's handiwork.

It was John Ray who did most to expound the argument from design in natural history (see chap. 6, "The Darwinian Revolution"; see also Greene 1959). His *Wisdom of God* (1695) used a host of examples drawn from the structure of the human and animal bodies to show that only a designing intelligence could account for the ways in which the structures functioned so effectively. The eye and the hand, both so important to human life, were favorite examples. Ray had no intention of arguing that all the other species were designed for our benefit (although some clearly were, like the horse). Each species was designed to function within its own environment, thus confirming the creator's benevolence as well as his intelligence. This focus on the utility or usefulness of structures was to play a major role in shaping attitudes to nature, establishing a fascination with adaptation that would survive (albeit massively transformed) in the Darwinian theory. We have also seen, however, that Ray's faith in the existence of a divine plan of creation functioned as a motivation in his search for a rational system by which the immense variety of species could be classified. The foundations of modern taxonomy emerged from the belief that the human mind can comprehend and represent the order implicit in the divine plan of creation.

Over the following century, the radical thinkers of the Enlightenment challenged the argument from design by reviving the materialist vision of a world thrown together in a haphazard way by the blind laws of nature. Not everyone saw it in this way—Erasmus Darwin's theory of evolution saw the laws themselves as creative and, hence, the whole universe designed to achieve a purpose through its progress to higher states. But this was too much for the conservatives, especially after the trauma of the French Revolution. The argument from design was revived, especially in Britain, along with a more biblical view of the earth's history. Erasmus Darwin was one of the targets in William Paley's classic restatement of the argument from design, his *Natural Theology* (1802). Here the mechanist foundations of the argument were made explicit in the comparison with the watch and the watchmaker: complex mechanical systems adapted to a purpose require intelligent design.

The resulting wave of enthusiasm for the design argument has sometimes been dismissed as a blind alley in the development of science. It encouraged the accumulation of endless examples of adaptation, all uncritically presented as evidence of design, as in the eight contributions to the *Bridgewater Treatises* of the 1830s. But as we have seen in our discussion of the Darwinian Revolution, natural theology was not entirely static. Paleontologists such as Buckland used the concept of adaptation to understand the lifestyles and environment of the fossil species they described and postulated a series of creations, each adapted to the climate of a particular geological period. More imaginative uses of the design argument came from naturalists such as Louis Agassiz and Richard Owen, who looked for patterns uniting the whole of creation into a comprehensive whole. Owen's concept of the archetype unfolding into a variety of differently-specialized forms provided useful evidence for Darwin and brought Owen himself close to the idea of evolution. Robert Chambers's anonymous *Vestiges of the Natural History of Creation* (1844) placed the concept of evolution as the gradual unfolding of a divine plan very firmly in the public domain. Chambers linked everything from the nebular hypothesis to the expansion of the human brain into a vast law-bound system of progress, all brought about by the laws impressed by the creator onto nature at the very beginning of the universe.

The Challenge of Darwinism

Darwin's theory was different (see chap. 6, "The Darwinian Revolution"). It certainly depended on a complex interaction of many law-bound processes, but it was difficult to imagine how the ensemble could be seen as an expression of divine purpose. His theory seemed to revive the old materialist challenge to the argument from design, especially as the variation on which natural selection feeds was supposed to be "random" in the sense that it produced many different modifications with no apparent purpose. Could natural selection itself be the source of the creator's purpose? This was difficult to believe seeing that it worked through the death and suffering of an untold myriad of unfit individuals. In the end, many chose to believe that evolution must work more along the lines suggested by Chambers. There must be some pattern built into the laws of variation ensuring that evolution is pushed in the right direction. But to invoke God's design as the only explanation for why evolution moved toward progress seemed increasingly out of date to scientists hoping to understand the world solely in terms of natural law. Rather than build the supernatural into the laws themselves, it might be better to look for overarching trends in evolution that—while only indirectly the products of complex lawlike interactions—nevertheless could be seen as the expressions of a divine plan. The element of design became much less explicit, and the resulting theories were sometimes hard to distinguish from those of the materialists, many of whom also took it for granted that the universe would inevitably generate progress.

Darwin himself began as a religious man and was charmed when he read Paley's *Natural Theology*. Even when he first devised the theory of natural selection, he seems to have thought that it was a process consistent with divine benevolence because the suffering of the few brought about adaptation and hence happiness for the whole species over future generations (Ospovat 1981; see also Gillespie 1979). This changed when he began to realize the full implications of Malthus's principle of population, which would ensure that many would have to die even if the species were already well-adapted. He gradually became more aware of the cruelty of nature and, hence, less inclined to see selection as the agent of divine providence—although he never became an outright atheist. He also remained convinced that in the long run, and despite many blind alleys, evolution did generate higher forms of life, including humans. It was thus not entirely cynical for him to end his *Origin of Species* with a hymn to progress through suffering and with the implication that this was all part of the creator's purpose.

Despite these efforts at reconciliation, the materialist implications of the theory were all too obvious, and the resulting debate was at first highly emotive. One of the classic confrontations was between "Darwin's bulldog," Thomas Henry Huxley, and Bishop Samuel Wilberforce at the 1860

FIGURE 15.3 Cartoons of Bishop Samuel Wilberforce (*left*) and T. H. Huxley (*right*), the two antagonists in the 1860 confrontation over Darwinism at the British Association, from the magazine *Vanity Fair* in 1869 and 1871. Later accounts of the event by those favorable to science created the myth that Huxley had vanquished the bishop by exposing the shallowness of his appeal to popular sentiment.

meeting of the British Association (fig. 15.3). Although now widely assumed to have been a victory for Huxley, we now know that it was a much less conclusive debate. Over the next decade or so, many educated people converted to evolutionism, although very few accepted natural selection as an adequate explanation (Durant 1985; Ellegård 1958; Moore 1979). The difficulty of reconciling Darwin's theory with design was clearly uppermost in the minds of many, and to be fair, the advocacy of Huxley and Herbert Spencer was bound to inflame their fears. Both supported a philosophy of "scientific naturalism" in which only processes governed by law could be used to explain the world, all elements of the supernatural being discounted, even in the form of an original plan imposed at the creation. A purely naturalistic theory of evolution was an integral part of such a philosophy, and even if Huxley and Spencer had their reservations about the adequacy of natural selection, they were bound to support it as an example of the kind of theory their philosophy required. To more conservative thinkers such as Wilberforce, this total repudiation of design was exactly what made the theory unacceptable. The respected astronomer Sir J. F. W. Herschel dismissed the theory as the "law of higgledy-piggledy" and called for evolution to be seen as a process working under divine superintendence. Richard Owen—often assumed to be an outright opponent of evolution because he wrote a critical review of the *Origin*—made a similar appeal, as did his disciple, the Catholic anatomist St. George Jackson Mivart. They were calling for what has sometimes been called "theistic evolutionism"—evolution working with an element of supernatural design built into the very laws by which it operates, guaranteeing that the process will work toward a predetermined goal.

The tensions created by the theory can be seen in the reaction of a figure who, ostensibly, counted himself as one of Darwin's supporters. This was the American botanist Asa Gray, a deeply religious man who nevertheless saw the advantages to the scientist of a theory based on a process of adaptation. In the papers collected in his *Darwiniana* of 1876, we see Gray grappling with the question of whether natural selection can be accepted as a divinely instituted process designed to bring about complex adaptive structures. He tries to argue that it doesn't really matter how the end is achieved—as long as it is achieved, God's purpose is fulfilled. But further reflection forces him to concede that a process that requires the endless production of useless variations (the "scum of creation," born only to perish, in his picturesque language) lays itself open to difficulties. In the end, he advises Darwin to assume that variation is not random but "has been led along certain beneficial lines" (Gray 1876, 147–48). As Darwin himself protested, this would make selection unnecessary. More seriously, it reintroduces the supernatural in a way that many scientists would find disconcerting because the effect is built into the laws of nature and cannot be distinguished from them.

One way out of Gray's dilemma was to opt for the only other available mechanism of adaptive evolution, the inheritance of acquired characteristics, now known as Lamarckism. The wave of enthusiasm for Lamarckism in the late nineteenth century was an integral component of the eclipse of Darwinism and was certainly generated in part by religious and moral concerns about the consequences of the selection theory. Lamarckism allowed a species to adapt itself to changes in its environment through the collective efforts of all its members in response to a new habit (such as the giraffes feeding off the leaves of trees). It was a perfectly natural process, still plausible in the period before the emergence of Mendelian genetics, and it worked without the elimination of the unfit because all members of the species learned the new habit and began adapting themselves to the new way of life. For the neo-Lamarckian paleontologist Edward Drinker Cope, in his *Theology of Evolution* (1887), the ability of the animals to direct evolution through their own efforts could be seen as God's creativity delegated into the life force that animates them. Similar views were articulated from a moral rather than a theological perspective by the novelist Samuel Butler, who became one of Darwin's chief opponents. For Butler, natural selection represented a soulless materialism in which animals lived or died according to the luck of the draw. Lamarckism thus became the preferred alternative to many who had reservations about the selection theory, even though there was little direct evidence for the actual inheritance of acquired characters.

Those who wanted to see evolution as the expression of a divine purpose also stressed its progressive character and the implication that the human mind or spirit was its intended product. Such an interpretation was still popular in the early twentieth century, when it became part of a concerted effort made by some scientists and theologians to argue that the hostility of the Victorian era had been overcome (Bowler 2001; Livingstone 1987; Turner 1974). In the 1920s the British biologist J. Arthur Thomson wrote a popular book called *The Gospel of Evolution*. Like many of his contemporaries, Thomson was inspired by the French philosopher Henri Bergson's vision of a "creative evolution" driven by a life force struggling to overcome the limitations of matter. On this model, the exact course of evolution was not predetermined, only the overall character of what counted as progress, that is, the ascent toward mind. The psychologist Conwy Lloyd Morgan promoted the idea of "emergent evolution," in which new qualities such as life, mind, and spirit appeared suddenly at key points in the ascent toward increasing complexity. To many liberal Christians, these ideas seemed to make the basic concept of evolution acceptable. But there was a strong element of opposition both to Darwinian selectionism and to the mechanist view of life built into this attempted synthesis. As the twentieth century progressed, however, it became increasingly clear that Darwinism and mechanism were becoming dominant forces in biology. Modern theologians are still trying to work out the implications of these developments.

Even at the height of the eclipse of Darwinism by Lamarckism, there were conservative Christians who remained suspicious of the compromise based on the supposedly purposeful nature of evolution. The problem with

FIGURE 15.4 The Scopes trial, 1925. Clarence Darrow, in his shirtsleeves because of the heat, argues the case for the defense to the jurors.

the idea of progress was that it undermined the traditional belief that humans are fallen, sinful creatures in need of salvation through Christ. These concerns began to manifest themselves more strongly in early twentieth-century America, where in the South especially there was a fear that modern ideas and values were undermining the foundations of Christian society. The Fundamentalist movement (named after a series of pamphlets entitled *The Fundamentals*) gained considerable support, and there were increasing calls to limit the teaching of Darwinism because it served as a key plank in the modernist platform. Some states began to pass legislation forbidding the teaching of evolution, leading to a notorious episode centered on the 1925 "monkey trial" of John Thomas Scopes for flouting the act passed in Tennessee (fig. 15.4). The myths surrounding this trial depict simpleminded creationists making fools of themselves before the world's press, but the true story is much more complex (Larson 1998; Numbers 1998). The Fundamentalists were not biblical literalists (some even accepted a form of evolution), and their concerns focused more on genuine fears about the materialist implications of Darwinism. There was indeed a revival of "young earth creationism" at this time, as a few figures such as

George McCready Price renewed the old idea that all the fossil-bearing rocks were laid down in Noah's flood (Numbers 1992). But the movement remained largely isolated until renewed fears sparked by the success of the modern Darwinian synthesis triggered a wave of support for its position in the 1960s. Efforts to have this form of "creation science" taught in public schools were blocked, in part because the young earth position is so obviously linked to the Genesis story. Creationists' attention has now focused on "intelligent design" theory, which revives Paley's old argument from design by arguing that some biological processes are so complex that they cannot have been formed by gradual evolution.

Materialism and Human Nature

The Fundamentalist response reminds us that there is another aspect to the problem: evolution not only raises questions about how God governs the universe, it also threatens the traditional concept of the human soul. Christianity had always assumed that humans are distinct from animals in that they possess an immortal soul that can be judged by its creator. By deriving humans by a natural process from the animals, evolution threatens this belief and encourages us to think of human nature as merely an extension of the mental powers already possessed by animals. It thereby links itself with a more general materialist philosophy that claims that the mind is at best only a by-product of the physical activities of the brain. Bigger brains mean higher mental powers, but those powers are still produced by a material system governed by natural law—they are completely determined (undermining the concept of free will) and vanish when the brain is destroyed at death. Religious thinkers were deeply concerned by these consequences. Many were able to convince themselves that the human mind might be produced by evolution, but they rejected the materialist position and were thus inclined to argue that evolution was driven by mind, perhaps by the willpower of the animals themselves.

Descartes had applied the mechanical philosophy to animals, claiming that they were just complex machines, but he had insisted that a human being combines a material body with a nonphysical soul. It was the Enlightenment materialists who boldly advanced the claim that the human mind is a by-product of physical processes in the brain. J. O. de La Mettrie's *Man a Machine* (1748) made the point quite explicit. In the early nineteenth century, the movement known as phrenology taught that each mental function is generated in a particular part of the brain and maintained that an individual's character could be inferred from the shape of his or her skull. Phrenology was soon dismissed as a pseudoscience, yet in the later nineteenth century there were major developments in neurophysiology (the study of the functioning of the brain and nervous system) that indicated that the proper working of the brain was indeed necessary for mental functions to be exhibited. The prospect of a totally naturalistic explanation of the mind emerged and became deeply troubling to many religious thinkers (see chap. 18, "Biology and Ideology").

In his *Vestiges of the Natural History of Creation,* Chambers had used phrenology to argue that the expansion of the brain produced by progressive evolution generated a consequent increase in mental powers leading up to the human mind. Darwin took a materialist view of the mind for granted and used his theory to explain how and why particular mental functions had been generated in the course of human evolution. For Darwin, our moral values are a consequence of social instincts built into us by natural selection. Huxley went even further: he had little interest in the process by which we had evolved, but he championed the claim that animals are essentially automata and made no effort to conceal the fact that he thought the human mind could be explained in a similar way. The materialist position was widely promoted in Germany and was linked to evolutionism by Ernst Haeckel. Haeckel was ostensibly a monist, holding that mind and matter are just parallel manifestations of a single underlying substance. But he made no secret of his contempt for the traditional view of the soul: humans are just parts of nature, governed by the same natural laws. The mind is a product of the brain and disappears at the point of death, so there can be no such thing as an immortal soul. Haeckel's *Riddle of the Universe* (translated into English in 1900) was a widely read expression of this philosophy and was seen as a powerful challenge to religion. Significantly, though, Haeckel's denial of any supernatural creator did not prevent him from believing that evolution is necessarily progressive. It was the laws of nature, not a divine plan, that ensured the ascent toward humankind.

Opposition to the materialist position led many religious people to favor both scientific theories and philosophical concepts that seemed to offer an alternative view of life and mind. The Lamarckian theory drew much of its support from the belief that if it were true, living things had the power to choose new habits and thus direct their own evolution. Bergson's creative evolution built on the same antimaterialist vision. There was a temporary wave of opposition to mechanism in physiology at the end of the nineteenth century, led by Hans Driesch, followed by a further wave of support for holistic and organismic theories that held that complex systems can exhibit properties that cannot be deduced from the behavior of their constituent parts. But the theologians who endorsed these movements in science ran the risk of paralleling the materialists' refusal to admit any clear distinction between humans and animals. This was why the theory of emergent evolution became popular, because Lloyd Morgan postulated several distinct steps in which new properties had emerged, generating life, mind, and spirit, with the latter property being characteristic only of the last stage in the evolution of humanity.

Physics against Materialism

Efforts to promote an antimechanist biology were less than successful in the early twentieth century, and further advances in neurophysiology and cognitive science have only compounded the problem for those wishing to defend the traditional notion of the soul. But some relief came from an unexpected quarter: physics itself now turned its back on the mechanical view of nature, leading some philosophers and theologians to hope that it had reopened the way for the mind to be an independent entity. Whether the physicists had ever seriously endorsed the simpleminded billiard-ball model of reality attributed to them by the materialists is doubtful—Newtonianism itself had endowed matter with the almost mystical power of attraction at a distance. But by the end of the nineteenth century a self-conscious alternative to mechanism had emerged in the theory of the ether, a subtle fluid that was supposed to pervade the universe to serve as the medium through which light and other forms of radiation were transmitted. Perhaps the ether might offer a means by which the mind interacted with the coarser forms of matter. The revolutions in physics that marked the early twentieth century saw the ether discredited, but at the same time the emergence of quantum mechanics seemed to undermine the traditional materialist vision of a self-contained and entirely law-bound universe existing apart from the minds that perceive it.

The ether dominated the theoretical outlook of some of the most creative late nineteenth-century physicists, including Lord Rayleigh and J. J. Thomson. To them, the existence of this tenuous medium was self-evident, since without it there would be no means by which energy could be transmitted. But the ether had a wider philosophical, theological, and ultimately ideological role in their thinking. It challenged the materialists by suggesting that the world was a unified, interlocking cosmos rather than a collection of atoms moving at random through space, thus bringing physics back into line with natural theology. In the hands of Oliver Lodge, however, it also made the mind and the spirit appear real once again because it offered a place where their activity could be understood to take place independent of, yet linked to, the material body. Lodge was one of a small group of eminent scientists who took a serious interest in spiritualism and the paranormal, and he wrote a series of books arguing that the spirit survived the death of the material body on the ethereal plane (Oppenheim 1985). He also exploited the idea of progressive evolution both in the organic and the spiritual worlds.

By the 1920s, relativity made Lodge's ether physics out of date, but another revolution in physics seemed to drive this area of science even further away from materialism. Quantum mechanics and the uncertainty principle undermined the mechanistic viewpoint by showing that the behavior of particles was governed by statistical laws and could never be predicted with absolute accuracy (see chap. 11, "Twentieth-Century Physics"). Even if the mind were a product of physical activity in the brain, that activity was not rigidly predetermined, allowing some religious thinkers to argue that the freedom of the will was no longer compromised. Moreover, it seemed that the final state of a system was only resolved when it was actually observed, so that the conscious observer played a role in the creation of reality—he or she was not merely a passive onlooker. This meant that the human mind was integral to the physicists' new vision of reality and raised the prospect that the whole universe was in some sense dependent on a Mind that somehow transcended all the individual acts of observation. As A. S. Eddington observed in his immensely popular *The Nature of the Physical World,* "religion first became possible for a reasonable scientific man about the year 1927" (1928, 350). James Jeans went further in his *Mysterious Universe* (1930), declaring that under the new physics, the universe was best pictured as a thought in the mind of a mathematical creator. It was as though Kepler's Platonic version of natural theology had been revived. Small wonder that theologians rushed to hail the new physics as the foundation for a new reconciliation between science and religion, although not all the physicists welcomed their interpretation.

Jeans and Eddington were both cosmologists as well as physicists and were aware of the latest developments showing that our galaxy is only one among many. The universe is immense almost beyond comprehension—but does this mean that it must contain other inhabited planets? Jeans spearheaded an attack on the nebular hypothesis, arguing instead that planetary material was drawn from the sun in a near miss by a passing star (almost a revival of Buffon's theory). He argued that since such near collisions would be extremely rare, ours is one of the very few planetary systems to exist in the whole universe. This meant that the human race once again became central to the whole of creation, in the sense that we are probably the only conscious observers of the system that has created us. The cosmologists were also aware of the immense age of the universe and of the evidence suggesting that it was expanding outward from a point of origin that was later dubbed the Big Bang. The congruence between this model and the story of an original act of creation was not lost on liberal theologians. Efforts to understand the nature of the Big Bang would eventually lead to the claim that it was "fine tuned" so as to guarantee the emergence of a universe in which intelligent life could evolve. Thus liberal theologians found both physics and cosmology a fertile source of inspiration, although the renewed threat of Darwinism in biology encouraged Fundamentalists to reject both cosmology and geology as guides to the history of the world.

Conclusions

A historical survey of the relationship between science and religion reveals that they cannot be seen either as natural allies or as natural enemies. The "warfare" model breaks down in the face of the long tradition of natural theology and the evident fact that such a theology has often given positive support to scientists' thinking. But any attempt to argue that science can always be harmonized with religion has to confront the many episodes where religions have dug their heels in over points of doctrine that they will not yield in the face of scientific advance. For every religious liberal, willing to accommodate his or her thinking to the latest trends in science, there is a conservative who regards certain beliefs about nature or human nature as articles of faith that cannot be abandoned. There is no single, natural form of the relationship of the two entities, because there are many religions (including many different forms of Christianity) and many different areas of science, each posing its own problems. Even within the same debate, it is often possible for different interpretations to be put on a theory or a theological principle that will encourage either reconciliation or conflict. To the historian, the interesting question is: Who chooses which policy, and why?

Rather than encouraging those who call for a single policy of amity or enmity, the history of science shows that the interaction is a contingent and local one, different in different countries and communities, and constantly changing through time. The historian's job is to understand the scientific, theological, and cultural factors that determine the outcome in each situation. If there are lessons to be learned from such study, they include the need to be aware of the diversity within our modern belief systems and the need to recognize the values embedded in the strategies that conflicting parties bring to their interpretations of the past. By stressing a carefully chosen selection of episodes, each side can make its own position seem to be in tune with a historical trend. Comprehensive study suggests that a less dogmatic and more nuanced approach is required.

References and Further Reading

Bowler, Peter J. 2001. *Reconciling Science and Religion: The Debate in Early-Twentieth-Century Britain.* Chicago: University of Chicago Press.

Brooke, John Hedley. 1991. *Science and Religion: Some Historical Perspectives.* Cambridge: Cambridge University Press.

Burnet, Thomas. [1691] 1965. *The Sacred Theory of the Earth.* Edited by Basil Willey. London: Centaur.

Cohen, I. Bernard, ed. 1990. *Puritanism and the Rise of Science: The Merton Thesis.* New Brunswick, NJ: Rutgers University Press.

De Santillana, Giorgio. 1958. *The Crime of Galileo.* London: Heinemann.

Durant, John. 1985. *Darwinism and Divinity: Essays on Evolution and Religious Belief.* Oxford: Blackwell.

Eddington, Arthur Stanley. 1928. *The Nature of the Physical World.* Cambridge: Cambridge University Press.

Ellegård, Alvar. 1958. *Darwin and the General Reader: The Reception of Darwin's Theory of Evolution in the British Periodical Press, 1859–1871.* Göteburg: Acta Universitatis Gothenburgensis. Reprint, Chicago: University of Chicago Press, 1990.

Gillespie, Neal C. 1979. *Charles Darwin and the Problem of Creation.* Chicago: University of Chicago Press.

Gillispie, Charles C. 1951. *Genesis and Geology: A Study in the Relations of Scientific Thought, Natural Theology, and Social Opinion in Great Britain, 1790–1850.* Reprint. New York: Harper & Row.

Gray, Asa. 1876. *Darwiniana: Essays and Reviews Pertaining to Darwinism.* New York: Appleton. Reprint edited by A. Hunter Dupree. Cambridge, MA: Harvard University Press, 1963.

Greene, John C. 1959. *The Death of Adam: Evolution and Its Impact on Western Thought.* Ames: Iowa State University Press.

Harrison, Peter. 1998. *The Bible, Protestantism, and the Rise of Natural Science.* Cambridge: Cambridge University Press.

Jaki, Stanley. 1978. *The Road of Science and the Way to God.* Chicago: University of Chicago Press.

Larson, Edward J. 1998. *Summer for the Gods: The Scopes Trial and America's Continuing Debate over Science and Religion.* New York: Basic Books; Cambridge, MA: Harvard University Press.

Lindberg, David C., and Ronald L. Numbers, eds. 1986. *God and Nature: Historical Essays on the Encounter between Christianity and Science.* Berkeley: University of California Press.

———. 2003. *When Science and Christianity Meet.* Chicago: University of Chicago Press.

Livingstone, David N. 1987. *Darwin's Forgotten Defenders: The Encounter between Evangelical Theology and Evolutionary Thought.* Grand Rapids, MI: Eerdmans.

Merton, Robert K. 1938. *Science, Technology and Society in Seventeenth-Century England.* Bruges: St. Catharine Press. Reprint, New York: Harper, 1970.
Moore, James R. 1979. *The Post-Darwinian Controversies: A Study of the Protestant Struggle to Come to Terms with Darwinism in Great Britain and America, 1879–1900.* Cambridge: Cambridge University Press.
Noble, David F. 1997. *The Religion of Technology: The Divinity of Man and the Spirit of Invention.* New York: Knopf.
Numbers, Ronald L. 1992. *The Creationists.* New York: Knopf.
———. 1998. *Darwinism Comes to America.* Cambridge, MA: Harvard University Press.
———. 2003. *When Science and Christianity Meet.* Chicago: University of Chicago Press.
Oppenheim, Janet. 1985. *The Other World: Spiritualism and Psychical Research in Britain, 1850–1914.* Cambridge: Cambridge University Press.
Ospovat, Dov. 1981. *The Development of Darwin's Theory: Natural History, Natural Theology, and Natural Selection, 1838–59.* Cambridge: Cambridge University Press.
Redondi, Pietro. 1988. *Galileo Heretic.* London: Allen Lane.
Turner, Frank Miller. 1974. *Between Science and Religion: The Reaction to Scientific Naturalism in Late Victorian England.* New Haven, CT: Yale University Press.
Webster, Charles. 1975. *The Great Instauration: Science, Medicine and Reform, 1626–1660.* London: Duckworth.
Westfall, Richard. 1958. *Science and Religion in Seventeenth-Century England.* New Haven, CT: Yale University Press.

ʌR SCIENCE

THE WORDS "SCIENCE" AND "POPULAR" seem to our modern eyes to sit rather incongruously together. We often take science to be the antithesis of popular—a highly esoteric and expert activity requiring years of dedicated training. When we think of popular science at all it is probably in terms of "gee whiz" science programming on TV or episodes of *Star Trek*. Breathless exclamation over the latest technological gadgetry seems rather far removed from what we know about the actual practice of science. In that respect popular science, insofar as it exists at all, might seem rather peripheral to what scientists themselves do—a matter of mere dissemination of watered-down facts, theories, and applications to a passive public rather than the real thing. Science and scientists often seem rather disconnected from the popular as well. Scientific spokespersons worry publicly about the "public understanding of science" but that often only seems to mean that the public should know enough to let the real scientists get on with their job rather than being any serious call for engagement. Whenever science appears to be becoming part of popular culture it is often castigated for having been reduced to triviality. Scientists' engagement with the popular seems to be a distraction from their proper task.

Looking at things from a historical perspective, this view of science as being totally disengaged from popular culture is deeply flawed. Science has always had a public face and still has it now. If only to defend their own turf, scientists have always cultivated an audience beyond the immediate one made up of their own peers and fellow researchers. In any case, the perception that science is, or ought to be, the particular province of a relatively small and culturally isolated group of highly trained experts is a comparatively recent one. Until well into the nineteenth century at least, a knowledge of and some level of engagement with the latest science was widely considered to be a mark of culture. Literary magazines and journals carried reports of the latest discoveries and reviews of the latest scientific bestsellers as routinely as they discussed Dickens or Dostoyevsky. The cultural critic C. P. Snow famously described the breakdown of this common cultural context in a controversial essay, *The Two Cultures* (Snow 1959). The extent of this common cultural context still needs to be taken with a pinch of salt, however. Popular engagement with science was never a mass activity. We also need to bear in mind that Snow's common culture masks a variety of different ideas about what science was, how it should be practiced, and just what the relationship between science and popular culture ought to be.

When historians have looked at popular science, they have often done so as if it were something external to science proper. The usual model has been one of diffusion. Science is produced by experts and then disseminated to a popular audience through a variety of media including books, lectures, museum exhibitions, and, more recently, television. From this perspective the process of dissemination has no effect on the science itself or the way in which it might be practiced. More recently, however, historians have started to rethink the relationship between science and popular culture and between scientists and their audiences. We now think of audiences as having an active rather than just a passive relationship to the production of scientific knowledge. Not only does the way in which the scientist chooses to present his or her work to different audiences and the context within which that work is presented have important consequences for the way science is understood, but audiences themselves actively interpret and redefine the knowledge they receive as they go along. From this perspective, studying popular science does indeed engage with the actual content of science and the process of knowledge making.

Historians study popular science in a number of contexts. They look at the kinds of places where science is done in public, such as lecture theaters and exhibition halls. They look at the different media through which scientific communication takes place, including books, journals, and television shows. They look at the wide variety of ways in which scientific knowledge is transmitted to the public and the ways in which different audiences receive it. Historians also study the various ways in which particular sciences have themselves, at different times, been popular. Examples we shall examine in more depth here include the sciences of mesmerism and phrenology during the first half of the nineteenth century. From a modern perspective these activities might appear to be pseudo-sciences—not real sciences at all—but to large numbers of people during their heyday they were taken very seriously indeed. Their supporters argued vociferously that they were genuine scientific practices and that the efforts of opponents to deny them scientific status was evidence that they wanted to keep science out of the hands of the common people. Looking at popular science in all its various aspects plays an important part in helping us understand how science has come to be demarcated from other aspects of culture and how, at different times and places, the boundary between science and culture was differently drawn.

Lecture-Room Culture

As we saw in a previous chapter, one important feature of the so-called Scientific Revolution of the sixteenth and seventeenth centuries was the shift in focus of much natural philosophical activity from the universities to a more civic, often courtly context. Philosophers such as Francis Bacon argued that natural philosophers needed to be men of the world rather than cloistered academics (see chap. 2, "The Scientific Revolution"). In keeping with this new ethos of science as part of civic culture, natural philosophers actively sought out new audiences for their activities. In England, France, and Italy, such scientific societies as the Royal Society London, the Académie des Sciences, and the Accademia dei Lincei were founded with the express purpose of integrating science into civil society (see chap. 14, "The Organization of Science"). Carrying out public experiments in the presence of eminent witnesses was an important part of the ritual surrounding the consolidation of novel matters of fact. As new audiences for natural philosophy developed among the upper and middle classes, public lecturing came to be a new source of potential income and prestige for new generations of natural philosophers. English natural philosophers in the Newtonian tradition quite explicitly characterized themselves as "priests of nature," charged with the responsibility of spreading the Newtonian gospel far and wide. For these men of science, lecturing was a moral responsibility as much as a financial necessity.

By the beginning of the eighteenth century, the main venues for natural philosophical lectures were the increasingly ubiquitous coffeehouses. By 1739, according to one survey, London had 551 of them. Coffeehouses had developed as centers for the rapid informal dissemination and exchange of information (often financial information) largely during the second half of the seventeenth century. Their patrons ranged from bankers and merchants through entrepreneurs of all kinds to the rapidly expanding breed of literary hacks. They were the places people went to get the latest news and financial gossip or to persuade potential patrons of the benefits of some new invention or other novelty. Working men might drop by to read the newspapers. This motley clientele turned out to be the ideal audience for the new vogue for scientific lecturing, anxious for new information of all kinds (Porter 2000). Natural philosophers doing the circuit typically offered lecture courses of between a dozen and two dozen lectures on the rudiments of Newtonianism and the mechanical philosophy, enlivening their performances with demonstrations and experiments using the latest philosophical instruments, such as air pumps and electrical machines. Lectures on the mechanical philosophy and demonstrations of their own experimental dexterity could also prove a good way to display their credentials to potential patrons and to gain financial backing for some new invention or project (Stewart 1992).

John Theophilus Desaguliers is a good example of an experimental natural philosopher who made his name through popular lecturing. An avid Newtonian, Desaguliers presented his lectures on electricity and other powers of nature as demonstrations of Newton's claims about the relationship between God and nature. Making the powers of nature visible was a way of making God's immanence in the universe visible as well. Desaguliers made full use of the latest experimental technologies to impress his coffeehouse audiences, administering shocks, showing off the powers of electrical attraction and repulsion, drawing off sparks from electrical machines. Not only did these kinds of spectacular displays make Desaguliers's reputation as a natural philosopher, they also helped bring him to the attention of potential patrons, such as the Duke of Chandos. Lecturers across Europe vied with each other to produce ever more spectacular displays of nature's powers. In France, the prominent Parisian public lecturer Jean Antoine Nollet used Leyden jars to great effect, applying the shocks they delivered to make lines of Carthusian monks and palace guardsmen jump in unison. The German electrician Georg Matthias Bose, as well as the English coffeehouse lecturer Benjamin Rackstrow claimed to be able to produce an effect they called beatification—literally making a member of the audience glow in the dark. Displays like these brought audiences flooding to scientific lectures in every European metropolis (Heilbron 1979).

In the British Isles, the vogue for popular scientific lectures rapidly spread outside London. Fashionable towns such as Bath soon acquired their own local scientific lecturers as well as attracting metropolitan performers following the well-heeled crowds who flocked there. The popular performer James Graham started his career as a lecturer and philosophical showman there with spectacular demonstrations of electricity's mysterious powers.

By the 1780s, he was one of London's best-known philosophical performers, charging £50 a night for use of the "celestial bed" at his Temple of Health and Hymen. In Newcastle, James Jurin, the local grammar school master and, later, secretary of the Royal Society, advertised lectures on natural philosophy from 1712 onward aimed at local industrialists. Desaguliers himself advertised lectures there for similar audiences during the 1740s. Itinerant lecturers like Benjamin Martin traveled from town to town, advertising lectures in the local papers and targeting their material to local needs. For the real stars of the lecture circuit like Desaguliers, there was even the possibility of international bookings, and in the 1730s, for example, he lectured in the Netherlands. As the century progressed, lecturers' claims grew more extravagant and their demonstrations more spectacular as they sought new audiences. There was an increasing emphasis as well on the utility of natural philosophy, particularly in popular lectures aimed at hard-nosed northern industrialists (see chap. 17, "Science and Technology").

By the end of the eighteenth century—as James Graham and his celestial bed illustrate—popular lecturers went to ever greater lengths to attract their audiences. Another example was the fashionable astronomical lecturer Adam Walker, who lectured at the Haymarket Theatre in London from the 1770s on. During the 1780s, the main attraction at his shows was the Eudouranion—a huge twenty-foot orrery featuring luminous globes representing the planets of the solar system. By the beginning of the nineteenth century, the lecture circuit in places like London was well-established. Increasing numbers of scientific establishments, such as the Surrey or the London Institutes, offered popular lecture courses to the paying public. Out in the provinces, the new vogue for literary and philosophical societies provided lecturing opportunities, too. In North America during the eighteenth century—both before and after the Revolution—there was a similar taste for scientific lecturing. The American Philosophical Society developed from the self-styled junto of philosophical enthusiasts around Benjamin Franklin in 1749. Philadelphia's Franklin Institute, founded in 1824, aimed popular lectures at the working man. Similar schemes developed across the British Isles with the burgeoning Mechanics' Institutes movement. Establishments like these provided a basic income for impecunious men of science, as well as catering to the public's appetite for scientific lecturing (Hays 1983).

In England, at least, the doyen of popular scientific institutions was the fashionable Royal Institution in Albemarle Street off Piccadilly. It was founded in 1799 by the exiled American royalist Benjamin Thompson, Count Rumford. Under its star performers Humphry Davy and later Mi-

FIGURE 16.1 Michael Faraday delivering one of his famous Christmas lectures to children at the Royal Institution (Wellcome Medical Library, London). Sitting in the front row of the audience, facing Faraday, are Albert the prince consort and the young prince of Wales. Notice the number of women in the audience.

chael Faraday, the Royal Institution developed a formidable reputation as a purveyor of scientific knowledge to the rich and famous. Davy made a name for himself at the Royal Institution with flamboyant lectures showing off his mastery of the newly invented Voltaic battery, impressing his audiences with spectacular shows of sparks and electrical conflagrations (Golinski 1992). Faraday followed in his master's tradition. During the 1820s, he established the institution's (still on-going) series of Christmas lectures for children (fig. 16.1). He also set up the institution's famous Friday evening discourses, which rapidly became a feature of the fashionable London season. Each Friday during the season, Faraday or an invited speaker would deliver lectures and demonstrations on the latest scientific discoveries and inventions to captivated crowds of metropolitan socialites (Berman 1978). Out in the provinces, meetings of the British Association for the Advancement of Science (BAAS)—established in 1831 and held each year in a different provincial town or city—attracted crowds of thousands to its lectures (Morrell and Thackray 1981).

Throughout the nineteenth century, popular scientific lecturers were public figures. Faraday, for example, was certainly as well (if not better) known for his bravura lecture performances as he was for his electrical theories. Another good example is T. H. Huxley—Darwin's bulldog—now best remembered for his fiery encounter with "Soapy Sam" Wilberforce, bishop of Oxford, at the 1860 meeting of the British Association for the Advancement of Science (see chap. 6, "The Darwinian Revolution"). Huxley was particularly known for controversial lectures to the working classes. Huxley had started delivering regular lecture series aimed at working men from the 1850s onward, continuing a tradition started by the geologist Henry de la Beche at the Museum of Economic Geology on Piccadilly. By the 1860s, he was drawing in nightly crowds of hundreds, or even thousands, for his courses (Desmond 1994). Huxley did not confine himself to the metropolis. He toured the country, offering his radical lectures at mechanics' institutes and working men's halls. By 1868 he was principal of his own working men's college in south London. Of course Huxley's lectures, however populist, had a serious political agenda. He was trying to persuade his audiences that science rather than religion was the authority they should turn to (see chap. 15, "Science and Religion").

Huxley's lecture activities were not confined to the merely national sphere either. Traveling to the United States in 1876, he was the latest in a long line of British popular scientific lecturers to tour North America in this fashion. The geologist Charles Lyell lectured across the states in the 1840s. The physicist Sir William Thomson, later Lord Kelvin, lectured there in 1884. The vogue for popular scientific lecturing was not just a British peculiarity. In the rest of Europe and the United States, crowds flocked to such events, and the best-known lecturers were significant public figures. Huxley and Thomson are probably the best British examples from the second half of the nineteenth century. Hermann von Helmholtz in Germany and Louis Pasteur in France were of a similar stature and would have been similarly recognizable to the public at large. What this indicates is the extent to which science easily crossed boundaries into other areas of culture. From the seventeenth and eighteenth centuries onward, practicing natural philosophers regarded this kind of public performance as something intrinsic to the practice of their science. Public lectures formed one of the main avenues of communication between natural philosophers and scientists and their audiences. It was more than a matter of making a living—though that was certainly an issue—it was what men of science did.

At the Exhibition

Collections of scientific instruments and artifacts have a long history. From the Renaissance onward, cabinets of curiosities were increasingly popular. Wealthy patrons collected specimens of strange and unusual natural or manmade objects and put them on show to amaze and impress (see chap. 2, "The Scientific Revolution"). Scientific instruments were themselves often designed for display, as highly ornate surviving examples of microscopes or telescopes from the seventeenth and eighteenth centuries vividly demonstrate (Morton 1993). By the beginning of the nineteenth century, the business of collecting and displaying specimens and artifacts was increasingly commercialized. Cabinets were no longer solely the preserve of those privileged enough to enter the private houses and institutions where they were kept. Anyone willing to spend a few pennies at the door could gain access. From the middle of the nineteenth century, the science museum and the scientific exhibition were ubiquitous. These kinds of collections had—and still have—a crucial impact on the way people have viewed science and the natural world. The way in which objects are arranged in a museum, be they dinosaur fossils, scientific instruments, or steam engines, has a profound effect on the way those objects are understood. It is through these kinds of displays that audiences from the Victorian period onward have learned to make sense of large parts of science.

The artist Charles Willson Peale's Philadelphia Museum at the beginning of the nineteenth century catered to the interests of an American public already fascinated by the curious and fantastic. Peale's museum featured natural historical curiosities such as the bones of a mastodon unearthed in New York State, his own historical paintings, antiquarian curiosities, and new mechanical inventions and contrivances. Even the showman P. T. Barnum's extravagant exhibitions of the exotic played on his public's fascination with science. In many ways, the secret of his success lay in his ability to challenge his audience's capacities to tell the difference between fakes and the genuine article. The Philadelphian entrepreneur and inventor Jacob Perkins might well have had Peale's Museum in mind when he opened his National Gallery of Practical Science on Adelaide Street off the Strand in London in 1832. His gallery certainly featured a similar combination of natural historical specimens, mechanical and scientific contrivances, and exotic curiosities of all kinds. The public who paid their shillings at the door could view the latest scientific and technological wonders, listen to scientific lectures along with musical performances, and even watch as feeding time for the electric eels took place. The Adelaide Gallery soon had a com-

FIGURE 16.2 A view of the Main Hall at the Royal Polytechnic Institution in London. This was one of the city's main centers for popular science. You can see one of their main attractions, the diving bell and diver, in the background.

petitor, the Royal Polytechnic Institution on Regent Street, with a similar array of offerings (fig. 16.2; Morus 1998).

Places like the Adelaide Gallery and the Polytechnic Institution played a key role in defining science for the early Victorian London public. These kinds of places, rather than the august precincts of the Royal Institution, were where most of them who were interested in such matters were most likely to encounter science. At the galleries, it was the material culture of science, not its theoretical abstractions, that mattered most. Science for audiences there was about machines, technological ingenuity, and entertainment. These places also found themselves competing with the rest of London's exhibition industry. They vied with theatrical productions, panoramas, and magic lantern shows for the public's attention. Other shows incorporated natural philosophy into their own offerings. The Coliseum in Regent's Park advertised itself as owning the world's largest electrical machine. The exhibitions offered employment to natural philosophers. The electrician William Sturgeon lectured at the Adelaide Gallery while the chemist William Leithead supervised the Department of Natural Magic at the Coliseum with its gargantuan electrical machine. They were a vital resource for hopeful inventors too. Competitors trying to market rival telegraphic systems during the 1840s, such as Edward Davy, put their inventions on show at the galleries to attract investors' attention as well as finance their inventing activities. It was as a piece of showmanship as much as a new communications system that the Victorian public first got to know the electric telegraph (see chap. 17, "Science and Technology").

The opening of the Great Exhibition of the Works and Industry of All Nations in London's Hyde Park in 1851 was a watershed for Victorian scientific showmanship. The exhibition, organized by the Royal Society of Arts with Prince Albert, the prince consort, playing a central role, was designed to demonstrate the superiority of British industry and technological ingenuity. The exhibition building itself—the Crystal Palace—was a tour de force of Victorian architectural and engineering know-how. Designed by the landscape gardener Joseph Paxton, it was in effect a gigantic greenhouse of cast-iron girders and plate glass (fig. 16.3). The British public along with thousands of foreign visitors flocked inside to wonder at more than 100,000 separate exhibits. The latest science and technology was prominently visible. Visitors could keep time with the electric clock in the Great Transept. A wide variety of competing electric telegraph equipment was on show. The Danish inventor Sören Hjorth won a prize for his electromagnetic motor. The Birmingham firm of Elkingtons displayed a wide range of electroplated silverware. British and foreign instrument makers put on show a wide variety of batteries, electromagnets, photographs and photographic equipment, telescopes, and other scientific apparatus. The photographs included particularly impressive ones of the surface of the moon by the Harvard astronomer William Cranch Bond.

The Great Exhibition's success and the seemingly endless appetites of the public for scientific and technological exhibition helped stimulate, and

FIGURE 16.3 The Crystal Palace housing the Great Exhibition of the Art and Industry of All Nations in Hyde Park, London, 1851.

in some cases even finance, a new vogue for scientific museums. Some of the profits from the Great Exhibition were poured into developing a new "city of science" in South Kensington. The jewel in the new site's crown by the end of the 1860s was the Natural History Museum, whose curator Richard Owen used it as a vehicle to impose his own vision of the natural world past and present. Because Owen—inventor of the word "dinosaur"—was in a position to determine how the collection's ancient fossils were displayed he also had a built-in advantage in selling his own ideas of the ways these antediluvian creatures looked and behaved to the museum's visitors. Museums like this were increasingly popular during the second half of the nineteenth century. Being in possession of a good museum was an important source of civic pride. In towns and cities across Europe and North America, museums, both in external appearance and internal arrangement, stood for progressive, scientific values and, by implication, the role played by local communities and their leaders in the march of progress.

The Crystal Palace's success also inaugurated a whole series of national and international exhibitions throughout the second half of the nineteenth century and into the twentieth. In 1853, Dubliners organized their own Great International Exhibition of All Nations, anxious not to be outdone by their colonial masters. The French followed with an international exhibition in Paris in 1855 that attracted more than 5 million visitors. Others followed in 1862 and 1867, by which time British manufacturers were increasingly worried that the exhibitions showed the degree the rest of the world was catching up with British industry. London hosted its own follow-up of the Great Exhibition in 1862, and New York had in 1853 attempted an international industrial exhibition as well. The first really successful American effort was, however, Philadelphia's Centennial Exhibition of 1876. Among its many claims to fame was the first public demonstration of Alexander Graham Bell's telephone. Australia celebrated its own centennial (of Captain Cook's discovery) in 1888 with an international exhibition in Melbourne. By the beginning of the twentieth century, these were truly massive affairs. The Pan-American Exposition in Buffalo, New York, in 1901 used power from the newly opened Niagara Falls generating station to drive the exhibits at its Palace of Electricity as well as more than 200,000 electric lights throughout the grounds (Beauchamp 1997).

Late Victorian pundits reckoned that the nineteenth was the century of exhibitions. They seemed to symbolize the confident and progressive public face of science and technology. Electricity and exhibitions seemed particularly made for each other. By the end of the century, massive and spectacular electrical displays of the sort featured at Buffalo were a common characteristic of international exhibitions. Large electrical companies such as Westinghouse and Edison in the United States (Westinghouse was responsible for the Buffalo installations) or Siemens in Europe competed energetically with each other to put on the most spectacular displays. They put on extravagant shows of electric lights, they featured the latest experimental systems of electrical locomotion, and they boasted electrical generators of mammoth proportions. Chicago's Columbian Exposition of 1893 gloried in 90,000 arc and incandescent electric lights. The laurels went to Edison's tower of light standing eighty-two feet high in the center of the Electricity Building (Marvin 1988). Exhibitions were showcases through which late Victorian science and technology could be sold to its publics. Not only were these events far and away their most visible and spectacular face, they also provided the context for prize giving, mutual backslapping, and international scientific congresses. Standard electrical units were established by electrical congresses at such gatherings in the 1880s.

International world fairs continued the tradition of scientific exhibitionism on a grand scale into the twentieth century. Hosting such events were the focus of massive local civic and international pride. Cities competed with each other for the opportunity of putting on such shows. Chicago celebrated a "century of progress" with a world fair in 1933, the centenary of its foundation as a city. The New York World Fair ran through 1939 to 1940 on the eve of American entry into the Second World War. Britain celebrated its survival and rebuilding in 1951 with the Festival of Britain, not coincidentally the centenary of the triumphalist Great Exhibition of 1851. The festival's "dome of discovery" embodied contemporary hopes about the ways in which progressive science and technology would be the driving force behind the country's economic and social recovery. The festival's organizers made concerted efforts to bring science and art together. Visitors could buy shirts and neckties made of fabrics printed with designs based on the patterns of crystalline materials. Scientific exhibitionism remains big business today at places like the Smithsonian's National Air and Space Museum or the Science Museum in London. Innovative centers like San Francisco's Exploratorium use technologies that are still surprisingly similar to the philosophical toys on show in places like the Adelaide Gallery more than a century and a half previously.

Science in Print

The beginnings of the Scientific Revolution closely coincided with a revolution in print (see chap. 2, "The Scientific Revolution"). Indeed, some historians have identified the printing revolution as one of the harbingers of the Scientific Revolution (Eisenstein 1979). By the eighteenth and nineteenth centuries, books and journals bringing natural philosophy to a broad and popular audience proliferated. Again, as we shall see, popular scientific publishing was not simply a matter of disseminating a pre-established body of knowledge to a docile audience. Scientific authors and publishers had a whole range of interests and motives for producing books and journals of all kinds. Making money was clearly one interest. Some nineteenth-century popular science texts, such as Robert Chambers's notorious *Vestiges of the Natural History of Creation* (1844),were, by the standards of the time, bestsellers (see chap. 6, "The Darwinian Revolution"). But authors had particular visions of science to place before their publics as well. Indeed, a book such as Chambers's *Vestiges* only sold so well because its message chimed so well with what its middle-class readership wanted to hear (Secord 2000). Audiences were by no means docile receivers of scientific knowledge either. The reading public during the nineteenth century, for example, had its own ideas about what good science should be. The latest science book was as likely as the latest George Eliot novel or Macaulay history to get a mauling in the prestigious quarterly reviews if it failed to live up to expectations.

Print culture for much of the seventeenth and eighteenth centuries was fluid and malleable (Johns 1998). The powerful Stationers' Company, for example, governed the printing trade in seventeenth-century England. The company had been incorporated under the Tudors to oversee and regulate the production of printed material in London. Only it and a limited number of other bodies, such as the universities and, interestingly, the Royal Society, could license printing. Authors had little if any power over their own works. Booksellers and printers (effectively the same at this period) could vary their texts at will. The situation was much the same in other European capitals, such as Paris. By the beginning of the eighteenth century, writing for pay was coming to be a viable way of making a living in such large cities as London with its emerging Grub Street culture. The seventeenth century saw the genesis of magazines and newspapers. Literary hacks churned out journalistic pieces of variable reliability along with dramas, philosophical essays, novels, and anything else that might attract the paying customer. There was a steady market for pornography (Newton's publisher was a pornographer) and political sedition. By the middle of the eighteenth century, science was enough of a marketable commodity for there to be a flourishing trade in popular scientific literature as well.

One of the particular features of early seventeenth-century natural philosophical such as Galileo's *Dialogues on the Two Chief Systems of the World* and his *Discourse on Two New Sciences* was that they were written in the vernacular rather than in the academic and Church language, Latin. This in itself was an indication that Galileo wanted to be read by the laity. In keeping with the seventeenth-century ethos of science as the province of the cultured, civic-minded gentleman of parts, natural philosophers increasingly commonly aimed their books at more than just a scientific audience (if indeed such an audience can properly be said to have existed at all during the period). The early eighteenth century saw a swathe of publications aimed at explaining and supporting the esoteric truths of Newtonianism to that vast majority of the literate population unable to comprehend Newton's mathematics. Texts such as Joseph Priestley's *History and Present State of Electricity* (1767) or his *Experiments and Observations on Different Kinds of Air* (1776) carried his version of the Newtonian message to a dissenting middle-class audience eager for Priestley's argument that Newtonian science paved the way for moral and social revolution. Science was at the core of radical French philosophers Diderot's and d'Alembert's grand plan for an universal *encyclopédie* classifying all knowledge.

Not only scientific books, but scientific journals proliferated. Even august periodical publications like the Royal Society's *Philosophical Transactions* were not aimed at a purely scientific audience. Its contributors wanted to be read by leisured gentlemen as well. Scientific academies across Europe produced similar publications with the same kind of audience in mind. More significantly, science was part of the staple diet churned out by the new genres of magazines that flourished during the eighteenth century. A gentleman reading the *Gentleman's Magazine* (founded in 1731) or a lady reading the *Lady's Magazine* (founded in 1770) could hope to be kept up with the latest in science and scientific gossip. These journals and a host of others like them were aimed at a rapidly expanding market of (relatively) leisured, literate, and mainly urban middle-class readers, and science was part of what that audience expected to receive in its literary diet. In France and in the Americas, too, science was part of the literary culture built around more-or-less short-lived magazines and journals discussing the latest politics and society scandal of the day. Natural philosophy and some of its practitioners' pretensions were enough a part of this day-by-day literary culture to become the subject of satirists' wit in such novels as Jonathan Swift's *Gulliver's Travels*.

By the nineteenth century, scientific publishing was well-established, with authors aiming their works at a variety of audiences. Such writers as Jane Marcet published for children with books like her *Conversations on Chemistry* (1806). Mary Somerville's *Connexions of the Physical Sciences* (1834) dutifully portrayed the findings of gentlemen of science to a polite middle-class audience. The gentlemen of science themselves aimed their books at popular audiences as well. Texts such as Charles Lyell's *Principles of Geology* (1830) or William Robert Grove's *Correlation of Physical Forces* (1846) aimed for a broad readership. Such diverse organizations as the Society for the Diffusion of Useful Knowledge, their rivals at the Anglican Society for the Promotion of Christian Knowledge, and, later in the century, the evangelical Religious Tract Society published series of popular scientific books aimed at the working and lower middle classes. The publishing phenomenon of the first half of the century was Robert Chambers's anonymously published *Vestiges of the Natural History of Creation* (1844), which both became a best-seller and caused huge controversy with its outspoken defense of progressive development in nature and society (Secord 2000). Texts such as Alexander von Humboldt's five volume *Cosmos* (1845–62) were widely read across Europe and America. For much of the century, American publishers depended on republishing the works of European scientific authors. By the end of the century, however, such American writers as Edward Livingston Youmans were making names for themselves in their own right. Youmans was a key figure in establishing the International Scientific Series of popular science books during the early 1870s.

Popular scientific publishing continued throughout the twentieth century and into the twenty-first. Scientists such as the astronomer Arthur Eddington and the physicist James Jeans wrote popular books bringing the claims of Einstein's theories of relativity to a broad general audience. Books such as Eddington's *The Nature of the Physical World* (1928) or Jeans's *The Universe around Us* (1929) had a crucial role in shaping early public perceptions of the philosophical implications of the new physics (see chap. 11, "Twentieth-Century Physics"). As the sciences became increasingly professionalized and esoteric, scientists often turned to popular scientific writing to make claims or express views that they were unable to maintain in the context of professional journals. Discussions about the relationship between science and religion, for example, were often played out in popular rather than professional publications (Bowler 2001). Physicists like Sir Oliver Lodge whose views were increasingly at odds with those of mainstream physics also turned to the popular press as a vehicle for their ideas. During the middle of the twentieth century there was also a distinct trend toward popular scientific writing with a pronounced socialist tendency. Such authors as Lancelot Hogben, in *Science for the Citizen* (1938), argued for the central role that science and scientific planning should play in a progressive society.

Scientific journals proliferated throughout the nineteenth century. Again, even such elite publications as the French *Académie*'s *Comptes Rendus* or the science magazine *Nature,* toward the end of the century, aspired to more than just a professional audience. In the United States, the *Scientific American* quite explicitly aimed to make itself a mouthpiece for populist science. Both in Europe and North America, a wide variety of scientifically oriented magazines and journals, including the Adelaide Gallery's *Magazine of Popular Science,* tried to take advantage of the market for popular science. Such publications as the *Inventor's Advocate* or, more successfully, the *Mechanics' Magazine* tried to market themselves as the organs for those excluded from mainstream scientific discourse. Middle-class British weeklies like the *Literary Gazette* or the *Athenaeum* included news about scientific meetings and the latest science gossip in their columns. Similarly, such prestigious quarterly journals as the liberal *Edinburgh Review* or the more conservative *Quarterly Review* provided reviews of the latest science in their pages. In France, the Jesuit priest Frédéric Moigno was science correspondent for *La Presse* as well as editing his own weekly popular science journal, *Cosmos.* A steady diet of scientific news and information was a standard part of the journalistic repertoire offered by popular magazines such as the *Penny Cyclopaedia* and other periodicals produced by such organizations as the Religious Tract Society, keen to sell their own vision of science to their readers. Throughout the century, big scientific events, among them the British Association for the Advancement of Science's annual meetings, attracted acres of newsprint in the big daily papers.

Science poked its nose into fiction increasingly as well. The way Charles Dickens poked fun at the "Mudfog Association for the Advancement of Everything" in the *Pickwick Papers* assumed a familiarity with the activities of his target (the BAAS) on the part of his readership. George Eliot could crack scientific jokes in her novels. By the second half of the nineteenth century, scientific speculation was starting to become a literary genre (what we would call science fiction) in its own right. Jules Verne's novels—*Around the World in Eighty Days* (1873), for example—played with the possibilities of contemporary science and technology while his *From the Earth to the Moon* (1865) speculated about future possibilities. H. G. Wells used fictionalized science in the *Time Machine* (1895) to mount a critique of late nineteenth-century industrialized society's social divisions, as did Edward Bulwer Lytton in the *Coming Race* (1871). Books like these were at the more popular end of a growing vogue for scientific utopian and dystopian novels speculating about the moral and social consequences of scientific progress (Fayter 1997). By the beginnings of the twentieth century, H. G. Wells in particular had established himself as a scientific speculator with the *War of the Worlds* (1895) and as social prophet with *The Shape of Things to Come* (1933).

Science fiction became an increasingly important and popular literary genre during the first half of the twentieth century. In the United States in particular, such science fiction magazines as *Amazing Stories* brought science fiction short stories to dedicated audiences of fans while providing a living for such budding authors as Isaac Asimov and Robert Heinlein. By the 1950s, science fiction was appearing on the relatively new genre of television, with space operas like *Flash Gordon.* As the Cold War got chillier during the 1950s, science fiction both in the movies and on television was a way of playing out fears of invasion and evil empires in a (seemingly) depoliticized setting. From the late 1960s on, Gene Roddenberry used his *Star Trek* series to break new boundaries, criticizing the Vietnam War and introducing the first televised interracial kiss safely far away in space and time on the starship *Enterprise.* Shunned and ridiculed by mainstream literary critics, science fiction authors maintained (and still maintain) a steady cult readership. Beginning in the late 1970s, the success of the movie blockbuster *Star Wars* brought about a renewed wave of Hollywood space operas as well as reviving Roddenberry's *Star Trek* franchise. As with many early nineteenth-century science exhibitions, the success of much science fiction as a literary, movie, and television genre depends on playing with and extending their audiences' knowledge and expectations of contemporary science.

Alternative Science

Popular science has never been wholly under the control of those defining themselves as mainstream or professional practitioners of science. In lectures, exhibitions, books, and latterly, television, it has always been a focus for reappropriation and for redefinition of what it might mean to be scientific. The audiences for science have never been entirely passive. Far from it, they constantly engage with scientific questions and concerns, turning them to address their own issues and preoccupations. From that perspective, popular science, broadly understood, has always been a battleground between competing groups' perceptions of what kind of activities might legitimately be described as science in the first place. We have already seen that in whatever genre popular science has historically been presented, its presenters always try to form it in their own image, as it were. Some producers of popular science have produced versions of scientific activity that mainstream practitioners have violently rejected. In many ways, what we now consider to be orthodox, respectable science is the result of these kinds of debates in the past. The late eighteenth and nineteenth centuries were particularly ripe grounds for these kinds of "alternative" sciences and we will consider two of them—mesmerism and phrenology—here (fig. 16.4).

Mesmerism or animal magnetism had its origins in the work of the late eighteenth-century Viennese physician Franz Anton Mesmer (1734–1815). Mesmer believed he had found a way to manipulate the magnetic fluid inherent in human and animal bodies in order to effect a number of distinct physical effects on his experimental subjects and patients. By passing his hands over particular parts of the body or by staring into his patients' eyes, Mesmer was able to produce a range of bodily and mental sensations and behaviors on the part of his subjects. Their limbs might move involuntarily or become paralyzed, they might become hysterical or comatose, or Mesmer might be able to command their bodies to carry out particular actions without their being aware of them. Having fled Vienna to escape state persecution for his philosophical and religious beliefs, Mesmer settled in Paris where animal magnetism rapidly became a fashionable craze. Crowds flocked to mesmeric salons to be mesmerized by Mesmer or one of his disciples. Enthusiasts hailed animal magnetism as a revolutionary new science of the mind. Critics accused Mesmer of being a charlatan and even

TOWN HALL, CHARD.

Mr. DAVEY

Who has just concluded a Course of Nine Lectures at Tiverton, respectfully announces to the Gentry and Public generally of CHARD, and its Vicinity, that he will deliver a Course of

THREE EXPERIMENTAL LECTURES,

ON THE UTILITY OF

MESMERISM

PHRENOLOGY,

SYMPATHY & MINERAL MAGNETISM,

AT THE ABOVE HALL, KINDLY LENT FOR THE OCCASION,

On Tuesday 8*th*, *Thursday* 10*th* & *Friday* 11*th December*, 1846,

TO COMMENCE AT SEVEN O'CLOCK, P. M.

That all persons may have an opportunity of witnessing the greatest wonder of the age, the price of Admission will be reduced one-half, viz. RESERVED SEATS, 1s.; SECOND CLASS, 6d.; BACK SEATS, 3d.

THE LECTURER

Will explain in his preparatory Lectures the locality, use, and abuse of the Organs, and the application of Mesmerism to human welfare, and exhibit a number of Busts, whose characters are before the Public. He will then undertake to produce Mesmeric Sleep, Rigidity of the Limbs, Power of Attraction and Repulsion, and the Transmission of Sympathetic Feelings. He will also demonstrate Phrenology, by exciting the Organs while in a state of Coma. The sleepers will perform Vocal and Instrumental Music, Dancing, Talking, Nursing, Eating, Drinking, and other feelings of mirth, imitation and independence, even up to the highest manifestations of benevolence, veneration and sublimity, while in the Mesmeric Sleep.

Mr. D. will be accompanied by Miss Henly, daughter of the late Capt. Henly, from Newton-Abbot, born Deaf and Dumb; and he feels confident of bringing into action those faculties which have been dormant from her birth, by the aid of Phreno-Mesmerism.

"FACTS ARE STUBBORN THINGS."

FIGURE 16.4 A poster advertising a series of popular lectures on mesmerism and phrenology in 1846. Lectures like these were crucial in the spread of new popular ideas.

drew connections between mesmerism and the French Revolution. A royal commission, established by the Académie des Sciences and including Benjamin Franklin among its members, roundly condemned Mesmer as an outright fraud.

Mesmerism underwent a major revival in the British Isles from the 1830s onward as some of Mesmer's disciples arrived in London to try their luck in a new environment. Practitioners traveled the country offering mesmeric lectures and performances in which members of the audience would be invited onstage to be magnetized. Middle-class women conducted mesmeric séances in the privacy of their own homes, mesmerizing their servants, their daughters, and their neighbors. The journalist and author Harriet Martineau caused a scandal when she announced that animal magnetism had cured a long-standing illness. Mesmerism became a cause célèbre in the hands of radical doctor John Elliotson at University College Hospital in London during the 1830s. Elliotson thought that mesmerism could provide the foundations for a new and materialist science of the mind, showing how all mental states were simply the results of the physical state of the body. He carried out mesmeric experiments on his patients at the hospital before invited scientific witnesses such as Michael Faraday. His experiments on a young working-class girl, Elizabeth O'Key, were particularly notorious. His advocacy of mesmerism led to a rift between Elliotson and fellow radicals, including Thomas Wakley, editor of the *Lancet,* and led to his eventual dismissal from his post at University College (Winter 1998). During the 1840s, mesmerism was even tried out as a form of anesthetic.

Elliotson's advocacy provides one clue to mesmerism's widespread popularity both in Europe and North America. It was a radical science providing an alternative account of mind and action to that on offer from many orthodox scientists and from established religion (see chap. 15, "Science and Religion). This also helps explain the virulence with which scientific conformists greeted the upstart contender. Mesmerism offered to turn the science of mind into a physical science. The way people behaved and their position in society could be explained on the basis of the flow of magnetic fluid through their brains rather than on the grounds of divine providence or heredity. Another clue to mesmerism's success was the egalitarian nature of its practice. While gentlemen of science increasingly argued that only the highly trained few could engage in real science, mesmerists argued that anybody could join their ranks. Women and working-class men were just as likely to make good mesmerists as middle-class gentlemen. Working-class political radicals looking for a materialist basis for their social theories supported mesmerism for these kinds of reasons. More prosaically, despite the frantic opposition of middle-class scientists, large swathes of the middle classes were fascinated by mesmerism simply as a source for speculation and entertainment. Mesmerism was a science that, to some degree at least, crossed class and gender boundaries (see chap. 21, "Science and Gender"). It provided an avenue through which working-class men and women could approach middle-class men as intellectual equals.

Like mesmerism, the science of phrenology had its origins in late eighteenth-century efforts to establish a materialist science of mind. It was the result of efforts by another Viennese physician, Franz Joseph Gall (1758–1828), to understand the relationship between the brain's physical structure and different mental states. Gall's new science was based on a number of seemingly straightforward and uncontroversial principles: that the brain was the organ of the mind; that the mind was composed of a number of distinct faculties; that each faculty was associated with a distinct organ in the brain; that the size of each organ determined the relative power of the corresponding faculty; that the size and shape of the brain was determined by the size and shape of the respective organs; and that the contours of the skull were determined from the size and shape of the brain. All of this meant that it was possible to read off the shape of the brain and, therefore, the size of the respective organs and the power of the corresponding faculties from the size and shape of the skull. Gall and his disciple J. C. Spurzheim (1776–1832) toured extensively across Europe throughout the early years of the nineteenth century lecturing on the science of phrenology and its implications.

The New Science first came to prominence in the British Isles following a savage review of Gall's work published in the *Edinburgh Review* in 1815. Spurzheim's spirited defense of his master's work before a largely hostile audience of medical men at Edinburgh University attracted considerable sympathy and focused attention on phrenology's underlying principles. Phrenology rapidly became a popular science (Cooter 1984). Its main supporter in Edinburgh—and in the country more generally—was George Combe, whose phrenological *Constitution of Man* (1828) was a huge bestseller, with 350,000 copies sold by the end of the century. Combe's book placed phrenology in the context of other efforts to establish a naturalist science of man's place in nature and society as resulting from the workings of natural law. Combe was also instrumental in founding the Phrenological Society in 1820, along with its *Phrenological Journal*. Such societies flourished in the British Isles, the rest of Europe, and North America during the 1830s. As with mesmerism, popular performers offered phrenological readings to lecture audiences and popular books offered "do-it-yourself" guides to phrenological science. The American phrenologist L. N. Fowler's phrenological lecture tours across the United States and Europe during the second half of the century did much to revive phrenology's flagging popularity.

As with mesmerism, one of the key reasons for phrenology's popularity was its populism. On the one hand, it was a science that anyone could practice. Its guiding principles were relatively straightforward and easy to understand. Once they had been mastered, all that the budding practitioner needed was a map of the position of the various phrenological organs of the brain and the corresponding bumps on the surface of the skull, and they were in business. Phrenology's implicit materialism made it attractive to political radicals. Even more so than in the case of mesmerism, phrenology carried an egalitarian and antihierarchical message. If character and aptitude were determined by the size and shape of the organs of the brain, then those were the characteristics that should determine social status and opportunity, not inherited position and family wealth. It was a deeply attractive proposition to the lower-middle-class audiences that flocked to phrenological lectures or devoured Combe's book. It seemed to provide a scientific basis for their claims to political and social reform on meritocratic grounds. Political and social power should be given to those who were phrenologically fit for the job. Popular phrenological lecturers offered readings to parents about the likely occupational aptitudes of their offspring or to worried household heads about the trustworthiness of potential servants (figs. 16.4 and 18.1, p. 418).

Conclusions

Looking at the cultural dynamics surrounding such alternative sciences as mesmerism or phrenology shows us just how important popular science can be in understanding the relationships between science and society. Both of these sciences flourished through the full range of historical media, by means of lectures, exhibitions, popular books, and journals. Enthusiasts established societies to promote these new sciences just as they established societies to promote astronomy or geology. Their histories show just how important the public and popular face of science can be in determining how, throughout history, science as a practice has been defined. It is through interaction with a broader public audience like this, at least as much as through interaction with their peers, that scientists define what science is. In the cases of mesmerism and phrenology during the nineteenth century, their advocates struggled hard to establish these practices as sciences and in so doing redefine what science was, how it ought to be practiced, and what kind of people its practitioners should be. These were debates that by their very nature had to be carried out in the public domain and through popular as much as professional media. This was simply because in cases like this, at least, it was the public that ultimately decided what science was—or at least what science they wished to pay attention to.

But having a popular face also matters to other sciences and scientists that we might more readily recognize as being orthodox. Throughout most of the period covered in this book, scientists and natural philosophers have sought to engage with their publics not as something peripheral to their main concerns but as a central feature of what their science was about. As we have seen, they have engaged in a whole array of different venues toward that end. For most of this period, too, the public has found nothing unusual about scientists' efforts to engage them. On the contrary, they have been keen enthusiasts for science in one form or another. In this respect C. P. Snow is quite right that the breakdown into "two cultures" is a modern phenomenon. That is not to suggest that what we had in the past was a common context. Different groups, classes, and genders had different experiences and expectations of popular science and the uses to which it might be put. The various genres appealed to a range of different constituencies and producers of popular science, be they scientific practitioners, literary hacks, showmen, political or religious partisans, or sci-fi authors and television producers, have all had their different goals and aspirations—and competing ideas about what science was.

References and Further Reading

Beauchamp, Ken, 1997. *Exhibiting Electricity.* London: Institution of Electrical Engineers.

Berman, Morris, 1978. *Social Change and Scientific Organization: The Royal Institution, 1799–1844.* London: Heinemann.

Bowler, Peter J. 2001. *Reconciling Science and Religion: The Debate in Early Twentieth-Century Britain.* Chicago: University of Chicago Press.

Cooter, Roger. 1984. *The Cultural Meaning of Popular Science: Phrenology and the Organization of Consent in Nineteenth-Century Britain.* Cambridge: Cambridge University Press. .

Desmond, Adrian. 1994. *Huxley: The Devil's Disciple.* London: Michael Joseph.

Eisenstein, Elizabeth. 1979. *The Printing Press as an Agent of Change: Communications and Cultural Transformations in Early Modern Europe.* Cambridge: Cambridge University Press.

Fayter, Paul. 1997. "Strange New Worlds of Space and Time: Late Victorian Science and Science Fiction." In *Victorian Science in Context,* edited by Bernard Lightman. Chicago: University of Chicago Press.

Golinski, Jan. 1992. *Science as Public Culture: Chemistry and Enlightenment in Britain, 1760–1820.* Cambridge: Cambridge University Press.

Hays, J. N. 1983. "The London Lecturing Empire, 1800–50." In *Metropolis and Province: Science in British Culture, 1780–1850,* edited by Ian Inkster and Jack Morrell. London: Hutchison, 1983.

Heilbron, John. 1979. *Electricity in the Seventeenth and Eighteenth Centuries: A Study of Early Modern Physics.* Berkeley: University of California Press.

Johns, Adrian. 1998. *The Nature of the Book: Print and Knowledge in the Making.* Chicago: University of Chicago Press.

Marvin, Carolyn. 1988. *When Old Technologies Were New.* Oxford: Oxford University Press.

Morrell, Jack, and Arnold Thackray. 1981. *Gentlemen of Science: The Early Years of the British Association for the Advancement of Science.* Oxford: Oxford University Press.

Morton, Alan. 1993. *Public and Private Science: The King George III Collection.* Oxford: Oxford University Press.

Morus, Iwan Rhys. 1998. *Frankenstein's Children: Electricity, Exhibition and Experiment in Early Nineteenth-Century London.* Princeton, NJ: Princeton University Press.

Porter, Roy. 2000. *Enlightenment: Britain and the Creation of the Modern World.* London: Allen Lane.

Secord, James. 2000. *Victorian Sensation: The Extraordinary Publication, Reception, and Secret Authorship of Vestiges of the Natural History of Creation.* Chicago: University of Chicago Press.

Snow, C. P. 1959. *The Two Cultures and the Scientific Revolution.* Cambridge: Cambridge University Press.

Stewart, Larry. 1992. *The Rise of Public Science: Rhetoric, Technology, and Natural Philosophy in Newtonian Britain, 1660–1750.* Cambridge: Cambridge University Press.

Winter, Alison. 1998. *Mesmerized: Powers of Mind in Victorian Britain.* Chicago: University of Chicago Press.

CHAPTER 17

SCIENCE AND TECHNOLOGY

FROM A CONTEMPORARY POINT OF VIEW, science and technology seem to be inextricably connected. The relationship, if anything, seems to be becoming increasingly closer. Even a decade ago policymakers and scientists themselves would have routinely assumed boundaries between pure and applied research, between science and engineering, or between theory and application that seem increasingly tenuous today. Now they are more likely to reject such distinctions as at best artificial and at worst positively misleading in their assumptions of how both scientific and engineering disciplines work in practice. In a more popular context, science and technology are usually represented as being largely coterminous as well. Science on television, for example, is usually presented as the sum of its technological applications. This perception is, however, relatively recent. Historically, philosophically, and sociologically, the relationship between science and technology has been a matter of considerable debate and remains contested today. The dominant philosophical view has been that the relationship between science and technology has been straightforwardly hierarchical. Scientists following the scientific method produce new theories that engineers and technologists then employ to find applications that they use to solve practical problems like building bridges or exploding nuclear bombs. As we shall see, this view of the relationship has its own history.

Contemporary sociological and social historical views of the relationship between science and technology tend toward seeing them as being closely intertwined, if not indistinguishable activities. The sociologist Bruno Latour, for example, brings the two together under the label "technoscience" and treats them as identical (Latour 1987). He suggests that from the point of view of the sociologist (or historian) trying to understand science and technology, there is no practical difference between them. There are three points to be made here. First, there can be little question following the work of historians of science and of technology over the past fifty years that the historical relationship between science and technology is far more complex than the hierarchical view supposes. Second, it is also clear that disciplinary boundaries between science and engineering that may have existed a few decades ago are now becoming increasingly porous. Third, historians and sociologists of science now tend to think of science itself as being (at least in part) a practical activity rather than a theoretical abstraction. They are therefore more likely to focus their attention on those features of science that bear most resemblance to technological practice. The cultural turn in the history of science more generally, of course, also means that historians are more interested in looking for connections between science and other areas of culture than they might previously have been.

Historians are now increasingly aware that not only is the historical relationship between science and technology constantly changing but that historical figures themselves have had a range of conflicting views about just what that relationship was or should be. Understanding those conflicts is an important part of understanding the relationship between science and technology. Conflicts of this kind were usually part and parcel of broader debates about the nature of science, how it should be pursued, and by whom. Arguments about the close relationship between science and industry, for example, were, as we shall see, commonly put forward during the Victorian period by those advocating increased state funding of scientific activity. It was in their interest to argue that science made a tangible contribution to economic productivity. Those opposed to increased state funding tended to deny the link. In earlier periods, arguments about science and utility have often been put forward by those seeking patronage for their activities. As well as professional interests, other political concerns might be at stake in disputes like this, too. Arguments about the relationship between science and technology have often been arguments about whose cultural property science should be considered to be. We shall see that many contemporary historiographical positions concerning the relationship between science and technology have their equivalents in the claims made by historical figures themselves about that relationship.

We shall start this chapter with an overview of some contemporary historical perspectives on the relationship between science and technology, along with the claims that some historical figures themselves have made about that relationship. From at least Francis Bacon onward, natural philosophers have periodically put forward a range of arguments concerning the utility of their sciences. As far as Bacon was concerned, for example, its utility was one of the important features that distinguished the New Science from the old scholasticism (see chap. 2, "The Scientific Revolution"). France, during the revolutionary and Napoleonic era, made concerted efforts to harness science for national benefit. British commentators —Charles Babbage and William Whewell, for instance—quarreled over the relationship between science and the arts. Early twentieth-century left-wing scientists such as J. D. Bernal wanted to turn science to the national good. We shall then look at two case studies—the steam engine and the telegraph—in an effort to get some sense of the dynamics of the relationship between science and technology during the eighteenth and nineteenth centuries. There are, of course, a number of other case studies, such as the nineteenth-century chemical industry or twentieth-century electronics, that could be used to make the same kinds of points. Finally, we shall look at the ways in which debates concerning the relationship between science and technology—whether technology is a product of science or science a product of technology—have often been intertwined with broader debates concerning the social identity of the scientist.

The Chicken and the Egg

For much of the last century, most of its practitioners regarded the relatively new discipline of the history of science as entirely distinct from the history of technology. The history of science was to be about ideas and their origins, not technological applications. Still less was there any sympathy for the idea that science itself might have its roots in practical technological activity. The historian George Sarton, who founded *Isis,* one of the first specialist history of science journals in 1912, regarded technological applications as largely irrelevant distractions for the historian of science. Science was about the production of truths, not technologies (Sarton 1931). Science might indeed have beneficial applications, but this was an incidental by-product of the search for truth, motivated by "disinterested curiosity." Sarton's view was not unusual. The French historian of ideas Alexandre Koyré put forward similar arguments. The great figures in the history of science, such as Galileo or Newton, had nothing to do with engineers or craftsmen. Their science was the product of theory rather than practice (Koyré 1968). A similar point of view was put forward by the British historian Herbert Butterfield and his students. Science clearly had technological applications, but these were incidental and lay outside the purview of the history of science that should concern itself primarily with ideas (Butterfield 1949).

To some extent, at least, the self-conscious idealism of many of the history of science's founding fathers and their reluctance to engage with the relationship between science and technology was a reaction to an emerging strand of Marxist history that regarded science as being an offshoot of economic and technological development. This position was most famously articulated by the Soviet economic historian Boris Hessen in his 1931 essay "The Social and Economic Roots of Newton's 'Principia'" ([1931] 1971). According to Hessen, the mathematical science produced by Newton and his contemporaries was simply the consolidation in theoretical language of the practical knowledge generated by artisans, craftsmen, and engineers engaged in economic and technological activity. The driving force behind the emergence of modern science was its potential utility. It was the science of ballistics, fortification, navigation, and shipbuilding. This was a classically Marxist line of argument, privileging economic activity as the ultimate source of all historical development. A similar position was adopted by the sociologist Edgar Zilsel (1942). According to Zilsel, the rise of science was inseparable from the rise of modern capitalism. Like Hessen, he argued that the emergence of science should be understood as the appropriation by academic scholars of the craft skills and technological knowledge of manual laborers like carpenters, instrument makers, and miners.

Some historians of science outside the idealist tradition have avoided engagement with this debate (often characterized as the externalism-internalism debate) by focusing their attention on the development of applied science, particularly in the context of the Industrial Revolution and its aftermath (Cardwell 1957). Cardwell, for example, while acknowledging that economic and technological developments during the sixteenth and seventeenth centuries clearly played some role in the Scientific Revolution of those centuries, denied the economic and technological determinism implicit (and occasionally quite explicit) in arguments such as those put forward by Hessen and Zilsel. At the very least, he suggested, such a thesis was ultimately untestable. Cardwell's main interest lay in charting the development of connections, particularly institutional connections, between science and engineering from the late eighteenth century. Having a background in both "pure" and "applied" science (as he characterized them), he was anxious to insist on the benefits of fruitful interaction between science and technology for both enterprises. In this respect his work is typical of much recent history of the relationship between science and technology. Historians have often taken the categories for granted and tried to characterize what the relationship between the two activities was at a particular historical juncture. One way of questioning this approach is to start looking at what historical figures themselves have had to say about the relationship.

Francis Bacon was only one of many early modern commentators on the New Science of the sixteenth and seventeenth centuries to insist that utility was one of its defining features. As we saw earlier, Bacon was insistent that natural philosophy should be cultivated by worldly gentlemen rather than reclusive scholars (see chap. 2, "The Scientific Revolution"). One of his reasons for this insistence was that science ought to be at the service of the commonwealth. One of the benefits to be expected from the systematic pursuit of knowledge was an increased capacity to manipulate nature for utilitarian ends (fig. 17.1). As Bacon expressed it in probably his most famous dictum, "knowledge is power." Seventeenth-century natural philosophers across Europe in search of state or private patronage happily repeated Bacon's dictum to their potential patrons. One of the motives underlying the foundation of both the Royal Society of London and the French Académie des Sciences was that the institutions would foster science's utility (see chap. 14, "The Organization of Science"). Some historians have identified this utilitarian turn as one of the defining features of the Scientific Revolution (Merton 1938; Webster 1975). Eighteenth-century popular lecturers such as J. T. Desaguliers highlighted the potential technological potential of their natural philosophical knowledge as they toured eighteenth-century London's coffeehouses in search of audiences and patrons (Stewart 1992; see also chap. 16, "Popular Science").

The late eighteenth-century revolutionary French state was one of the first to make a concerted institutional effort to harness what many of the revolution's backers regarded as the self-evident technological potential of the sciences. Even before the revolution, many French military officers were increasingly interested in applying science to the improved design and production of weaponry (Alder 1997). Following the systematic reorganization of French educational and scientific institutions in the immediate aftermath of the revolution, new institutions like the École Polytechnique were explicitly designed to deliver an education in natural philosophy (to army cadets in particular) that was fully expected to result in technological and engineering expertise. Such advocates of scientific education as the revolutionary general Lazare Carnot argued that scientific tools like geometric analysis were ideally suited for engineering problem solving. Under the revolutionary regime, Gaspard Monge, the École's main instigator, also argued that geometry was the foundation of engineering knowledge.

FIGURE 17.1 The frontispiece of Francis Bacon, *Insturatio magna,* showing the ship of discovery about to set out through the Pillars of Hercules to sail the seas of knowledge. The Latin tag translates as "many set forth and knowledge is increased."

Following reforms of the École under Napoleon and particularly the rise of power of the mathematical physicist Pierre-Simon Laplace, physics came to occupy an ever greater portion of the curriculum. Reforms like this were not, however, an indication of any loss of faith in science's capacity to deliver technological benefits. The dispute was about what kind of science would deliver the best results.

British sympathizers with the Napoleonic regime in France looked enviously at the French state's support for science and largely agreed with French natural philosophers' assumption that natural philosophical education had tangible industrial benefits. Charles Babbage's *Reflexions on the Decline of Science in England* (1830) was a vituperative attack on the Royal Society's leadership and on the English government's laissez-faire attitude to the sciences that drew highly unflattering comparisons between English and French support for science on precisely such grounds. Babbage and fellow sympathizers, such as John Herschel, were adamant that science was an indispensable tool for industrial progress. Nothing other than the systematic application of scientific principles—not only to the process of technological invention but to industrial organization, too—could guarantee steady progress (Ashworth 1996; Schaffer 1994). Babbage returned to the attack in his *Exposition of 1851* in which he took the Great Exhibition's organizers to task for failing to plan the exhibition on proper, scientific principles. Babbage's position was supported by others, such as the chemist Lyon Playfair, who were equally adamant that chemical science was a prerequisite of chemical industry. As we shall see, however, Babbage and his supporters were deeply unsympathetic to any idea that the craft skills and knowledge of artisans and mechanics might contribute to science. On the contrary, though, technological innovation depended on replacing craft with hard science.

Few British scientists during the first half of the nineteenth century shared Babbage's enthusiasm either for the Napoleonic regime or for the French model of state funding for scientific institutions. There was, however, a strongly utilitarian bent in most British accounts of scientific progress. The Welsh natural philosopher William Robert Grove was quite typical in his claim that Britain's greatness depended on her industry and commerce and that industry and commerce depended on science. There were dissenting voices, however. The polymathic William Whewell, Master of Trinity College Cambridge, was deeply suspicious of claims about science's utility. Whewell denied outright that science was a necessary prerequisite of industrial progress. The industrial arts and the sciences each progressed according to their own internal principles of development. If there was any relationship between the two, then it was the reverse of what Babbage and his allies suggested according to Whewell. The relationship between art (i.e., technology) and science was equivalent to that between poetry and criticism. Technology, in other words, came first. Science might seek to understand the natural process through which technology operated, but it was not as a result a secure source of technological innovation. Whewell, like his friend George Bidell Airy an opponent of state funding for the progressive sciences, saw no merit in Babbage's argument that state funding for science was essential to ensure continued technological progress.

Other commentators elsewhere were equally suspicious of the utilitarian argument for science. The American physicist and first secretary of the Smithsonian Institute Joseph Henry, for example, was anxious to keep American science respectable, was worried about his countrypeople's tendency to prefer inventors to discoverers, and was dubious about the connection between science and technology. By the beginning of the twentieth century, however, proponents of the utilitarian argument in favor of state funding were increasingly vociferous. In Britain, liberal and socialist scientists were adamant that scientific progress was indispensable to continued economic prosperity. This meant that science needed to be subject to centralized state control and funding and that scientists needed to be actively involved in planning economic policy. By the 1930s, the Marxist scientist J. D. Bernal was arguing along with Boris Hessen that science and technology were locked in a symbiotic embrace and that economic forces were the key to their development. Progress in science and technology needed centralized planning to achieve the common good. As Bernal put it, "Science consciously directed, rather than left to blind chance, can transform almost without limit the material basis of life" (1954). Bernal's scientific and technological utopianism was shared by others, particularly following the seeming success of wartime scientific planning (see chap. 20, "Science and War"). In fact, it was largely in response to what they regarded as the hijacking of science's history by Marxists and technological determinists like Bernal or Hessen for such purposes that idealist historians of science tried to disentangle science from technology.

Steam Culture

Someone like Bernal certainly assumed that the steam engine was a classic example of science's role in the production of technological innovation. In fact, from the late eighteenth century onward the steam engine has been

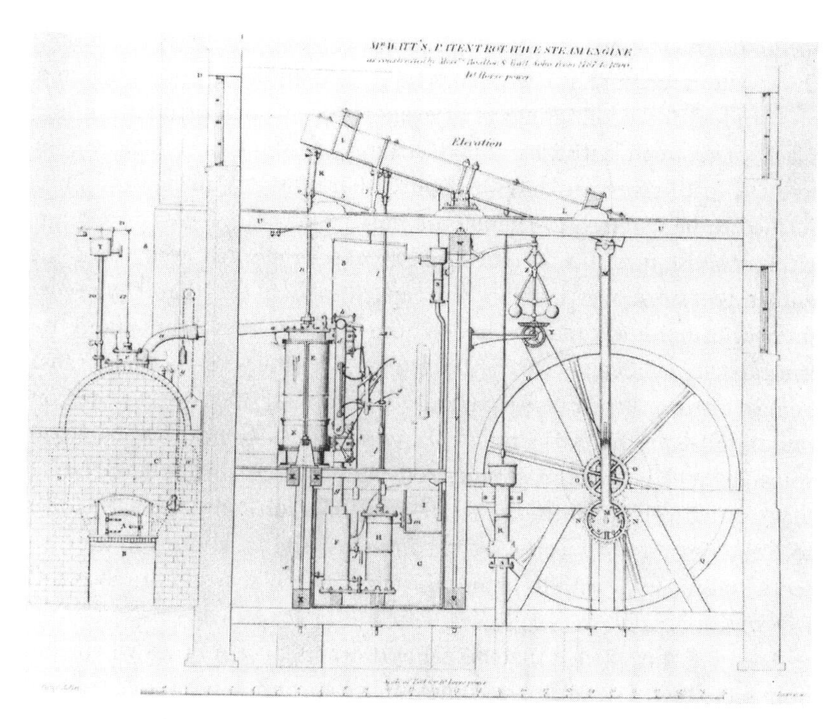

FIGURE 17.2 An illustration of James Watt's steam-engine improvements.

one of the stock examples used by proponents of science as a vehicle of industrial progress. Nineteenth-century advocates of state funding for scientific activity—and scientific education in particular—pointed, in particular, to James Watt and his contribution to the development of steam technology as evidence of a direct link between scientific principles and technological innovation. Watt's improvements to the steam engine were held to be the direct outcome of his understanding of the science of heat (fig. 17.2). Twentieth-century Marxist historians of science such as Bernal took the argument a stage further and argued that not only was Watt's innovation the outcome of applied science, but the eighteenth-century science of heat he had applied to improving the efficiency of steam engines was itself the product of steam technology. Bernal suggested that it was late seventeenth- and early eighteenth-century developments in the use of steam engines for pumping water from mines by Newcomen and others and the increasing scale of industrial activity involving heat that had focused scientific attention on the problems of heat transfer in the first place.

Major developments in the science of heat took place in Scotland during the second half of the eighteenth century. William Cullen, who taught medicine and chemistry at both Edinburgh and Glasgow, drew attention to the apparent relationship between pressure and the boiling point of liquids and studied the cooling effects of evaporation. Cullen's student, Joseph Black, made further strides in his efforts to understand heat. Black was interested in the way in which it seemed that different quantities of heat seemed to be required to change the state (solid, liquid, gas) of different kinds of substances. Black argued that his experiments showed that different kinds of substances had different capacities for heat—that they needed more or less of it in order to change from one state to another—and that this capacity seemed to be something inherent to the substance rather than being a feature of their density. He also argued that, since the change from one state to another takes time, a certain amount of heat was being absorbed by substances during this period without increasing its temperature. Black called this quantity the latent heat of a substance and developed techniques to try to measure it by comparing the time taken to heat water to boiling point with the time taken for it to boil it away entirely (Cardwell 1971).

This was the science that Bernal and others had in mind when they pointed to Watt's steam engine innovations as examples of applied science (Bernal 1954). James Watt, as a young Glaswegian instrument maker, had been hired by Joseph Black to repair the university's model Newcomen steam engine. According to the traditional account, Watt found that the problem lay in the fact that steam was condensing in the cold piston cylinder, and Black explained the phenomenon to him in terms of latent heat. It was this observation, according to the legend, that led Watt to realize how much more efficiently a steam engine would run if the steam were condensed in a separate condenser. The source of this tradition can be found in a late eighteenth-century article on steam engines by John Robison, one of Black's students, in the *Encyclopaedia Britannica*. Watt himself, in later life, denied the story but throughout the nineteenth century and into the twentieth century it remained in place as the classic example of science's contribution to technology. Cardwell has pointed out that the legend as it stands is particularly implausible since it fails to correspond to what is known about the chronological sequence of Watt's inventions (Cardwell 1971).

What does seem clear, however, regardless of the specifics of the legend, was that for many mid-eighteenth-century inventors, natural philosophers, and entrepreneurs, there was little practical difference among their activities—they all inhabited the same culture. Both during Watt's early career as an instrument maker in Glasgow and in later life as an engineering entrepreneur in England, he moved in circles that drew little practical distinction between natural philosophy and technological innovation. Within the Glasgow circles in which Watt moved during his early career, men like Joseph Black or his student friend John Robison shifted easily between practical problems and natural philosophical abstractions. The presence of the model Newcomen engine that Watt was asked to repair in the classroom is an indication of the practical nature of the curriculum. The solution to the question of the origins of Watt's knowledge—Did science or technology come first?—in this case is simply to appreciate that, in the context in which Watt worked, little practical distinction was made between the two. The same can be said of the circles in which Watt moved in later life in England. Watt was a member of the Lunar Society—an informal gathering of like-minded natural philosophical enthusiasts in the Birmingham area that included his business partner Matthew Boulton, the industrialist Josiah Wedgwood, the doctor and early exponent of evolution Erasmus Darwin, and the radical chemist Joseph Priestley. It is an indication of this group's lack of interest in drawing distinctions between what we think of as science and what we think of as technology that Priestley's own response to Joseph Black's isolation of "fixed air" (carbon dioxide) was to develop a new industrial technique for producing carbonated water.

As we saw in our earlier discussions of the origins of the conservation of energy, Sadi Carnot's efforts to develop a theory of ideal heat engines during the 1820s was also an effort to find practical ways of increasing the efficiency of steam engines (see chap. 4, "The Conservation of Energy"). Carnot was educated by his Republican engineer father and at the École Polytechnique—an institution dedicated as we have seen, to applying natural philosophy to France's technological, military, and economic advancement. With such a background it is not surprising that Carnot fils would turn to natural philosophy in an effort to further technological improvement. During the 1830s and 1840s, the search for ways of improving the efficiency of steam engines was a preoccupation of French engineers and natural philosophers alike. They were looking for ways to transform French industry so that it could compete with that of their old enemy Britain. The engineer Marc Séguin's influential 1839 treatise *De l'influence des chemins de fer* discussed the matter of steam engine efficiency and its improvement at some length. Victor Regnault, one of the rising stars of French physics during the 1830s, was appointed by the French Ministry of Public Works to carry out experimental research on steam engine efficiency. The results of his endeavors were not published in full until 1870. In the meantime, however, Regnault's Paris laboratory was increasingly recognized as one of Europe's most prestigious centers for systematic experimental physics.

It was to Regnault's laboratory that William Thomson came in 1845 following his graduation from Cambridge, hoping for training in the latest experimental techniques to match the mathematical education he had acquired at the university. Given Thomson's Glasgow background and established interests in natural philosophy this was not a surprising choice of laboratory. In Glasgow, Thomson, along with his father and brother, moved in circles that took the utility of science for granted. At institutions such as the Glasgow Philosophical Society, university academics and industrialists mixed easily and shared an ethos that regarded science as an agent of economic improvement. It was not so much the case that Thomson and his brother regarded their researches into the science of heat as being of potential use in improving the design of steam engines but that they regarded the two enterprises—natural philosophical investigation and technological improvement—as being two sides of the same coin (Smith 1999). Along with other mid-nineteenth-century British engineers and natural philosophers, including W. J. M. Rankine, they simply did not make a systematic distinction between the two. As in the case of Watt and Carnot, rather than looking for specific examples of natural philosophical principles directly being used to produce a technical innovation or of a technical improvement yielding a new scientific principle, it is more fruitful to look at the cultural context within which individuals developed their work as both science and technology.

There is little question that in early and mid-nineteenth-century Britain, the steam engine was widely regarded as the primary vehicle of economic progress. It stood for Britain's industrial supremacy. Political economists and other commentators worked to try to explain the role the steam engine and industrial machinery more generally played in powering the national economy. Popular scientific writers wrote books and articles explaining the principles underlying these new machines to their readership. Enthusiasts for the new "factory system," such as the chemist Andrew Ure in his *Philosophy of Manufactures* (1835) or Charles Babbage with his *Economy of Machinery and Manufactures* (1832), extolled the virtues of industrial machinery. Steam would replace human and animal labor as the main source of power. Machinery would keep the workforce in line as well. Ure looked forward to a future in which humans and machines would work together harmoniously, all regulated by a central engine. Commentators like this took it for granted that science was the ultimate source of technical innovation. It was the harnessing of natural philosophy to practical ends that they saw as being primarily responsible for British industrial progress during their century. Even natural philosophers like Michael Faraday, who disagreed with the view that the main goal of science was technological improvement, agreed that science inevitably delivered such benefits.

Networks of Power

Even during the 1830s, when the steam engine was just coming into more common use as a means of locomotion following the 1829 Rainhill trials on the Liverpool & Manchester Railway when Stephenson's *Rocket* triumphed, some commentators were already predicting its decline. The competition came from electricity, that other great symbol of Victorian progress. By the early 1830s, concerted efforts were already being made to use electricity as a source of motive power for locomotion and other purposes. Its promoters were optimistic that it was just a matter of time (and not much time either) before electricity superseded steam as the main source of industrial and locomotive power. They looked forward to the day when it would be possible to cross the Atlantic with only a few gallons of acid and a few pounds of zinc (for the battery) as fuel. By the end of the 1840s, these commentators had some practical success stories they could point toward as evidence that the nineteenth century would be the age of electricity. They could point to the development of the electrometallurgical industry using electrochemical techniques to plate metals. More spectacularly, they pointed to the rise of the electric telegraph as evidence of the way in which natural philosophy could deliver previously unimaginable technological progress.

Despite seemingly widespread unanimity concerning the electric telegraph as an example of the way natural philosophy could contribute to technological innovation, the origins of the telegraph on both sides of the Atlantic were surrounded by controversy over the question of just what the relationship between scientific discovery and technological invention might be. In England, the first electromagnetic telegraph was patented successfully by Charles Wheatstone, professor of natural philosophy at London's King's College, and William Fothergill Cooke in 1837. While studying anatomical modeling in Heidelberg, Cooke happened to come across the possibility of using electricity for long-distance signaling. Having attempted unsuccessfully to produce a working model, he got into contact with Wheatstone for advice, who informed him that he, too, had been working on the problem of long-distance electrical communication. The two joined together and, following their successful application for a patent, worked to persuade the directors of various railway companies to adopt their telegraph as a signaling system. By the mid-1840s, when Cooke along with a number of other investors established the Electric Telegraph Company,

Wheatstone had already sold his share of the partnership to Cooke in exchange for regular royalties. Their partnership had fallen apart over the question of their respective rights to describe the electric telegraph as their own invention.

The positions laid out by Cooke and Wheatstone as they tried to persuade each other and, eventually, a mutually agreed arbitration panel, of their respective claims to the telegraph's invention show how difficult it could be to distinguish between science and technology. In many respects, the problem was that neither of them agreed over what the basic criteria for deciding between their claims should be. Cooke argued that the original idea for the device was his, that he had already produced working models before he approached Wheatstone, that he had worked out a comprehensive system (or "projection" as it was described) of how the telegraph should work in practice, and that he had been responsible for its practical implementation. Wheatstone countered that he, too, had come up with the idea of a telegraph and experimented with the possibility before meeting Cooke. Crucially, however, he insisted that Cooke's working models could never have been made to work over long distances without his own superior knowledge of the scientific principles of electricity. The final arbitration carefully divided their claims to invention. Cooke was acknowledged as the first "projector" of the system and Wheatstone was recognized as having provided the knowledge of scientific principles that made the invention possible. In effect, the arbitrators tried to distinguish between invention and discovery, making Cooke the inventor and Wheatstone the discoverer of the telegraph (Morus 1998).

A similar controversy raged in the United States concerning the origins of the telegraph. As far as Americans were concerned, the inventor of the telegraph was neither William Fothergill Cooke nor Charles Wheatstone, it was Samuel Morse. Morse, an impoverished artist, had become enthusiastic about the possibility of transmitting information over long distances by means of electricity after watching electrical demonstrations during a trip to Europe in the early 1830s. Back in the United States, he set out to build a working model of a telegraphic system and eventually—following advice from Leonard Gale, professor of chemistry at New York and later one of his business partners, and from Joseph Henry, professor of natural philosophy at New Jersey College in Princeton—he succeeded. Morse applied for a U.S. patent for his invention in 1837 and set out to exhibit his invention and attract potential patrons (fig. 17.3). In 1843, he succeeded in persuading Congress to award him a grant of $30,000 to develop his system, which included his code for transmitting the alphabet as a sequence of dots

FIGURE 17.3 Samuel Morse posing heroically with his hand resting on his electromagnetic telegraph.

and dashes on a strip of paper. A year later, he transmitted the first telegraphic message ("what hath God wrought") between Baltimore and Washington, DC.

As was the case with Cooke and Wheatstone, the disputes between Morse and his erstwhile partner Leonard Gale, as well as with the Princeton professor (and later secretary of the Smithsonian Institute) Joseph Henry, hinged on the question of just what contribution science had made to this new technology. The debates came about at least partly in the context of attempts to overturn Morse's patent on the grounds that the invention was based on previously known principles of natural philosophy. Both Gale and Henry insisted that the scientific advice they had given Morse was indispensable to the telegraph system's operation. Morse, of course, disagreed saying that the advice given was incidental to the telegraph's successful working. The advice offered to Morse by the two natural philosophers had hinged on the best methods of winding the wire coils around the electromagnets used in the telegraph. Henry had first established his name as a natural philosopher with experiments to determine the best kind of coils to produce different kinds of magnetic effects. He also claimed that he had been using relays of the kind employed by Morse to boost the electrical signals periodically over long distances in his classroom before Morse appropriated it for his telegraph. What both series of disputes show is that despite the apparent unanimity surrounding the telegraph as an example of the application of science to technological progress, deciding just what the details of the contribution might be was fraught with difficulty.

Later in the nineteenth century, tension remained between the new profession of electrical engineering and physics over the question of just who had the right kind of expertise to understand the workings of the telegraph system. Practical telegraph engineers, such as the head of the British Post Office's Telegraph Service, William Henry Preece, argued that it was the experience of men like him, who had long practice in dealing with the peculiarities of electrical systems, that were best suited for dealing with the everyday operations of the telegraph network. Physicists such as Oliver Heaviside, Oliver Lodge, and the American Henry Rowland argued that, on the contrary, it was their knowledge of the theories of electromagnetism developed by the Scottish physicist James Clerk Maxwell that made them the experts on telegraphy (see chap. 4, "The Conservation of Energy"). The debate came to a head in 1888 when the theoreticians took advantage of the German physicist Heinrich Hertz's discovery of electromagnetic waves to argue that the Maxwellian theory of electrical and magnetic fields worked better than the telegraph engineers' robustly empirical view of electricity in a wire as a fluid running through a pipe (Hunt 1991). Again it was at least in part a dispute about the relative role of scientific and technical skills in developing technology. It was a debate made even more important by the increasingly critical role the telegraph played in sustaining late nineteenth-century imperialism (Headrick 1988).

Throughout the nineteenth century it was often difficult to distinguish between the discoverer and the inventor. Even at the end of the century, as the reputation of a figure like Thomas Alva Edison suggests, there was little practical distinction, at least as far as the wider public was concerned.

Edison carefully cultivated a public image for himself as a self-made, self-educated man whose success was the result of his own inventive genius rather than any scientific training (Millard 1990). Behind this image, however, Edison took full advantage of the scientific credentials of the employees beavering away at his Menlo Park laboratories. Edison was nevertheless an icon of individual scientific and inventive genius to the late nineteenth- and early twentieth-century public. He went out of his way to identify himself with his company's inventions as publicly and flamboyantly as possible (Marvin 1988). As the historian of technology Thomas Hughes suggests, however, by the end of the nineteenth century, at least, developing industries of the kind with which Edison was associated, like the rapidly expanding electric power industry, needed a "seamless web" of all kinds of different expertise in order to compete successfully (Hughes 1983). As far as the twentieth century's development of large-scale technological systems was concerned, any distinction between science and technology was simply irrelevant.

Invisible Technicians

Debates about the respective roles of science and technology have as a rule been built around the relationship between head work and hand work. Scientists work with their heads; engineers, technicians, and craftspeople work with their hands. Debates like these have often had an important political dimension as well since, more often than not, they are debates about respective social status. Traditionally—probably since the time of Greek civilization—those who work with their hands have been regarded as socially inferior to those who work with their heads. In early slave-owning societies, such as Greece or Rome, there was a clear social stigma attached to bodily labor of any kind—it was something that slaves did. During the Middle Ages, the view that working with the hands was beneath the dignity of a gentleman, on the one hand, coexisted with a monastic tradition, on the other, that increasingly exalted manual labor as a route toward personal salvation. By the early modern period, a range of positions concerning the dignity of manual labor could be found. It remained the case, however, that gentlemen did not perform manual labor. Given that early modern natural philosophers modeled themselves, as discussed earlier, on gentlemanly norms of behavior, they, too, had an ambivalent attitude toward working with their hands and toward those who did so (see chap. 2, "The Scientific Revolution").

As the historian of science Steven Shapin has pointed out, technicians (those who carried out the manual labor involved in doing experiments) usually remained invisible during the early modern period unless something went wrong. Experiments such as those Robert Boyle carried out with the air pump, for example, required a great deal of technical skills and manual labor if they were to work correctly. The air pumps themselves needed to be constructed by highly skilled craftsmen. Little if any of this behind-the-scenes technical activity appeared in finished published accounts of Boyle's experiments, however. In most cases the reader of such a text could have been forgiven for thinking that Boyle had carried out the experiments entirely on his own. Certainly no technician (or "laborant" as they were usually called during this period) would have been credited with any authorial role in the production of scientific knowledge. Even in rare cases (such as one involving a named assistant, Denis Papin) where Boyle made explicit the fact that a technician had actually done most of the work, the experiment remained firmly Boyle's property. In principle, seventeenth-century advocates of experimentalism argued that natural philosophers themselves should get their hands dirty and carry out even the most degrading manual labor themselves. In practice it is clear that this was rarely done (Shapin 1994).

The ambivalent relationship between natural philosophical experimentation, technical craft skill, and manual labor is clearly illustrated in the career of Boyle's contemporary, Robert Hooke. During his early career, Hooke is known to have himself been one of Robert Boyle's laboratory technicians and was responsible for improving the original version of his air pump. In 1662, following his appointment as curator of experiments at the Royal Society he was, by his own estimation, well on the way to becoming a natural philosopher in his own right. His difficulties in this respect show how hard making the transition from technician to philosopher could be. As far as his employers at the Royal Society were concerned, Hooke remained a mechanic whose task it was to carry out experiments at their instruction rather than conduct his own autonomous researches. His status as someone who worked with his hands (and in particular his status as someone who did so for pay) made it difficult for him to be accepted as a natural philosopher. The ways in which craftsmen and mechanics were expected to behave with regard to their work was simply different to the ways in which natural philosophers were expected to behave. As far as gentlemanly natural philosophers were concerned, there was a question as to what extent a technician could be trusted.

Early nineteenth-century natural philosophers shared some of these concerns about the differences between the ways technicians and scientists might be expected to behave. John Herschel, for example, drew a distinction between natural philosophers' habits of openness and transparency and craftsmen's tendencies to cloak their activities with an air of mystery. Artisans and mechanics could only expect to be regarded as men of science, he argued, if they abandoned their secretive practices and started acting like natural philosophers. Artisans and mechanics, by way of contrast, often regarded natural philosophers' efforts to draw a line between science and technology as illegitimate efforts to deny them the fruits of their labor. Such commentators as Joseph Robertson and Thomas Hodgkin, the founders and editors of the *Mechanics' Magazine* (fig. 17.4), argued that artisans' intimate familiarity with machines and with natural processes gave them unique insights into the operations of nature. Technicians, rather than gentlemen like Herschel, were the real scientists. In the *Mechanics' Magazine*'s view, natural philosophers more often than not made their scientific discoveries simply by stealing artisanal knowledge and claiming it as their own. One of the reasons that the magazine's editors supported the early Mechanics' Institutes movement was because they hoped it would help mechanics develop into men of science in their own right rather than having others exploit their knowledge.

The magazine represented itself as the artisan-inventor's champion against the gentlemen of science. On several occasions the editors mounted public campaigns to protect particular individuals whose rights of credit to a discovery or invention they felt to be under threat by unscrupulous natural philosophers. When the Scottish clockmaker Alexander Bain, for example, claimed that Charles Wheatstone had stolen his ideas for making an electric clock, the *Mechanics' Magazine* came out in his defense. They represented Wheatstone as a straightforward plagiarist who had attempted to take advantage of the prestige of his status as a professor of natural philosophy to deny a working man his due claim to invention. In the case of a similar controversy surrounding electrometallurgy, one side argued that there had been no invention: electrometallurgy was simply the straightforward application of known natural philosophical principles to an industrial process. The *Mechanics' Magazine,* in contrast, was prepared to compare the "discovery" to Newton's discovery of gravity. They certainly did not recognize the kind of distinction between the practices of scientists and mechanics that John Herschel or Charles Babbage suggested. As far as they were concerned it was all just a matter of who had the social prestige to be able to claim what.

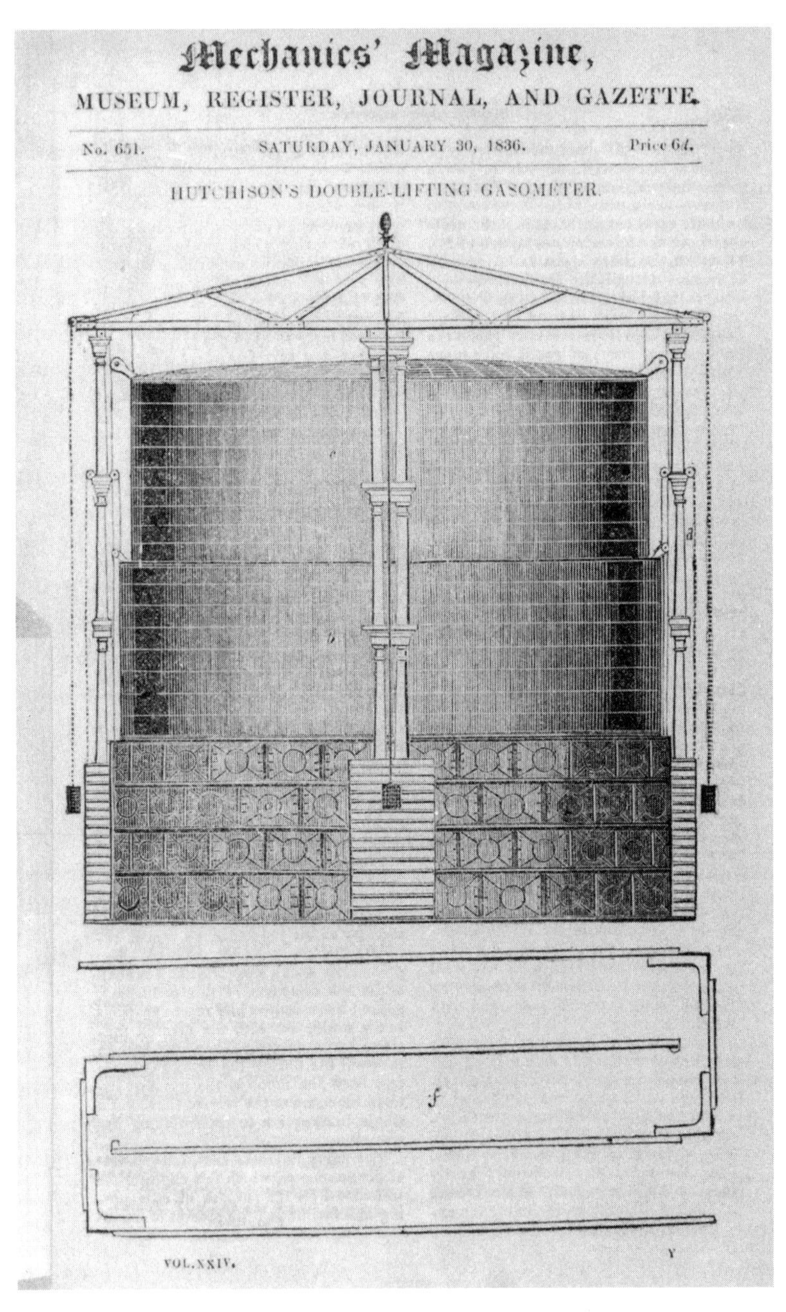
Mechanics' Magazine,
MUSEUM, REGISTER, JOURNAL, AND GAZETTE.
No. 651. SATURDAY, JANUARY 30, 1836. Price 6d.
HUTCHISON'S DOUBLE-LIFTING GASOMETER.
VOL. XXIV. Y

FIGURE 17.4 The first page of an 1830s issue of the *Mechanics' Magazine*. Journals like this played an important role in spreading new technical information to a wide audience.

Babbage or Herschel, conversely, were prepared to argue that not only was science different in principle from the working practices of mechanics and artisans but, also, scientific principles needed to be applied to those practices if sustained economic and technological progress was to happen. In order to make progress a certainty, rather than a mere matter of happenstance, it was essential that the way in which people worked was subjected to detailed and constant scientific scrutiny. The relationship between science and technology from this perspective was very much a hierarchical affair, with science and scientists firmly in control. As scientists throughout the nineteenth century and into the twentieth century argued for state support for science and education, this was the form their argument usually took: science was the only sure source of technological progress and sustaining that progress required the maintenance of a strict hierarchy between science and technology. A similar ethos was at play in new managerial philosophies such as Taylorism and Fordism that appeared at the beginning of the twentieth century. Making work more productive was a matter of applying strict scientific principles. This meant replacing workers' standards and expectations of how to carry out particular tasks with those of scientifically trained managers, just as Herschel had insisted that craftsmen's traditions of secrecy needed to be replaced by scientific transparency.

It should be clear from these examples that there has been more at stake historically in deciding the relationship between science and technology than just philosophy or epistemological nicety. Traditionally in Western societies, those who work with their heads—scientists in this context—have been regarded as culturally superior to those who work with their hands. As we have seen, head work is also often taken to be epistemologically prior to hand work—in other words, it is assumed that mental labor has a higher rank, so to speak, than manual labor. This is what Herschel or Babbage would have had in mind in defining the practical relationship in hierarchical terms as they did. It is also how Boyle might have justified his position with regard to his experimental technicians. That philosophical hierarchy, therefore, carried with it some important cultural and political implications. It is clear that for Boyle, for example, the kind of work one did (or did not do) carried important implications for social status. The same is true for the nineteenth-century case as well. The reason that the editors of the *Mechanics' Magazine* and others argued so strongly that science and technology were ultimately interchangeable activities was because they wanted to reorder the traditional hierarchical distinction. In other words, defining the boundary between science and technology was (and is) ultimately about defining the social position of practitioners as well.

Conclusions

Debates about the nature of science and technology and the proper relationship between them still rage today. One of the things that looking at the history tells us is that there are no right or wrong answers to debates like this. At different times, different people have seen the connections between science and technology in a whole variety of ways. Francis Bacon and other exponents of the New Science of the seventeenth century argued that natural philosophy could, if properly organized, be a source of useful discoveries and inventions. Arguing like this was one way of trying to differentiate their science from that of academic scholastics. As men of the world they wanted their science to matter in the world. In the nineteenth century, advocates of state scientific funding in Britain argued that science was essential to industrial progress. According to this view, there was a straightforward hierarchical relationship between science and technology. Scientists made discoveries and those discoveries could be exploited for economic gain. Only scientific management of the process of invention could guarantee progress according to Charles Babbage. Opponents of state funding for science, such as William Whewell, denied there was any kind of link between science and technology. Neither was necessary for the other's continued progress. As science became both an academic and an industrial profession during the twentieth century, many university scientists regarded the activities of their counterparts in industry as something less than "pure" science.

We have also seen that historians own views about the nature of the relationship between science and technology in the past has often been informed by their perceptions of contemporary issues as well. Such historians of science as George Sarton, Alexandre Koyré, or Herbert Butterfield were anxious to dismiss connections between science and technology because they wanted to defend a particular image of modern science. Like at least some of their academic scientist colleagues, they regarded science as a purely intellectual achievement—something that was the proper concern of humanists not technicians or bureaucrats. They also wanted to defend science from the economic determinism of Marxist historians such as Boris Hessen. Hessen and J. D. Bernal argued that science and technology were intimately related because as Marxists they wanted to show that science was the product of particular economic conditions related to the development of modern capitalism. This kind of view of science as the product of particular historical conditions rather than individual genius was anathema to many intellectual historians who saw science (just as Whewell had) as progressing according to its own internal logic rather than in response to specific cultural developments. Protecting science from accusations of cultural contamination (as they saw it) meant separating it from technology as well. In the contemporary context, with scientists, engineers, and policymakers themselves increasingly calling the boundary between science and technology into question, historians are also developing a renewed concern for the historical relationship between science and technology.

References and Further Reading

Alder, Ken. 1997. *Engineering the Revolution: Arms and Enlightenment in France, 1763–1815*. Princeton, NJ: Princeton University Press.

Ashworth, Will. 1996. "Memory, Efficiency and Symbolic Analysis: Charles Babbage, John Herschel and the Industrial Mind." *Isis* 87:629–53.

Bernal, J. D. 1954. *Science in History*. London: Watts & Co.

Butterfield, Herbert. 1949. *The Origins of Modern Science, 1300–1800*. London: Bell.

Cardwell, Donald. 1957. *The Organisation of Science in England*. London: Heinemann.

———. 1971. *From Watt to Clausius: The Rise of Thermodynamics in the Early Industrial Age*. Ithaca, NY: Cornell University Press.

Headrick, Daniel. 1988. *The Tentacles of Progress: Technology Transfer in the Age of Imperialism, 1850–1840*. Oxford: Oxford University Press.

Hessen, Boris. [1931] 1971. "The Social and Economic Roots of Newton's 'Principia'" In *Science at the Cross Roads*, edited by N. I. Bukharin. London: Frank Cass.

Hughes, Thomas P. 1983. *Networks of Power: Electrification in Western Society, 1880–1930*. Baltimore: Johns Hopkins University Press.

Hunt, Bruce. 1991. *The Maxwellians*. Ithaca NY: Cornell University Press.

Koyré, Alexandre. 1968. *Metaphysics and Measurement*. Cambridge, MA: Harvard University Press.

Latour, Bruno. 1987. *Science in Action: How to Follow Scientists and Engineers through Society*. Milton Keynes: Open University Press.

Marvin, Carolyn. 1988. *When Old Technologies Were New: Thinking about Electric Communication in the Late Nineteenth Century*. Oxford: Oxford University Press.

Merton, Robert K. 1938. *Science, Technology and Society in Seventeenth-Century England*. Bruges: St. Catherine's Press.

Millard, Andre. 1990. *Edison and the Business of Innovation*. Baltimore: Johns Hopkins University Press.

Morus, Iwan Rhys. 1998. *Frankenstein's Children: Electricity, Exhibition and Experiment in Early Nineteenth-Century London*. Princeton, NJ: Princeton University Press.

Sarton, George. 1931. *The History of Science and the New Humanism*. New York: Holt & Co.

Schaffer, Simon. 1994. "Babbage's Intelligence: Calculating Engines and the Factory System." *Critical Inquiry* 21:203–27.

Shapin, Steven. 1994. *A Social History of Truth: Civility and Science in Seventeenth-Century England.* Chicago: University of Chicago Press.
Smith, Crosbie. 1999. *The Science of Energy.* Chicago: University of Chicago Press.
Stewart, Larry. 1992. *The Rise of Public Science: Rhetoric, Technology and Natural Philosophy in Newtonian Britain, 1660–1750.* Cambridge: Cambridge University Press.
Webster, Charles. 1975. *The Great Instauration: Science, Medicine and Reform, 1626–1660.* London: Duckworth, 1975.
Zilsel, Edgar. 1942. "The Sociological Roots of Science." *American Journal of Sociology* 47:245–79.

CHAPTER 18

BIOLOGY AND IDEOLOGY

IN THE MODERN WORLD, WE ARE ALL AWARE that biological knowledge can be applied to humans, but every attempt to create a biologically founded account of human nature has been marked by controversy. The suggestion that our behavior is determined by biological processes has been seen as an assault on human dignity and moral responsibility. If the mind is merely a reflection of physical changes in the brain, then perhaps it is the neurophysiologist, not the philosopher, to whom we should turn for advice on moral and even political issues. And if the brain is a product of natural evolution, then perhaps a study of the evolutionary process would explain why we are programmed to behave as we do or illustrate the best way to achieve social progress. Of necessity, these questions raise not only moral and theological questions but political or ideological ones, too. The scientist or philosopher may argue that the brain is the organ of the mind, but it is the ideologue who uses that assertion to justify social actions such as limiting the reproduction of persons alleged to have deficient mental powers or dangerous instincts. Liberals have come to see almost any attempt to apply biology to human nature as politically suspect, and they often use history to illustrate the dangers they perceive. By warning of the legacy of "social Darwinism" or of early efforts to provide scientific support for racism, they seek to brand modern forms of biological determinism as products of a conservative political agenda. History thus becomes a battleground for the competing ideologies, and historians of science have to work within the minefields that have been laid to defend modern positions.

Historians have focused a great deal of attention onto key areas in which biology has been applied to social issues and are well aware of the potentially controversial nature of what they do (for surveys, see Bowler [1993]; Smith [1997]). There is a vast literature on the attempts to show that human nature is dictated by the structure of the brain, by inherited limitations of intelligence or behavior patterns, or by the nature of the evolutionary process. Some of the older contributions to this literature may still attempt to present a conventional image of science as the source of objective, value-free knowledge, conceding only that the ideas and insights thus produced may be distorted by those seeking to apply them to the real world. On such a model, the Darwinian theory is the product of good science, but social Darwinism is a perverted application of concepts derived from it to social affairs. But more recently, historians have begun to interpret the scientific debates themselves in ideological terms. The old assumption that science offers objective knowledge has broken down in many areas, but in none more evidently than in the area of social Darwinism, where the human implications of scientific knowledge are so immediate. We have become increasingly sure that what passed for scientific knowledge at various points in the past was influenced (although not necessarily determined) by the social values of the time. As one influential voice in this movement has asserted, "Darwinism *is* social" (Young 1985a). It is not a question of Darwinism being applied to society, but of social images being built into the very fabric of science itself. The rise and fall of phrenology, an early theory of cerebral localization, was used as a case study by a pioneering member of the "Edinburgh school"—the most active proponents of the claim that scientific knowledge is socially constructed (Shapin 1979). Scientists often reject this claim as a challenge to their objectivity, but if history shows that earlier efforts to apply biology to the study of human nature were influenced by social values, the lesson should not be lost on those engaged in modern debates.

In this chapter, we will focus on topics that have attracted particular attention from historians, starting with the cerebral localization of mental functions. We shall also look at the complex area of social Darwinism, noting the importance of nonselectionist ideas of evolution in promoting social values often labeled as "Darwinian." Finally, we shall focus on theories alleging biological differences between the races and other applications of genetic determinism loosely grouped under the term "eugenics" (Francis Galton's term for a selective breeding program for the human species). But these areas are not as distinct as they sometimes appear. All depend on the assumption that the brain controls behavior, although this is often forgotten when attention focuses on the evolutionary origins of particular behavior patterns. Alleged mental differences between races are one manifestation of the more general claim that human character is controlled by heredity and cannot be modified by learning. Determinism itself often rests on assumptions about the role played by evolution in shaping the characters transmitted by heredity. The debate over the relative powers of "nature" and "nurture" in determining behavior raises a wide range of issues in the relationship between the biological and the social sciences. Modern theories such as sociobiology may thus combine influences from sources with separate origins in the development of biology. It may be worth noting that paleoanthropology, the science of human origins, has remained largely untouched by historians of science (an exception is Bowler [1986]), although the palaeoanthropologists themselves are aware of the history of their discipline and of the extent to which history reveals a tendency for scientific thinking to be influenced by prevailing social values (e.g., Lewin 1987).

Mind and Brain

The materialists of the eighteenth-century Enlightenment challenged the orthodox notion of the soul by declaring that the human mind was a by-product of the operations of the brain and nervous system. If Descartes could treat animals as merely complex machines, why should human beings be any different? Materialists such as J. O. de La Mettrie and Denis Diderot argued that changes in the brain, for instance, during illness, cause corresponding changes in the mind. A person with jaundice actually sees everything colored yellow. But despite their use of medical and other evidence, the materialists made no attempt to develop a detailed science of how the brain operates. Their program was developed more at the level of philosophy, although there was also a pronounced social agenda lying behind their attack on traditional religious beliefs—the Church was strongly identified with the political regime in prerevolutionary France.

A more focused attack on the notion that the mind exists on a purely spiritual level came in the early nineteenth century with the emergence of phrenology (Cooter 1984; Shapin 1979; Young 1970). This movement was pioneered by Franz Josef Gall and Johann Gaspar Spurzheim but attracted particular attention in Britain. Based on studies of cerebral anatomy and observed behavior, Gall and Spurzheim postulated a series of distinct mental functions, each located in a particular area of the brain. Individual behavior was, in effect, determined by the structure of brain—which was assumed to be detectable from the external form of the skull. So a person's character could be "read off" from a study of their head (fig. 18.1).

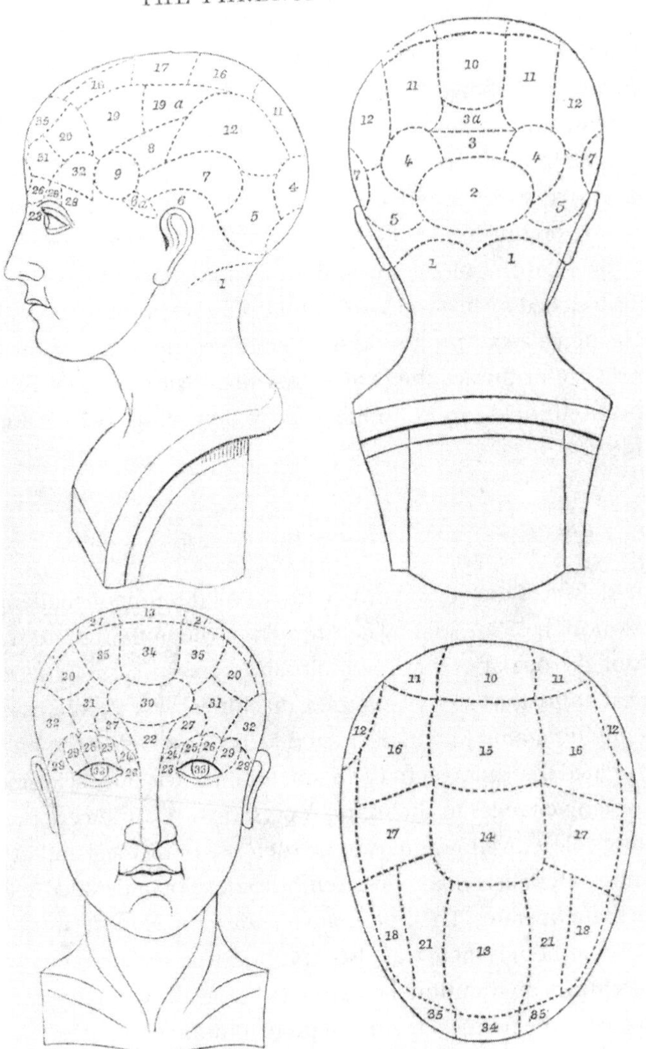

FIGURE 18.1 Phrenological head, the frontispiece to George Combe's *Elements of Phrenology* (Edinburgh, 1841). The head is divided into segments, each labeled with a particular mental faculty thought to be controlled by the area of the brain immediately below that section of the skull. The phrenologist "read" a personality by feeling the contours of the skull to see which areas had bumps showing that the underlying brain was well developed. Later critics dismissed phrenology as a pseudoscience because the skull does not, in fact, reflect the detailed contours of the brain.

Phrenology became popular in the 1820s and 1830s, despite the criticism leveled against it by both philosophers and anatomists. In Britain, the champion of phrenology, George Combe, linked it to a reformist social policy based on the claim that people could better control their lives if they knew their mental strengths and weaknesses. Combe's *Constitution of Man* (1828) was one of the early nineteenth century's bestselling books.

Phrenology is often dismissed as a pseudoscience because, as anatomists were quite right to point out, the structure of the brain is not reflected in the shape of the skull. Historians now see this easy dismissal as a product of hindsight that ignores the fact that phrenology's more fundamental claims were eventually endorsed by orthodox science. The study of cerebral localization later in the nineteenth century was able to show that some mental functions take place in certain regions of the brain because damage to that region affects the corresponding function. Under the circumstances, we need to ask a more sophisticated question: Who decides what is to be counted as scientific knowledge? Shapin (1979) and Cooter (1984) show that phrenology was accepted by those who gained from the reformist social philosophy linked to it by Combe and others. It was rejected by conservative thinkers who preferred the traditional view of the human soul as distinct from the body. Phrenology influenced many leading thinkers, including some who contributed to later developments in cerebral anatomy. Its initial elimination from academic science tells us more about the social processes that shape the attitudes of the scientific community than it does about the objective testing of theories.

Developments in neurophysiology eventually confirmed that some mental functions depend on the proper functioning of a particular part of the brain. In 1861, Paul Broca identified an area that, if damaged by a stroke, resulted in the loss of the ability to speak. This work was continued in the 1870s by David Ferrier and others. Ferrier had been influenced by the philosopher Herbert Spencer, whose 1855 *Principles of Psychology* adopted an evolutionary view of mental capacities and used this to argue that human nature adapts itself to changes in society. For Spencer, the individual mind was preshaped by the experiences of its ancestors: learned habits became instinctive behavior patterns transmitted by heredity. Spencer's psychology depended on the Lamarckian theory of the inheritance of acquired characteristics, but his assumption that learned habits could be transmitted depended on the belief that habits are determined by structures built up in the brain and transmitted by biological inheritance. Spencer's evolutionary psychology was also linked to his so-called social Darwinism (discussed below).

Ferrier's work was later extended into a comprehensive account of the actions of the nervous system by Sir Charles Sherrington. But Sherrington avoided discussion of mental states and thus kept neurophysiology distinct from psychology, which possibly held back the latter's development as a science in Britain (Smith 1992). A far greater impact was made by exponents of scientific naturalism, such as T. H. Huxley and John Tyndall, who argued that mental activity was merely a by-product of the physical activity of the brain. While accepting that the mental world could not be reduced to the physical, they nevertheless insisted that mind could not exert a controlling influence on the physical world. In a notorious address given in Belfast in 1874, Tyndall declared that science sought to explain everything in naturalistic terms, including the mind, thereby marginalizing religion. Twentieth-century developments in cerebral localization, confirming the real but very complex nature of the relationship between mind and brain, have gone largely unrecorded by historians, although they are now the center of intense public attention.

Phrenology also played a role in the evolution debates. Evolutionists naturally welcomed the implication that as animals acquired bigger brains, their mental powers were enhanced. This link was made explicit in *Vestiges of the Natural History of Creation,* published anonymously in 1844 by the popular writer Robert Chambers (Secord 2000). By the time Darwin had popularized the theory of evolution in the 1860s, it was taken for granted by many that the size of the brain was roughly proportional to the level of an animal's mental development. Darwin could exploit the obvious fact that the brain did indeed get bigger in the course of the history of life on earth, as displayed in the fossil record. But the link between evolution and cerebral localization had far-reaching implications when it was applied to the evolution of the human species itself.

Physical Anthropology and Race Theory

As early as the seventeenth century, anatomists such as Petrus Camper had compared the structure of human and ape bodies and had also argued that nonwhite races formed a kind of intermediate stage between them (Greene 1959). Camper defined the "facial angle" between the horizontal and a line joining the chin, nose, and forehead. Those individuals with a smaller facial angle had a more receding forehead, a classic sign of lower intelligence according to popular prejudices. Apes had a very small facial angle—but Camper and other physical anthropologists also portrayed nonwhite races as having the angle in between that of the ape and the white race. By the

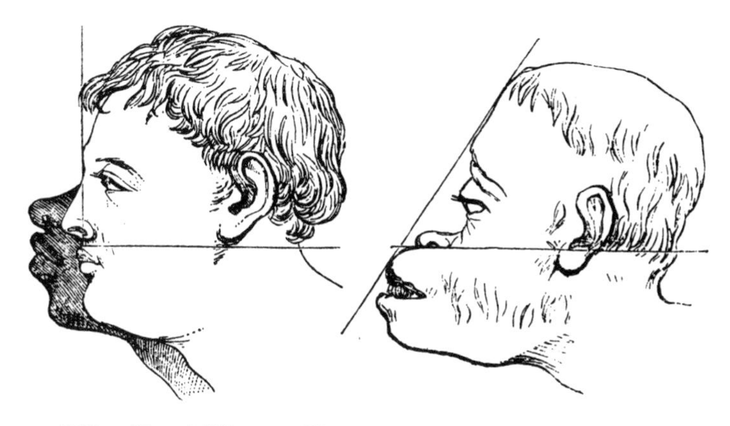

[*Profile of Negro, European, and Oran Outan.*]

FIGURE 18.2 The facial angle for a European, a Negro, and an orangutan, from Robert Knox's *The Races of Men* (London, 1851), 404. The facial angle is defined by the line from the forehead to the mouth and the horizontal, and a more acute angle implies a receding forehead, colloquially associated with a small brain and hence low intelligence. Knox clearly wishes his readers to believe that the Negro is intermediate in brain size, and hence in intellect, between the European and the ape.

end of the eighteenth century, anthropologists such as J. F. Blumenbach were dividing the human species into distinct races based on physical characters such as skull shape (Blumenbach had a famous collection of skulls from all over the world). All too often, these descriptions were manipulated to make it seem that the colored races were inferior to the whites (fig. 18.2). Phrenology only served to drive the point home: if the mind were the product of the brain, then individuals with bigger brains should have more intelligence. From this it was only a short step to arguing that some races had larger skulls than others and, hence, a higher level of intelligence.

In the early nineteenth century, physical anthropologists, determined to show that nonwhite races were less intelligent than whites, began to use craniometry (the measurement of cranial capacity) as a means of arguing their case (Gould 1981; Stanton 1960). Samuel George Morton used a volumetric technique, measuring the capacity of skulls with birdseed or lead shot. He claimed to find evidence that whites had the largest skulls—Gould notes how easy it would be for an unconscious bias to effect the results of such crude measurements. Broca, too, applied craniometry to physical anthropology and became convinced that the human race was divided into several distinct species, each with different levels of mental power. He founded an anthropological society in Paris dedicated to promoting this view of innate racial differences. In Britain, similar views were expounded by the anatomist Robert Knox, who had been discredited for buying corpses for dissection from the grave robbers—and murderers—Burke and Hare. Knox focused on what he perceived to be innate mental as well as physical differences between the races. In his *Races of Man,* first published in 1850, Knox declared: "With me race, or hereditary descent, is everything; it stamps the man." (1862, 6). He was particularly scathing about the character of both the black races and the Irish. Knox's disciple James Hunt soon founded a society in London with the same aims as Broca's group in Paris. By the time Darwin popularized the theory of evolution, it could be taken almost for granted that the "lower" races were relics of earlier stages in humankind's progress, their primitive character confirmed by their smaller brains and less highly developed intellectual powers. Darwin included figures allegedly confirming this in his *Descent of Man* (1871). This kind of physical anthropology continued to flourish through into the early twentieth century, often linked to evolution theory via the assumption that the lower races were relics of earlier stages in human evolution (see below and see also Haller [1975]; Stepan [1982]). These ideas have since been largely purged from science, at least on the surface, although their legacy continues to haunt popular debates.

A leading exponent of measurement applied to living human skulls was Francis Galton, the founder of the eugenics movement (discussed below; see fig. 18.3). Galton measured skulls as part of an effort to distinguish racial types—but he also introduced the systematic measurement of mental powers by testing large numbers of subjects. Early twentieth-century applications of intelligence testing, which were also held to confirm the inferior mental powers of nonwhite races, were founded on similar techniques of mass testing. By using questions that presupposed knowledge of a middle-class lifestyle, the tests used in America made it difficult for blacks or recent immigrants to show their full potential (Gould 1981).

Cultural and Biological Progress

The theory of evolution had an immense effect on the Victorians' view of human nature and society. Darwin avoided discussion of human origins in his *Origin of Species* because he realized how controversial the topic would be, although T. H. Huxley soon established the closeness of the relationship between humans and apes, especially in the structure of the brain. But far more than anatomical relationships were at stake. Did the relative increase

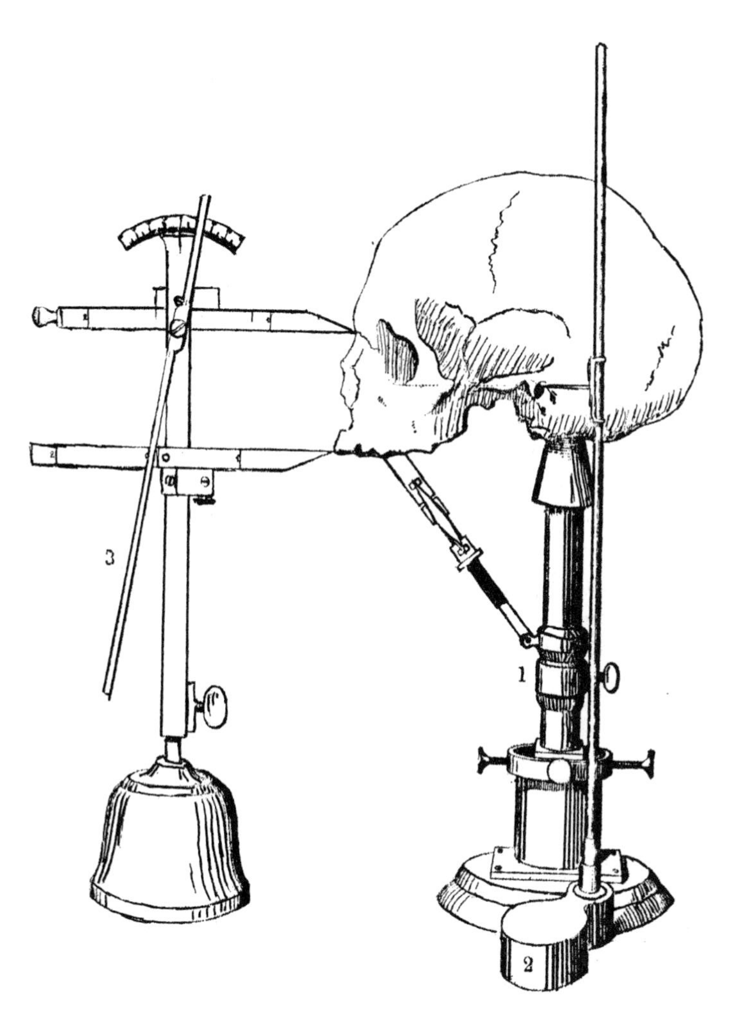

FIGURE 18.3 Anthropometric equipment for measuring the skull, including the facial angle, from Johannes Ranke, *Der Mensche* (Leipzig and Vienna, 1894), 1:393.

in the size of the human brain explain the emergence of the human mind, with the rational and moral powers that were once thought to distinguish us from the brutes? An evolutionary perspective on the mind had been developed, even before Darwin published, by the philosopher Herbert Spencer. Archaeologists and anthropologists were also developing the idea that culture and society had progressed from primitive origins. By the time Darwin published his *Descent of Man* in 1871, he could draw on a number of studies that had begun to explore the implications of evolutionism for the emergence of the human mind and the development of society. The later nineteenth century saw a flowering of interest in evolutionary models within the human sciences. Some of these models stressed the role of the struggle for existence as the motor of progress and have been widely labeled as "social Darwinism." But some evolutionary models contained elements that were not derived directly from Darwinism, and we must first explore the much more pervasive influence of the progressionist view of evolution.

Almost all the models of mental and social evolution assumed that development consisted in the ascent of a scale of increasing maturity. Such a model was proposed independently by late nineteenth-century anthropologists trying to understand the diversity of cultures and societies they saw in different parts of the world (Bowler 1989). Although histories of anthropology once assumed that this evolutionary perspective was stimulated by the Darwinian revolution, modern studies tend to see the two developments as parallel manifestations of the same cultural values. Evolutionary anthropologists such as Edward B. Tylor in Britain and Lewis H. Morgan in America assumed that modern "savages" were relics of the stage of cultural development that the ancestors of the white race had passed through in prehistoric times. Their inspiration lay in the new discoveries by archaeologists that, from the 1860s onward, confirmed the vast antiquity of the human race and created the notion of a primitive "Stone Age." The geologist Charles Lyell summed up this evidence in his *Antiquity of Man* (1863). The anthropologists assigned all living cultures a position on a scale of development, starting with Stone Age savages and culminating in modern, industrial civilization. Cultural differences were explained not by divergent evolution but as differences in the level of development along a single scale. At first the anthropologists resisted the claim that the culturally more "primitive" peoples were mentally inferior to the white race but the emergence of Darwinism made it increasingly difficult for them to separate mental from cultural development (see chap. 13).

Even before Darwin published, Herbert Spencer's philosophy of evolution firmly linked mental with cultural and social development (Richards 1987). Spencer's psychology stressed that there was no universal "human nature"—the human mind was shaped by its social environment, and the more stimulating the environment, the greater the level of individual mental development. Conversely, the greater the level of individual intelligence, the faster society would progress, creating a feedback loop between mental and social evolution. On this model it was inevitable that those races that preserved a primitive level of technology (assumed to mark a primitive level of social structure), must also be stuck at a lower stage of mental evolution. Savages were biological as well as cultural relics of the past, preserving a level of mentality little above the apes.

Darwin had adopted a materialist view of the mind from the start of his evolutionary researches. He was particularly interested in the origin of instincts, treating these as behavior patterns that had become imprinted on the brain by the process of evolution. Spencer adopted the Lamarckian view that learned habits could be transformed into hereditary instincts by the inheritance of acquired characters. But Darwin realized that natural selection could also modify instincts, as long as there was some variation within the behavior pattern. In the *Descent of Man,* he explained the origin of social instincts via both Lamarckism and the process of group selection (competition in which the groups with the strongest social instincts survive). For Darwin, it was human efforts to rationalize the instincts governing our social interactions that were the basis of all ethical systems.

Darwin accepted that, in the long run, evolution had steadily increased the level of animal intelligence—although he knew that many branches of the tree of life did not progress toward higher levels of development. He suggested that humans had developed a level of intelligence so much higher than the apes because our ancestors stood upright when they moved out from the forests to the open plains of Africa and had thus begun to use their hands for making primitive tools. The majority of evolutionists, however, had little interest in the possibility that there might have been a crucial turning point separating the paths of human and ape evolution. They sketched in an elaborate but totally hypothetical scale of mental development running through the animal kingdom up to humankind and, then, assumed that evolution would almost inevitably have advanced steadily up the scale (Richards 1987). This approach can be seen in the work of George John Romanes, who became Darwin's leading disciple in the area of mental evolution. In America, evolutionary models of the mind were proposed by James Mark Baldwin and G. Stanley Hall.

An important element within late nineteenth-century developmental theories was the concept of recapitulation: the belief that the evolutionary history of the species is recapitulated in the development of the individual organism (Gould 1977). In biology this was promoted by the German Darwinist Ernst Haeckel and by American neo-Lamarckians such as Edward Drinker Cope. The recapitulation theory offered a model of evolution in which progress toward the goal of increasing maturity seemed inevitable: evolution simply advanced up the scale, as did the embryo. Evolutionary psychologists were convinced that the development of the individual human mind passed through the phases of mental evolution that had marked the evolution of the animal kingdom. Romanes explicitly identified the mental capacity of the child at certain ages with various levels of animal mentality. This model encouraged the belief that savage races, assumed to be relics of the earlier stages in the advance from the apes, had minds equivalent to those of white children and little better than apes. In Italy, Cesare Lombroso proposed a system of "criminal anthropology" in which criminals had minds equivalent to those of savages—they, too, could be dismissed as relics of earlier stages in human evolution.

It was through the recapitulation theory that evolutionism had an impact on what soon became the most controversial view of the human mind: Sigmund Freud's analytical psychology (Sulloway 1979). Having begun by studying the nervous system, Freud abandoned this approach to the study of mental functions and began to treat mental illness as the product of purely psychological tensions. He visualized an unconscious level to the mind that was a relic of the animal stages of evolution, driven mostly by sexual impulses. But where earlier evolutionists had seen the later and hence more highly evolved mental functions as being in control of the whole personality, Freud saw the conscious mind as struggling to cope with the socially unacceptable drives arising from the unconscious. Here the optimistic progressionism of the nineteenth century gave way to a much harsher vision of the human personality that haunted twentieth-century thought. Freud insisted that he had rejected the role of biology, yet his theory was based on the notion that the mind consisted of distinct evolutionary layers. His anxiety to distance himself from his Darwinian roots parallels a more general trend in the human sciences of the early twentieth century whereby experimental psychology, sociology, and cultural anthropology all sought to proclaim their independence by insisting that biology did not predetermine human behavior (see chap. 13; Cravens 1978).

Social Darwinism

What was the driving force of mental and social evolution? In Darwin's theory of natural selection, change results from the elimination of the unfit in a struggle for existence, leaving the best adapted individuals to survive and breed. There were certainly many "social Darwinists" who proclaimed that struggle was the motor of progress. But to suppose that Darwin's theory was transferred from biology to society is—as far as some historians are concerned—to put the cart before the horse. We know that Darwin himself was directly influenced by Thomas Malthus's principle of population expansion, a classic product of free-enterprise economic thinking (see chap. 6, "The Darwinian Revolution"). This leads historians such as Robert M. Young (1985a, 1985b) to argue that ideological values were built into the heart of scientific evolutionism. Darwin's thinking certainly reflected the individualist social philosophy of the time, although he went far beyond Malthus in his willingness to see struggle as a creative force. But if the scientific theory itself reflected social values, it is hardly surprising that it was then used to legitimize the ideology on which it was built by arguing that society should be based on the "natural" principle of competition.

Much has been written on the vogue of "social Darwinism" in the late nineteenth century, with Spencer presented as the leading advocate of the claim that the free-enterprise system generated progress through struggle. Successful capitalists certainly justified the system by appealing to the metaphor of the survival of the fittest. The conventional view—supported in the classic study by Hofstadter (1955) and more recently by Hawkins (1997)—is that this claim was inspired by Darwinism. Some historians have urged caution, however, noting that the term "social Darwinism" was introduced by writers who themselves opposed the view that struggle should play a role in human affairs. It is also clear that there were many different social policies that could be justified on allegedly Darwinian principles (Bannister 1979; Jones 1980). Widespread use of the term "social Darwinism" by the movement's critics has highlighted the involvement of Darwin's theory, and there is no doubt that the selection theory was part of this ideology. But natural selection was by no means the only biological mechanism exploited in this way. Other theories, especially Lamarckism, were caught up in the enthusiasm for progress by struggle. Social Darwinism may be a convenient label for this whole movement, but it can be misleading if it is thought to imply that what modern biologists single out as Darwin's most important insight was the main inspiration of late nineteenth-century social thought.

The most widely discussed form of social Darwinism is the theory's application as a means of justifying the free-enterprise system lying at the heart of nineteenth-century capitalism. The parallel seems obvious enough: if natural evolution (assumed to be progressive) works by the selection of the fittest individuals in a struggle for existence, social progress will be ensured if a similar struggle is allowed to select out the best human individuals in every generation. The architect of this form of social Darwinism is supposed to have been Herbert Spencer, whose evolutionary philosophy gained wide popularity in Britain and especially in America. Many of the most successful, and most ruthless, of the American capitalists regarded themselves as his followers.

Spencer was certainly an advocate for unrestrained individualism as, since it was he, not Darwin, who coined the term "survival of the fittest" to describe natural selection, the link with biological Darwinism seems obvious enough. But there are several problems with an analysis that assumes that Darwin's selection theory (via Spencer) served to promote capitalism. For a start, we have seen that natural selection was not very popular among late nineteenth-century biologists—so why should it have been taken as a scientific endorsement for a social policy? Although he certainly supported a role for natural selection, Spencer himself was a lifelong Lamarckian who defended that theory actively when it came under fire from scientific critics. This theory, too, has parallels in his social evolutionism: for Spencer, the role of competition was not just to weed out the unfit but to force everyone to become fitter. When stimulated by the challenge of competition, most people learn how to improve themselves (although a few unfortunates may be unable to benefit and pay the penalty). And if the Lamarckian theory of the inheritance of acquired characteristics is valid, these self-improvements will be passed on to the next generation to benefit the race as a whole. Few supporters of free enterprise, then or now, argue that people must die to make social progress work—the argument has always been that state benefits make people lazy and unwilling to learn new skills to replace out-of-date ones. Thus, much of what has been described as social Darwinism may actually be a form of social Lamarckism. More appropriately, it might be better to regard Darwinism and Spencerian Lamarckism as parallel reflections of capitalist ideology in science. Even so, it may have been the Lamarckian component that gained most popularity, at least in the 1860s and 1870s when Spencer was at his most influential.

Partly as a result of the exaggerated emphasis placed on the Darwinian component of Spencer's thought, Lamarckism has gained a reputation as a theory that could more easily be used by the opponents of ruthless social policies. Lamarckians such as the American Lester Frank Ward believed that their theory offered a humane route to social progress: if children were taught appropriate social behavior, the resulting habits would eventually become inherited instincts. The human race itself would thus become more socialized. This was certainly one possible way of exploiting the theory, but the Lamarckian role in Spencer's endorsement of free enterprise should not be ignored. For Spencer, the "school of life" would always be more effective than anything provided by the state, because its lessons were backed up by the suffering that came as a penalty for failure. We should also note that Lamarckism played a major role in promoting the recapitulation theory, with its strong emphasis on the inferiority of "primitive" mentalities. The racial hierarchy that so many late nineteenth-century thinkers took for granted was based on a progressionist view of evolution that—as with Spencerianism—owed at least as much to Lamarckism as it did to Darwinism. Not that Darwin escaped the fascination with the idea of progress that was characteristic of his age. But he knew that, in most cases of evolution, "fitness" must be defined solely in terms of adaptation to the local environment. This does not imply the existence of an absolute scale of physical, mental, or cultural perfection.

The reminder that evolutionism was also applied to the race question points us toward another complexity in the nature of social Darwinism: the possibility of applying the concept of struggle at other levels besides that of individual competition within the same population. To the extent that late nineteenth-century thinkers accepted a role for natural selection, it was largely negative. They did not believe that it could create new forms of life—that was reserved for Lamarckism and other more positive mechanisms—but it might be able to weed out the less successful products of the drive toward progress. If evolution has produced several different forms of humanity (the supposedly distinct racial types), then these might be engaged in a competition among themselves to decide which was the most advanced. And the penalty for the losers in this competition would be extinction. There were few European or American scientists who doubted the superiority of the white race. The "lower" races were regarded as living fossils, relics of earlier stages in the ascent of humankind preserved in parts of the world that had been protected hitherto from the encroachment of the superior type. Now that the triumphant whites were colonizing the whole globe, the inferior forms of humanity must be swept aside in a racial struggle for existence. As the nineteenth century moved toward the age of imperialism, the Darwinian theory could be called in to justify the conquest and even the extermination of the native populations of the territories the whites coveted around the world. The Darwinist and imperialist Karl Pearson wrote that no one should regret "that a capable and stalwart race of white men should replace a dark-skinned tribe that can neither utilize its land for the full benefit of mankind, nor contribute its quota to the common stock of human knowledge" (Pearson 1900, 369). Only in the tropics would the better-adapted black peoples survive, to be ruled by their natural superiors in the new world order. By the early twentieth century, the decimation and potential elimination of the natives of America and

Australia was being compared with the wiping out of the Neanderthals by the Stone Age ancestors of modern humans—it was an unfortunate but necessary consequence of progressive evolution (Bowler 1986).

Even the rivalries between the European powers could be seen as a struggle for existence played out through the race for world domination. As early as 1872, the British political writer Walter Bagehot applied the logic of natural selection to national rivalries in his *Physics and Politics*. His message was that anything that supported the authority of the state was important in providing the national cohesiveness that would resist foreign threats. As international hostilities increased toward the end of the century, talk of a war to determine who should dominate Europe became commonplace (Crook 1994). Military writers in Germany insisted that a war to demonstrate the superiority of German culture was justified—and probably necessary. The inevitable outcome of these rivalries was the First World War, and when the American biologist Vernon Kellogg toured the German lines in Belgium he found that the officer corps was dominated by this ideology of nationalist social Darwinism. Ernst Haeckel's evolutionary philosophy played a key role here, and it has also been argued that Haeckel influenced the development of Nazi ideology in the next generation (Gasman 1971). This is a controversial claim, partly because Haeckel was articulating very widely held prejudices of the time, shared by many who were not Darwinians. He certainly endorsed the race hierarchy and envisaged a struggle between the races, but like Spencer, his evolutionism was as much Lamarckian as Darwinian.

The ideology of national competition was diametrically opposed to Spencerianism, which has been widely portrayed as the most fundamental form of social Darwinism. Spencer hated militarism and nationalism, regarding them as outdated relics of the feudal era in social evolution. They promoted an ideology of state control in the face of external threats, reversing Spencer's emphasis on the free competition between individuals in a society with only minimal government. The fact that mutually hostile ideologies could each justify themselves by appealing to different aspects of the Darwinian theory shows that social Darwinism was certainly not a unified movement and makes it difficult to see the selection theory as an active participant in the development of either society or political thought. Both the general idea of evolution and the specific theories of how evolution worked—Darwinian and Lamarckian—provided a rich vein of metaphor and rhetoric to be mined by the political writers of the time. And no doubt the biological theories of Darwin, Spencer, and many others were shaped by inspirations derived from cultural values. But to regard the various forms of social Darwinism prevalent in the nineteenth century as by-products of Darwin's selection theory is to credit the scientific community with far too much influence—it was the scientists who reflected the ideology of the time, and their ideas at best served to justify policies that were already being enacted. We must also be careful to note that the Darwinism prevalent in the late nineteenth century offered a general vision of progressive evolution in which natural selection played only a limited role. There was, in fact, a major transition still to take place at the intersection of biology and ideas about human nature, corresponding to the emergence of a far more rigid view of how heredity determines character.

Heredity and Genetic Determinism

Those nineteenth-century thinkers who argued that a person's level of ability was predetermined by his or her racial origin were advocating a form of biological or hereditary determinism. Liberal thinkers claimed in response that social background and education played the major role in shaping personality and ability, whatever a person's racial origin. This difference of opinion fed into an apparently never-ending debate over the relative significance of "nature" (heredity) and "nurture" (upbringing) in the determination of character. There was a major shift of emphasis toward heredity in the late nineteenth century. People had always been unwilling to admit that there was insanity "in the family." Now it was argued that all individual differences are predetermined by ancestry. Levels of ability, and perhaps even temperament, were transmitted by inheritance from parent to offspring, so anyone born with a "poor" heritage was doomed to inferiority whatever their education and upbringing. This development within social opinion coincided with a focusing of biologists' attention onto the topic of heredity, leading historians to ask about the role played by ideology in shaping scientific priorities, if not scientific knowledge itself.

Scientific support for hereditarianism was pioneered by Darwin's cousin, Francis Galton. While exploring Africa, Galton became convinced of the inferiority of the black races. He then began to argue that the hereditarian principle was applicable even among whites: intelligent people have intelligent children, and by implication stupid people have stupid children. His 1869 *Hereditary Genius* provided the scientific basis for a campaign to avert the social dangers that might flow from ignoring this alleged biological inequality. Galton argued that in modern society the "unfit" are no longer removed by natural selection because they can survive in the slums of the great cities. Here they breed rapidly and increase the level of poor heredity in the population as a whole. Galton coined the name "eugenics" for his program to improve the character of the race by restricting the breeding of the unfit and encouraging the fit to have more children (Kevles 1985; Mackenzie 1982; Searle 1976).

By the early twentieth century, Galton had become the figurehead of a powerful social movement. Eugenics flourished in most developed countries, buoyed up by fear of racial degeneration and enthusiasm for the idea that science offered the route to an efficiently managed society. In 1901 Galton's disciple Karl Pearson warned of the "degeneration" of the British population as revealed by the poor quality of the recruits for the army during the Boer War in South Africa (which the British won, but only at great cost). He argued that a eugenics program was necessary to improve the race and defend the empire. As noted above, he took it for granted that the white race was superior to the natives of the parts of the world that had been colonized. Support for eugenics coincided with the emergence of heredity as a major focus of biologists' attention. Pearson developed statistical techniques for evaluating the effect of selection on hereditary characters within a population, and the "rediscovery" of Mendel's laws came in 1900 (see chap. 6, "The Darwinian Revolution," and chap. 8, "Genetics"). Historians have linked these scientific developments with the changes in social opinion, and the most radical interpretations have argued that the structure of theories of heredity was determined by the uses to which they were put in supporting eugenics. As with the race question, it is comparatively easy to show that social pressures have focused scientists' attention onto particular issues but less easy to prove that the theories themselves reflect social values. The fact that rival theories were used to endorse the same social attitudes undermines the determinist interpretation, leaving room for the possibility that scientific questions shaped the details of thinking within a generally hereditarian framework.

Pearson supported Darwinian natural selection, and Darwinism has thus been seen as a model for eugenics: natural selection is replaced by artificial selection in the human population. Pearson laid the foundations of many modern statistical techniques, and his strong support for eugenics has led Donald Mackenzie (1982) to argue that those techniques were designed to highlight the effects of heredity in human society. A more recent study of Pearson's statistics suggests, however, that many of his techniques were motivated by biological problems; when he turned to human heredity, he introduced different methods of analysis (Magnello 1999). The link to Darwinism must thus be treated with care: Galton himself stressed the negative effects of removing selection pressure but did not believe that nat-

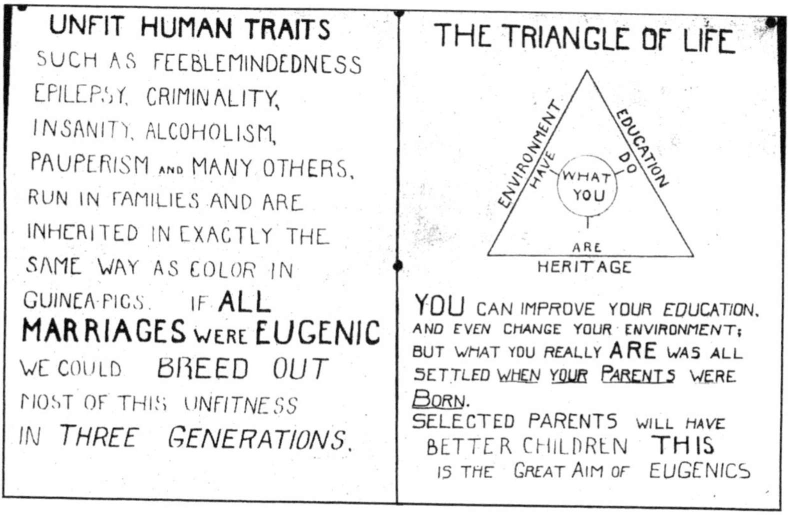

FIGURE 18.4 Eugenics displays at the Kansas Free Fair, 1929. Such displays were used to convince people that many physical and mental defects were inherited as unit characters and could thus be eliminated from the population by preventing individuals with the defects from breeding.

ural selection was the source of new characters in evolution. One of the most extreme British eugenists—he called for the compulsory sterilization of the Irish—was E. W. MacBride, one of the last defenders of Lamarckism.

The most characteristic product of the new wave of interest in heredity was, of course, Mendelian genetics. Although Gregor Mendel's laws of heredity had been published in 1865, they were largely ignored until rediscovered in 1900 by Hugo De Vries and Carl Correns. Soon Mendelism offered a powerful rival to Galton and Pearson's nonparticulate model of heredity, showing how rival forms of science could each be stimulated by the same social pressures. In America, especially, genetics was linked to the eugenics program via oversimplified assumptions about the genetic basis of human characteristics (Haller 1963). Every physical and psychological character was thought to be the product of a single gene (see fig.18.4). Charles Benedict Davenport argued that feeblemindedness, for instance, was a single Mendelian character that could easily be selected out of the population by sterilizing the carriers of the gene. There was no automatic link between Mendelism and eugenics, however. The leading British geneticist, William Bateson, did not support eugenics, while Pearson—Bateson's great scientific rival—distrusted genetics because he thought it an oversimplified theory that might undermine the credibility of eugenics.

The exact way in which enthusiasm for hereditarian thinking expressed itself in science thus depended on the circumstances of the scientists involved. One of the pioneers of population genetics, Ronald Aylmer Fisher, was deeply influenced by eugenics, although his work helped to show how difficult it would be to eliminate harmful genes from the human population. Similar work on the selection theory was done by J. B. S. Haldane, a socialist who was suspicious of the eugenics movement's efforts to limit the variability of the human population.

There were also important differences between the concerns expressed by eugenicists in different countries. In America, the movement became closely involved with opposition to the immigration of "inferior" races that would spread their characters into the general population. There were close links between American and German race scientists, links that were maintained after the Nazis came to power. In Britain, race was less of an issue (except perhaps for MacBride's fulminations against the Irish). Significantly, although some biologists endorsed eugenics and race theory, social scientists and anthropologists had abandoned the hereditarian position in the early twentieth century (Cravens 1978; see chap. 13, "The Emergence of the Human Sciences"). In Soviet Russia, there was ideological opposition to genetics encouraged by distrust of the claim that human nature could not be improved by social progress. In the 1940s and 1950s, T. D. Lysenko promoted a new form of Lamarckism and gained the support of the dictator Stalin to have geneticists purged from the Soviet scientific community (Joravsky 1970). Although Lysenko offered the hope (illusory as it turned out) of improving agricultural science, the hostility of the Marxists to genetics was generated by their hatred of genetic determinism. The Lysenko affair is often treated as an example of how attempts to impose ideological control on science are bound to backfire, but critics of determinism point to the Western biologists' enthusiasm for eugenics to illustrate that the ideological bias was not one-sided.

It was the excesses of the Nazis that eventually brought the hereditarian movement into disrepute in America and Western Europe. Their hostility to the Jews, culminating in the holocaust, was paralleled by draconian steps to eliminate "defectives" within the Aryan race. By the 1940s, a wave of revulsion against these excesses was forcing many people, including scientists, to reconsider their support for racism and eugenics (Barkan 1992). Scientific factors were also at work, however: the rise of the genetic theory of natural selection undermined the theories of parallel evolution that had been used to proclaim the distinct character of races and, at the same time, emphasized the genetic affinity between all modern humans. Advances in genetics undermined the claim that every character is the product of a single gene. Even so, some biologists have resisted the trend, and historians will continue to debate the extent to which science has contributed to, or has been driven by, social attitudes.

Conclusions

The horrors perpetrated in Nazi Germany led to a new wave of liberalism in the social sciences and generated support for the view that people can be improved by better conditions. In the 1970s, the debate over nature and nurture broke out again over the claims made by Edward O. Wilson for sociobiology (Caplan 1978). Wilson pioneered techniques for explaining many aspects of social behavior, especially in insects, in terms of instincts created by natural selection. When he suggested that human behavior, too, might be determined in this way, liberals reacted with outrage and claimed that a new wave of social Darwinism had begun. More recently, many neuroscientists have begun to support the view that genetic inheritance plays a role in determining the structure of the brain and, hence, both intellectual ability and instinctive behavior. There are renewed claims that different racial groups have different average levels of intellectual ability. The human genome project has encouraged the belief that there is a genetic "fix" available for every physical and emotional disorder. The latest developments in biotechnology have also increased fears that eugenics may reemerge—not through state control of reproduction but though parents being able to choose the characters of their offspring. There is once again considerable interest in the possibility that evolution and heredity may shape our personalities, and this inevitably focuses attention onto historical studies of earlier episodes when such ideologies were influential.

Historians have explored the ways in which science was used in an attempt to provide legitimacy for the assumption that nonwhite races and the lower classes in Western societies were mentally inferior. That science was used in this way is beyond question; the real issue confronting us is the extent to which these concerns shaped the development of science itself. The sociological perspective supposes that scientific knowledge reflects the ideological interests of those who produce it. Theories were constructed in a way that would maximize their ability to lend support to prejudices such as the assumption of white racial superiority. The wave of enthusiasm for theories of racial differentiation coincided with the age of imperialism, and this ideology almost certainly shaped the thinking of those scientists who dismissed other races as inferior. Historians have become wary, however, of adopting a determinist approach in which a particular ideology necessarily generates a particular scientific theory. Many different theories were adapted to the same social purpose, and this leaves the historian looking for other reasons why the scientists involved chose their particular theories. Most of the different evolutionary theories proposed in the late nineteenth and early twentieth centuries contributed to race science, Darwinian and non-Darwinian alike.

The fact that science became involved in such debates raises problems about the nature and objectivity of science itself. When we address the past, we are uncovering the origins of concepts and attitudes that still shape our rival visions of human nature. History is used as a means of labeling modern theories to highlight their alleged social implications, as in the identification of sociobiology with social Darwinism. Such appeals to the past show that history is still relevant today but also reveal the dangers that await any historian trying to delve into such controversial issues. We have a duty to warn about the misuse of history, including simpleminded claims that particular ideologies must necessarily be identified with particular scientific theories. But the historian has access to a wealth of information that can confirm the day-to-day involvement of past scientists with the social issues of their time. A socially informed analysis of history offers a valuable way of warning us all of the extent to which science may still be influenced by the same factors.

References and Further Reading

Bannister, Robert C. 1979. *Social Darwinism: Science and Myth in Anglo-American Social Thought.* Philadelphia: Temple University Press.

Barkan, Elazar. 1992. *The Retreat of Scientific Racism: Changing Concepts of Race in Britain and the United States between the World Wars.* Cambridge: Cambridge University Press.

Bowler, Peter J. 1986. *Theories of Human Evolution: A Century of Debate, 1844–1944.* Baltimore: Johns Hopkins University Press; Oxford: Blackwell.

———. 1989. *The Invention of Progress: The Victorians and the Past.* Oxford: Blackwell.

———. 1993. *Biology and Social Thought.* Berkeley: Office for History of Science and Technology, University of California.

Caplan, Arthur O., ed. 1978. *The Sociobiology Debate.* New York: Harper & Row.

Cooter, Roger. 1984. *The Cultural Meaning of Popular Science: Phrenology and the Organization of Consent in Nineteenth-Century Britain.* Cambridge: Cambridge University Press.

Cravens, Hamilton. 1978. *The Triumph of Evolution: American Scientists and the Heredity-Environment Controversy, 1900–1941.* Philadelphia: University of Pennsylvania Press.

Crook, Paul. 1994. *Darwinism, War and History: The Debate over the Biology of War*

from the "Origin of Species" to the First World War. Cambridge: Cambridge University Press.
Gasman, Daniel. 1971. *The Scientific Origins of National Socialism: Social Darwinism in Ernst Haeckel and the Monist League.* New York: American Elsevier.
Gould, Stephen Jay. 1977. *Ontogeny and Phylogeny.* Cambridge, MA: Harvard University Press.
———. 1981. *The Mismeasure of Man.* New York: Norton.
Greene, John C. 1959. *The Death of Adam: Evolution and Its Impact on Western Thought.* Ames: Iowa State University Press.
Haller, John S. 1975. *Outcasts from Evolution: Scientific Attitudes of Racial Inferiority, 1859–1900.* Urbana: University of Illinois Press.
Haller, Mark H. 1963. *Eugenics: Hereditarian Attitudes in American Thought.* New Brunswick, NJ: Rutgers University Press.
Hawkins, Mike. 1997. *Social Darwinism in European and American Thought, 1860–1945: Nature as Model and Nature as Threat.* Cambridge: Cambridge University Press.
Hofstadter, Richard. 1955. *Social Darwinism in American Thought.* Revised ed. Boston: Beacon Press.
Jones, Greta. 1980. *Social Darwinism in English Thought.* London: Harvester.
Joravsky, David. 1970. *The Lysenko Affair.* Cambridge, MA: Harvard University Press.
Kevles, Daniel. 1985. *In the Name of Eugenics: Genetics and the Uses of Human Heredity.* New York: Knopf.
Knox, Robert. 1862. *The Races of Man: A Philosophical Enquiry into the Influence of Race on the Destiny of Nations.* 2nd ed. London: Henry Renshaw.
Lewin, Roger. 1987. *Bones of Contention: Controversies in the Search for Human Origins.* New York: Simon & Schuster.
Mackenzie, Donald. 1982. *Statistics in Britain, 1865–1930: The Social Construction of Scientific Knowledge.* Edinburgh: Edinburgh University Press.
Magnello, Eileen. 1999. "The Non-Correlation of Biometry and Eugenics." *History of Science* 37: 79–106, 123–50.
Pearson, Karl. 1900. *The Grammar of Science.* 2nd ed. London: A. and C. Black.
Richards, Robert J. 1987. *Darwin and the Emergence of Evolutionary Theories of Mind and Behavior.* Chicago: University of Chicago Press.
Searle, G. R. 1976. *Eugenics and Politics in Britain, 1900–1914.* Leiden: Noordhoff International Publishing.
Secord, James A. 2000. *Victorian Sensation: The Extraordinary Publication, Reception and Secret Authorship of "Vestiges of the Natural History of Creation."* Chicago: University of Chicago Press.
Shapin, Steven. 1979. "Homo Phrenologicus: Anthropological Perspectives on a Historical Problem." In *Natural Order: Historical Studies of Scientific Culture,* edited by Barry Barnes and Steven Shapin. Beverly Hills, CA: Sage Publications, 41–79.
Smith, Roger. 1992. *Inhibition: History and Meaning in the Sciences of Mind and Brain.* London: Free Association Books.
———. 1997. *The Fontana/Norton History of the Human Sciences.* London: Fontana; New York: Norton.
Stanton, William. 1960. *The Leopard's Spots: Scientific Attitudes toward Race in America, 1815–1859.* Chicago: Phoenix Books.

Stepan, Nancy. 1982. *The Idea of Race in Science: Great Britain, 1800–1960*. London: Macmillan.
Sulloway, Frank. 1979. *Freud, Biologist of the Mind: Beyond the Psychoanalytic Legend.* London: Burnett Books.
Young, Robert M. 1970. *Mind, Brain and Adaptation in the Nineteenth Century.* Oxford: Clarendon Press.
———. 1985a. "Darwinism *Is* Social." In *The Darwinian Heritage,* edited by David Kohn. Princeton, NJ: Princeton University Press, 609–38.
———. 1985b. *Darwin's Metaphor: Nature's Place in Victorian Culture.* Cambridge: Cambridge University Press.

CHAPTER 19

SCIENCE AND MEDICINE

Medical breakthroughs and discoveries are nowadays regarded as being among the greatest achievements of modern science. We think of doctors and scientists as being much the same kind of people—sober, white-coated individuals working in and around laboratories of one sort or another. Science is widely regarded as being at the very heart of medical practice. Science provides doctors with a core of basic knowledge about the way the human body works and the ways in which diseases develop. Science provides new forms of therapy for hitherto incurable diseases by means of new drugs or improved understanding of the role genes play in health, for example. Science provides a constant source of new diagnostic technologies from X-rays at the end of the nineteenth century to magnetic resonance imaging (MRI) scanners at the end of the twentieth century. Substantial improvements in public health and longevity that have taken place over the last century—in the Western world at least—are laid at the door of scientific medicine. Scientists predict that they are on the verge of cracking genetic codes that will lead to an unprecedented revolution in the understanding and treatment of disease. We take this relationship between science and medicine as being self-evident to the point of banality. How else, after all, might medicine proceed?

The kind of relationship between medicine and science that we take for granted today is, however, of relatively recent historical origin (Porter 1997). Three hundred, or even a hundred and fifty years ago, the value of natural philosophy or science to medical practice was not at all self-evident. On the contrary, it was a matter of considerable dispute on the part of both doctors and their patients. Until comparatively recently, few medical practitioners were provided with anything resembling a scientific education. Medicine was regarded as a craft, to be acquired through apprenticeship to a skilled practitioner. Even physicians, the most elite of medical practitioners, were provided with only the most basic education in natural philosophy. What mattered to doctors was the hands-on knowledge and skill at diagnosis they developed through years of practice and their intimate knowledge of the foibles and idiosyncrasies of their individual patients. Natural philosophical advocates of the new sciences like René Descartes in the seventeenth century (see chap. 2, "The Scientific Revolution") might argue that the application of new understandings of the body would lead to a miraculous transformation in health and longevity but most doctors and their patients took such claims with a large dose of salt (Shapin 2000). Even during the second half of the nineteenth century as "scientific medicine" became increasingly established, many doctors still argued that it was hands-on knowledge rather than scientific book-learning that really mattered to medicine.

Forging the kind of link between medicine and science that we now take for granted should therefore be regarded as a major cultural accomplishment. Until relatively recently in historical terms the relationship between medicine and science was not at all self-evident. Making it so was a historically fraught and contingent process that needs careful historical attention. Doctors in the past had—by their own standards—very good reasons for being suspicious of science. They recognized, for example, that making medicine scientific would bring about major—and not necessarily beneficial—changes in the way they practiced medicine and the relationship between themselves and their patients. The relationship between medicine and science is not entirely uncontested today either. In fact it is probably more contested now than it has been for much of the past century. Advocates of various kinds of non-Western medicine accuse scientific medicine of being too materialistic, of focusing on the body at the expense of the soul. Scientific medicine is also reproached for paying too much attention to the human body as a collection of diseased parts and not enough to the body as a unified whole. Similar criticisms come from advocates of so-called New Age approaches to medicine. Social critics accuse scientific medicine of "medicalizing" the human body, turning perfectly normal aspects of the human condition and experience into diseases requiring medical intervention.

This chapter will commence with an overview of the state of medical practice during the early modern period, looking at the structure of the profession, the relationship between practitioners and patients, and knowledge of the body. It will focus on the impact of what some historians have described as the "birth of the clinic" at the end of the eighteenth century. We will then look at the rise of laboratory medicine during the nineteenth century and the increasing insistence of many advocates that medicine needed to become not just a science but an experimental science. Making laboratory-based scientific education an essential part of medical training was part of this process. Pioneers such as Louis Pasteur or Robert Koch argued that experiment lay at the core of their efforts to cure disease. We will then look at the therapeutic revolution that accompanied the introduction of new drugs during the twentieth century. The success of new antibiotic treatments like penicillin seemed to many to provide the final proof of the success of scientific medicine and to provide the blueprint for future therapeutic efforts. The twentieth century also saw the consolidation of physical medicine—the application of technologies like X-rays or radiation to the cure and diagnosis of disease. Across the board it seemed clear that science provided the only key to medical progress.

The Clinical Revolution

In one of the most important works in the history of medicine of the past fifty years, *The Birth of the Clinic,* the historian and social critic Michel Foucault described the transformation in the practice of medicine that took place at the end of the eighteenth century as crucial to the emergence of modern medicine (Foucault 1973). According to Foucault, modern medicine became possible with the establishment of the hospital as the central focus for medical practice. Another medical sociologist describes this moment as "the disappearance of the sick man from medical cosmology." The suggestion is that with the development of the hospital, doctors started playing less attention to the bodies of individual patients and started treating diseases as being entities in themselves (Jewson 1976). Jewson argues that until the end of the early modern period medical practice was focused around individuals' bodies while with the rise of hospitals and their large concentrations of patients, individual bodies were simply regarded as sites in which the symptoms of particular diseases manifested themselves. Medical practitioners increasingly regarded hospital patients as sources of information about the development of different diseases as much as being people to be cured. From this perspective, as Foucault argues, nosology—the classification of diseases—was the key medical science.

The eighteenth-century medical profession was broadly divided into three groups: physicians, surgeons, and apothecaries. Of these, only physicians, dealing with the internal ailments of the body, were expected to possess a university degree. Both surgeons, who dealt with external illness, and apothecaries, who prepared medicines, usually learned the necessary skills through apprenticeship to an established practitioner. The vast majority of these practitioners operated on an individual basis rather than as members of a larger institution such as a hospital. Patients wishing for treatment approached different doctors according to their availability, the nature of their illness, and their ability to pay. A wealthy patient unhappy with the treatment provided by one practitioner could easily move on to another. In this respect, the relationship between practitioner and patient was weighed heavily in favor of the patient—a feature to which many historians of medicine refer in explaining the patient-centered nature of eighteenth-century medicine (Porter and Porter 1989) For most people, however, the cost of visiting an accredited practitioner was usually prohibitive. These people made do with the services of a variety of local practitioners such as bonesetters, herbalists, or wise women. Most apothecaries, despite the fact that they were formally prohibited from prescribing medicine (the province of physicians) usually did so. Toward the end of the eighteenth century, apothecary-surgeons, later general practitioners, qualified in both branches of the profession, were increasingly common (Waddington 1984).

What did these practitioners know about the body and disease? Many doctors proscribed to a humeral theory of disease according to which the body was governed by four fluids or humors: blood, bile, black bile, and phlegm. In a healthy body, the humors were in balance. In an unhealthy body, they were out of balance and the doctor's task was to restore equilibrium. This was the rationale for early modern medical practices like bloodletting, for example. Natural philosophers argued over how they should understand the body. Enthusiasts for Newtonianism, such as the Dutch university professor Hermann Boerhaave, argued that the body should be regarded as being just like a machine made up of pumps, pulleys, and mechanical devices (fig. 19.1). Others, for instance, Georg Ernst Stahl, were animists, insisting that there was more to the human body than just the collection of its mechanical parts. Natural philosophers, including Albrecht von Haller, tried to classify the properties of different kinds of animal tissue, characterizing muscle tissue as irritable and nervous tissue as sensible, for example (Hall 1975). It is not clear to what extent these debates between mechanists and vitalists (as historians often characterize the two sides) had any discernible impact on the practice of medicine. Most doctors were probably too busy trying to cure their patients' ills to pay much attention (Bynum and Porter 1985).

For a number of reasons, as Foucault notes, hospitals became increas-

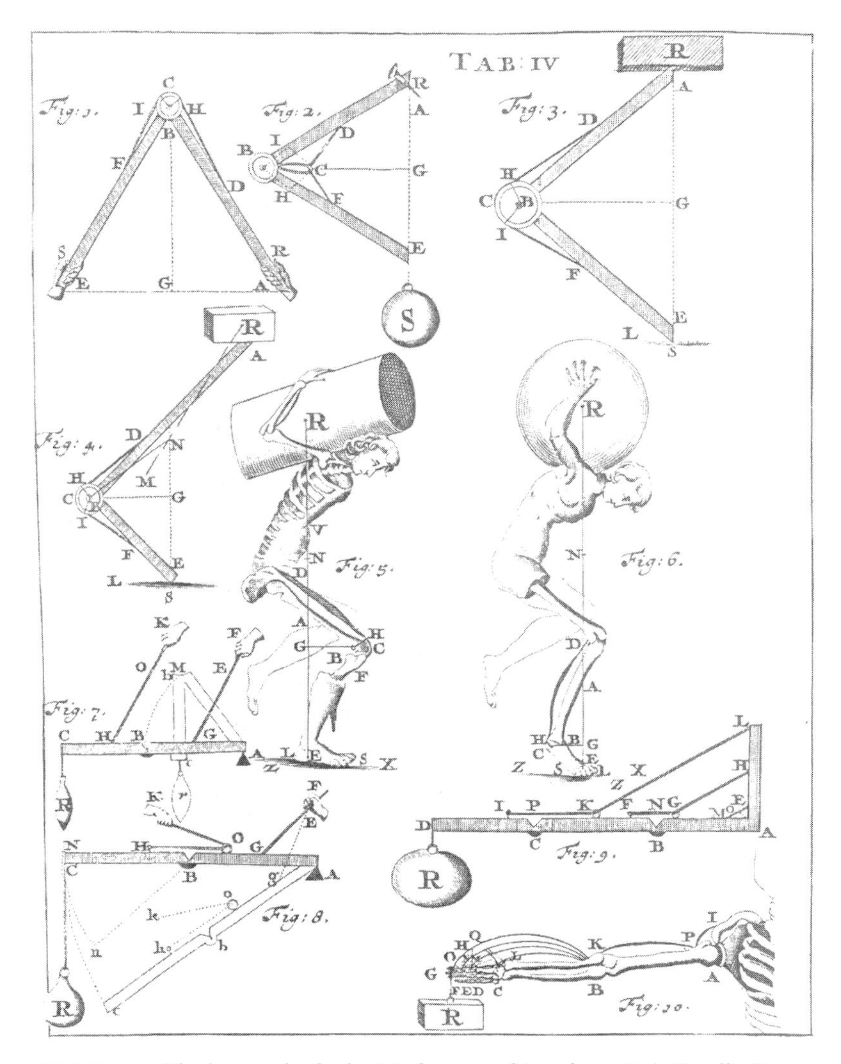

FIGURE 19.1 The human body depicted as a machine, from G. A. Borelli, *De motu animalium* (1680). The idea that the body could be understood as a mechanical system was increasingly common in early modern medicine.

ingly important as centers of medical practice and teaching during the course of the eighteenth century. Many eighteenth-century hospitals had histories stretching back to medieval times when they were founded as benevolent institutions, often under monastic direction, to provide care for the indigent poor. In eighteenth-century France, in particular, such institutions were increasingly brought under state control, particularly in the

FIGURE 19.2 A nineteenth-century Parisian hospital scene (Wellcome Medical Library, London).

aftermath of the French Revolution (fig 19.2). It is this assumption of state control and the reorganization of hospital practice that went along with it that Foucault had in mind as the "birth of the clinic." Hospitals came to play an increasingly important part in the career aspirations of ambitious doctors. In the process they helped transform the way in which doctors regarded patients and their diseases. Foucault's thesis is considerably less convincing outside France, in Britain or America, for example, where the state played little or no role in the establishment of hospitals. The development of the hospital as a center for medical learning and the changes in perspective that accompanied that process are evident in these countries as well however. Traditional healers such as midwives often found themselves displaced by new hospital-trained and accredited practitioners (like the new breed of *accoucheurs* or men-midwives, as they were called) who were often more attractive to the rising urban middle classes (Wilson 1995).

During the eighteenth century, nosology—the classification of diseases according to their particular characteristics and symptoms—became an increasingly central concern. It was in many ways the key Enlightenment medical science. New classificatory systems were very much an eighteenth-century trend, as the botanist Linnaeus's taxonomic system of natural history or even the French philosophers d'Alambert and Diderot's efforts to classify all knowledge in the *Encyclopédie* demonstrate. François Bossier de Sauvages (1706–67), physician and professor of medicine at the University of Montpellier was one of the first to embark on a systematic classification of diseases in his Nouvelles classes des maladies (1731). Sauvages identified ten different classes of disease, divided into 295 genera and 2,400 species. The Scottish surgeon and professor of medicine at Glasgow, William Cullen (1710–90), produced one of the eighteenth century's most influential efforts at disease classification in his *First Lines of the Practice of Physic* (1778–79). Increasingly, attention shifted from classification of diseases based on the subjective symptoms reported by patients and toward efforts to identify objective signs. Anatomical pathology came to be an increasingly important method of identifying diseases in an effort to identify particular organic lesions that corresponded to particular states of disease. Nosology came increasingly to depend on the existence of large numbers of patients for examination, as provided by large hospitals. In this way, hospitals started to be regarded as major centers of medical research as well as of teaching and medical care.

One of the most important features of the clinical revolution in medicine identified by Foucault is what he calls the emergence of the medical "gaze." Foucault argues that with the birth of the clinic doctors started to see their patients in literally quite different ways. Rather than regarding patients as individuals, each with their own unique needs and symptoms, doctors started to regard patients as sites for the manifestation of different diseases. Patients came to be regarded as experimental subjects. This was, in part at least, another aspect of the rise of large hospitals as major focuses of teaching and research. Hospital patients were by and large drawn from the poorer sections of society. They lacked the power that the middle and upper classes possessed to interact with doctors as social equals or even superiors. Poorer patients often feared hospitals as places of death and destitution. An example is the passing of the Anatomy Act in Britain in 1832. The act allowed doctors to use the unclaimed bodies of poor patients for dissection and experiment. They were the raw material for the new nosological systems that emerged during the early years of the nineteenth century as well as for the massive expansion of medical teaching that took place around the same period. The gaze, according to Foucault, submitted patients to the objectifying surveillance and control of modern medical authority.

Foucault paints a bleak picture of the emergence of scientific hospital medicine. From this perspective, the clinical revolution is to be regarded as the imposition of new forms of control and management on the bodies of patients rather than as an effort to improve health. It is possible, however, to appreciate many aspects of this analysis of the changes in medical practice at the end of the eighteenth century without necessarily regarding medical discourses and practices as just another form of power relationship. Changes in the way medical knowledge was organized, particularly the increasing interest in nosology, did develop alongside the emergence of new institutions and structures, such as the hospital as a space for teaching and research. Most historians of medicine would now also agree that Foucault's analysis is too quick to generalize from the particular example of France at the time of the French Revolution. Developments elsewhere in Europe or in North America did not take place at the same time or in the same way. There was little state involvement in hospital provision in Britain, for example, until the early twentieth century (Lawrence 1994; Peterson 1978). The state's involvement in hospital provision in the United States remains limited. The clinical perspective was nevertheless enthusiastically adopted in these countries, too, during the course of the nineteenth century.

Laboratory Medicine

Laboratories are now crucial sites of medical research. We look to laboratory science to produce new cures. More routinely, laboratory testing of drugs and samples now constitutes a veritable industry without which modern medicine could not continue at all. The contribution of laboratory work to medical practice is nevertheless of comparatively recent provenance. Only during the early years of the nineteenth century did the first systematic efforts to bring medicine into the laboratory take place. Proponents argued that only through the rigorous application of the methods of experimental science could medicine hope to progress and produce new and effective therapies. The application of laboratory science to medicine was by no means uncontested. Opponents throughout the nineteenth century (and continuing up to the present day) objected to the practice of vivisection, which seemed to be the inevitable accompaniment of scientific medical research. Many doctors also felt that scientific medicine was a distraction from the proper task of medical practice. They argued that medicine depended on the tact and experience that could only be gained through practice rather than on the technical knowledge and practices of laboratory science (Lawrence 1985). Treating the body as if it were a collection of mechanical parts distracted from a proper understanding of the body as a whole. Increasingly, however, proponents of scientific medicine were successful in making laboratory training a routine part of medical education.

Major therapeutic breakthroughs such as those of Louis Pasteur and Robert Koch were widely attributed to laboratory medicine.

In many ways, the model for early nineteenth-century laboratory-based medicine was developed by Justus Liebig (1803–73). Educated at Bonn, Erlangen, and Paris, Liebig was appointed professor of chemistry at Giessen in 1824 and established an institute of chemistry at the university there. Liebig is generally regarded by historians as the founder of one of the first research schools in chemistry (Brock 1997). He played a key role in establishing a coherent tradition of chemical and medical research, investigating biological functions as being the results of chemical and physical processes in the body rather than being the outcomes of innately vital activity (see chap.7, "The New Biology"). Claude Bernard, similarly, played a key role in promoting laboratory science as part of French medicine. Like Liebig, he is regarded as important as much for the contributions of his students and the powerful philosophical vindication he mounted for scientific medicine as he is for his own experimental work. Bernard's seminal book, *Introduction à l'étude de la médecine expérimentale* (1865) provided a powerful defense of the role of experimental science in medical training and research. Hospital observation was too haphazard and passive a process to provide reliable information about the way a disease progressed—its "pathophysiology" as Bernard called it. For this, experimentation on live animals in a controlled laboratory setting was essential (Holmes 1974).

Claude Bernard's reputation as France's most prestigious exponent of scientific medicine was soon eclipsed by the rising star of Louis Pasteur (1822–96). Pasteur graduated as a chemist from the École Normale Supérieure and moved through a succession of posts in provincial universities before he was appointed in 1854 to a university professorship in the French manufacturing center of Lille. It was there that Pasteur started out his research on the chemistry of fermentation, responding to the needs of Lille's industrial breweries. He succeeded in establishing that the fermentation process required the presence of microorganisms to take place. In the process, he also established the method for preventing beer (or milk) from going sour now known as "pasteurization." Pasteur made his reputation through a series of highly publicized debates with the radical doctor Felix Pouchet on the issue of spontaneous generation—whether organisms could be produced spontaneously from inanimate matter (Latour 1988). The radical materialist Pouchet argued that spontaneous generation was a reality. Pasteur, the conservative Catholic, argued that it was not. In a spectacular series of experiments, Pasteur showed that in all circumstances where the experimental apparatus was properly sterilized and environmen-

FIGURE 19.3 Louis Pasteur at work in his laboratory.

tal contamination prevented, organisms did not appear. In other words, the apparent observation of spontaneous generation was the result of contamination of the apparatus by extraneous microorganisms. His election to the French Academy of Sciences in 1862 sealed his reputation (Geison 1995).

Pasteur developed his work on microorganisms during the 1860s and 1870s, becoming a vociferous and energetic advocate of the germ theory of disease (fig. 19.3). He argued that diseases, as well as processes such as fermentation and putrefaction, were caused by the presence of microorganisms and that if the organisms responsible for causing a particular disease could be identified it should be possible to develop vaccinations against that disease. In 1879 he tested his theories by injecting chickens with "stale" cholera-causing microbes, showing that those birds already exposed to the stale version did not become infected when later exposed to a virulent form of the disease. He later carried out an even more stunning series of experiments at Pouilly de Fort in 1881 with anthrax—a major killer of both livestock and humans. He injected twenty-four sheep, six cows, and a goat with vaccine, and after repeating the treatment a few weeks later, he exposed the animals, along with a number of unvaccinated animals, to a live anthrax culture. The vaccinated animals' survival and the deaths of the unvaccinated ones were widely hailed as a triumphant vindication of Pasteur's theories. He mounted another spectacular in 1885 when he vaccinated nine-year-old Joseph Meister, bitten by a rabid dog. Meister survived. In 1888, the Institut Pasteur was established in Paris as a center for the kind of medical research that Pasteur had devoted his career to promoting.

Pasteur's experiments provided a major boost for the germ theory of disease (Geison 1995). Another significant and influential advocate of the germ theory was Pasteur's rival, the German experimenter Robert Koch (1843–1910). Koch had studied medicine at the University of Göttingen and after taking his MD in 1866 studied chemistry in Berlin with, among others, Rudolf Virchow (1821–1902). During his early career as the district medical officer for Wollenstein in the aftermath of the Franco-Prussian war, Koch made his reputation with a study of the transmission of anthrax. In 1880 he was appointed a member of the Imperial Health Bureau in Berlin (Brock 1988). There he continued to work on developing new methods of cultivating pure cultures of bacteria for study, including using the Petri dish, developed by his colleague Richard Julius Petri (1852–1921). Koch was particularly known for his famous four postulates, setting out the experimental procedures for establishing a link between a particular microorganism and a particular disease. They were (1) that the organism could be discoverable in every instance of the disease, (2) that, extracted from the body, the germ could be produced in a pure culture, maintainable over several microbial generations, (3) that the disease could be reproduced in experimental animals through a pure culture removed by numerous generations from the organisms initially isolated, and (4) that the organism could be retrieved from the inoculated animal and cultured anew.

The postulates were first put forward in a paper on the etiology (cause) of infectious diseases in 1879 and formalized in 1882. In that year Koch also announced a major breakthrough to the Berlin Physiological Society, identifying the bacillus *Mycobacterium tuberculosis* as the cause of tuberculosis, one of the biggest killers of the age. In 1883 Koch was dispatched to Egypt as part of the German cholera mission to investigate the outbreak of cholera there. Koch succeeded in identifying the "vibrio" (a form of bacteria) that caused the disease and brought pure samples of it back to Berlin for study. In 1885, Koch was appointed professor of hygiene at the University of Berlin and director of the newly established Institute of Hygiene at the university there. Throughout the 1880s and 1890s, Koch and his students at his prestigious Institute for the Study of Infectious Diseases in Berlin (of which he became director in 1891) proceeded to identify the organisms responsible for a train of the nineteenth century's worst killers, including diphtheria, typhoid, and pneumonia (Brock 1988). Not everyone was convinced, however. One German doctor actually drank a flask containing cholera microbes sent to him by Koch to demonstrate his contempt for the idea that disease could be caused by invisible organisms (Porter 1997). He survived, probably as a result of his stomach acidity being high enough to neutralize the microbes. Koch made several attempts to develop ways of curing tuberculosis on the basis of his discoveries. Despite his ambitious and optimistic claims, the cures proved largely ineffective, however. In 1905 Koch was awarded the Nobel Prize in Physiology and Medicine.

Breakthroughs like those of Pasteur or Koch that appeared to have a tangible and immediate therapeutic benefit in cases of hitherto intractable diseases did a great deal to make the case for laboratory medicine. As we have seen, however, even in these cases acceptance of the inevitable benefits of experimental science to medicine was by no means automatic. Vivisection was a major stumbling block as far as large sections of the nineteenth-century public, including many doctors, were concerned. Throughout the century, proponents of scientific medicine insisted that experiments on live animals were an essential feature of their practice. Claude Bernard, for example, argued that experimental animals had to be kept alive for the duration of an experiment to ensure that the whole process of disease could be properly monitored. Opponents argued in return that not only was the causing of pain to other creatures morally repugnant but that the responses of animals in extreme pain to particular stimuli provided no secure knowledge at all about their responses under normal circumstances. Antivivisection campaigns were particularly prominent in Britain. Following a scandal at the 1874 meeting of the British Medical Association when a French psychiatrist carried out public experiments on two unanesthasized dogs, a Royal Commission was convened to consider the situation. The result was the 1876 Cruelty to Animals Act, which prohibited animal experimentation without a license (French 1975). Despite this undercurrent of public discomfort and dissent, by the beginning of the twentieth century, it was increasingly coming to be accepted that laboratory science held the key to medical progress (Bynum 1994).

The Antibiotic Revolution

The name of Sir Alexander Fleming is probably one of the best known in the history of medicine. The story of his accidental discovery of penicillin in his laboratory at St. Mary's Hospital in London is frequently held up as an iconic moment for the relationship between science and medicine: all at once in his laboratory, the white-coated scientist Fleming made a single serendipitous discovery that changed the face of modern medicine. There is, of course, far more to the story than that (MacFarlane 1984). We have already seen that even the space in which Fleming worked—a laboratory in a hospital—did not appear simply by accident. The existence of such spaces was the end result of many decades of hard work and persuasive effort by enthusiasts for scientific medicine. Even granted the existence of hospital laboratories and researchers like Fleming trained to work in them, a great deal of work was needed to transform a single observation of a curious phenomenon into a successful drug. To produce penicillin and the drugs that followed it in useful quantities, scientific medicine had to become a mass-producing activity on an industrial scale. Pharmaceutical companies became one of the major industrial and scientific success stories of the twentieth century. They played a key role in turning medical research from being a relatively small-scale activity carried out by individuals and small groups with limited resources into a multibillion dollar industry employing hundreds of thousands.

Increasingly close links were already in place between medical researchers looking for drug cures and chemical industrial companies by the end of the nineteenth century. Paul Ehrlich (1854–1915), director of the Royal Prussian Institute for Experimental Therapy from 1899, had close links with German chemical industries. This was one reason he looked to chemical dyes (products of those industries) for therapies. Just as dyes seemed only to attach themselves to particular tissues, so drugs might be developed that attacked specific microbes. One of the early results of such research was salvarsan, a compound of arsenic that could be used to treat syphilis. Another German researcher, Gerhard Domagk (1895–1964), a research director at I. G. Farbenindustrie (another dye producer) from 1927 carried out research leading to the identification of sulfonamide as a means of curing streptococcal infections. By the 1930s, pharmaceutical companies were already producing the new generation of sulfa drugs, based on this work, in industrial quantities. Other researchers argued, however, that there were limits to what these kinds of chemical therapies could achieve. Pasteur and Koch's work in bacteriology had shown that biological, rather than simply chemical, agents might make the most effective cures. What was needed was a way of identifying so-called antibiotics—biologically derived drugs that would target the microbes causing particular diseases. This was the approach that Alexander Fleming thought most promising, particularly since he was aware of research by Frederick Twort (1877–1950) and Felix d'Hérelle (1873–1949) suggesting the existence of such bacteria-eating organisms.

As the popular myth has it, Fleming discovered penicillin by accident in August 1828. He had been working in his laboratory on the staphylococci, the bacteria responsible for a variety of conditions including septicemia and pneumonia. Returning from a holiday he discovered that a mold appeared to have destroyed a staphylococcus culture left growing in a Petri dish. Fleming promptly pursued the matter, identified the mold as penicillium and confirmed that it had a significant effect on a whole range of bacterial types while not affecting white cell functions. He published his conclusions the following year in the *British Journal of Experimental Pathology* (MacFarlane 1984). In reality, of course, matters were a little more complex. Fleming had been working on related research for several years. In particular, he had identified an enzyme (lysozyme) present in human tears that appeared to attack microbes. He was therefore already a believer that antibiotics, rather than chemicals, were the key to fighting disease. Penicillin, despite its success in attacking particular kinds of bacteria—so-called gram-positive bacteria—had no effect on gram-negative bacteria. It was also difficult to produce in large quantities and relatively unstable. As a result, most researchers felt that Fleming's discovery, while highly interesting, was unlikely to produce significant clinical benefits. Fleming himself made no effort to follow up his discovery, and for the next ten years neither did anyone else.

In 1938, however, the biochemist Ernst Chain (1906–79) rediscovered Fleming's paper in the course of his own researches into antibacterial agents. Along with Howard Florey (1906–79) at the Dunn School of Pathology at Oxford University, Chain set out to reproduce Fleming's experiments and to grow significant quantities of the penicillium mold. They achieved at least a partial clinical success when a patient suffering from septicemia was treated with their accumulated store of penicillin and improved for several days until the supply ran out. The patient's death underlined the difficulties of producing the drug in useful amounts. The resources of British pharmaceutical companies being taken up by the war effort, the Oxford group turned to the United States for help in producing penicillin on an industrial basis. By the early 1940s, both American and British pharmaceutical companies had perfected means of producing penicillin in large quantities. In 1945, Fleming, along with Chain and Florey, were awarded a Nobel Prize for the discovery of penicillin. The new wonder drug was widely hailed as having played a key role in saving Allied lives during D-day and the invasion of Europe in 1944. It had also demonstrated that not just scientific research but large-scale industrial production as well would be essential to make such drugs easily available to the public at large (fig. 19.4).

Penicillin's success prompted concerted efforts to look for other anti-

FIGURE 19.4 Early penicillin manufacturing equipment (Wellcome Medical Library, London). Notice the improvised milk churns.

biotics (Spink 1978). In 1939, the French-born bacteriologist René Dubos (1901–82), working at the Rockefeller Institute Hospital in New York, succeeded in isolating a crystalline substance he called tyrothricin from a culture of the soil organism *Bacillus brevis.* Tyrothricin acted as a powerful antibacterial agent, effective against a range of gram-positive bacteria. Following Dubos's success, the Russian émigré Selman Waksman (1888–1973) began research into the medicinal properties of soil microbes. In 1940 he isolated the antibiotic agent actinomycin. Unlike its predecessors, actinomycin was effective in attacking gram-negative bacteria such as those responsible for diseases like typhoid, dysentery, and cholera. It was too toxic to be used on humans, however. Four years later, Waksman isolated the antibiotic streptomycin, particularly effective in treating tuberculosis. Waksman was awarded the Nobel Prize in Medicine for his work on antibiotics in 1952. In 1948, Benjamin M. Duggar (1872–1956), recently retired professor of plant physiology and economic botany at the University of Wisconsin, isolated chlortetracycline from *Streptomyces aureofaciens.* Chlortetracycline, otherwise known as aureomycin, was the first so-called tetracycline antibiotic and the first broad-spectrum antibiotic. It was active against an estimated fifty disease-causing organisms. Antibiotics seemed to be on the verge of delivering a magic bullet to doctors, allowing them to combat a whole range of previously incurable diseases.

A notable feature of many antibiotic pioneers was the way their careers straddled academia and industry. Pharmaceutical companies played a critical role in the introduction of penicillin (Weatherall 1990). They provided the resources and the expertise that made its bulk production a feasible proposition. The American pharmaceutical company Merck was the first commercial producer of penicillin. One of Howard Florey's Oxford coworkers, Norman Heatley (1911–), who had come to the United States to investigate possibilities for large-scale commercial production, was released to Merck to help them develop the process. Selman Waksman also worked as a consultant for Merck and sent one of his students, H. Boyd Woodruff, to help with the production of penicillin. Unsurprisingly, Merck quickly started commercial production of Waksman's streptomycin as well. Duggar, who developed chlortetracycline, was a consultant for Lederle Laboratories, another pharmaceutical company. Arrangements like these, in which academic scientists also worked for commercial companies, were becoming increasingly commonplace as postwar pharmaceutical companies quickly came to recognize the profit-making potential of medical research into new drugs. Large pharmaceutical companies such as Burroughs Wellcome increasingly maintained their own large and well-funded laboratories for scientific research. The result was a transformation of medicine as massive resources were poured into medical research and new drugs and massive profits poured out.

Even from the beginning of the antibiotic revolution, however, some observers were warning that there were limits to the possibilities of the new miracle cures. René Dubos was one of the first to point out the danger of developing antibiotic resistance. In fact, Dubos eventually stopped researching into antibiotics because of his concerns that their indiscriminate use would lead to the appearance of microbial strains that were resistant to them. As early as 1940, Ernst Chain and fellow workers at the Dunn School in Oxford had identified a strain of the bacterium *Staphylococcus aureus* that could not be treated with penicillin. It was to be the first of many. By the 1950s, more strains of microbes that appeared to be resistant to antibiotic treatment were appearing, and by the end of the twentieth century it was increasingly recognized as a growing medical problem that new strains of "superbugs" resistant to all known antibiotics were becoming more common. There were increasing concerns that the antibiotic revolution would prove to be a lucky but short-lived aberration rather than a permanent achievement in the history of medicine. Some of the institutional features of the antibiotic revolution appear to be more resilient, however. By the second half of the twentieth century, the link between medicine and science (and big business, for that matter) seem to have become inextricable.

Physical Medicine

As well as drugs, we usually regard science as making major contributions to medicine in terms of the hardware it provides. Again, this is a relatively recent phenomenon. Many nineteenth-century doctors felt extremely uncomfortable at the prospect of introducing technology into medicine. They worried that instruments would quite literally get between them and their patients. A doctor's skill, from this perspective, rested with the hands-on nature of his relationship with those he treated. Even the introduction of the stethoscope by the French practitioner René Theophile Hyacinthe Laennec (1781–1826) was resisted by some for this reason. Nevertheless, in some quarters new technologies were regarded as holding the key to therapeutic progress. From the mid-eighteenth century onward, many enthusiasts argued that electrical machines and other paraphernalia could bring about a revolution in healing. By the middle of the nineteenth century, electromedical devices such as batteries, induction coils, portable magneto-electric generators, and electric belts were starting to become common. Electrotherapy quickly came to be accepted as a respectable form of medicine in nineteenth-century France. In Britain, acceptance was slower to come. Many doctors regarded any form of electrotherapy as nothing more than quackery. Nevertheless by the end of the century, electrical departments were relatively common features of large hospitals. Despite ongoing resistance, increasing numbers of doctors accepted that technology might be a valuable addition to the medical armory.

It was within such hospital electrical departments that some of the twentieth century's key medical technologies first became established. On 8 November 1895, the German physicist Karl Wilhelm Roentgen (1845–1923) made a startling discovery. Roentgen, a professor of physics at Wurtzburg since 1888, was interested in cathode rays—the strange glow given off when high-voltage electricity was passed through a hermetically sealed glass tube—and found that a nearby screen coated with barium platinocyanide glowed while his experiments were in progress. It seemed that some kind of invisible ray from the tube was affecting the screen. Further work established that the rays penetrated a wide variety of substances (see chap. 11, "Twentieth-Century Physics"). Eventually Roentgen even attempted—quite successfully—to use the rays to take a photograph of the inside of his wife's hand. He announced his results in a paper titled "Eine neue Art von Strahlen" (A new kind of ray) to the Physical-Medical Society of Wurtzburg on 28 December 1895 and published soon after in the *Sitzungs-Berichte der Physikalisch-medicinischen Gesellschaft zu Wurzburg*. The news was quickly picked up by the popular press and passed around the world. The new X-rays' diagnostic possibilities for medicine were quickly recognized. If they could be used to take photographs of the interior of living bodies, they could be used to identify bone fractures or the existence of solid masses inside the body.

What was probably the first X-ray photograph to be used for explicitly diagnostic purposes was taken by A. A. Campbell Swinton (1863–1930) in January 1896. Campbell Swinton was an electrical contractor who soon established himself as an X-ray consultant to the medical profession and established the first British X-ray laboratory at 66 Victoria Street, London. In Canada, John Cox, professor of physics at McGill University used X-rays to help remove a bullet from an injured man's leg in February 1896. Within a very few years, X-ray equipment was becoming a standard part of hospital technology as part of preexisting electrical departments. This was where X-rays remained until electricity started going out of fashion as a form of therapy in the 1920s and 1930s (Burrows 1986). X-rays were used for therapeutic as well as diagnostic purposes. Doctors were soon using X-rays to treat conditions like skin diseases, cancers, and tuberculosis. The first systematic practitioner of X-ray therapy was Leopold Freund (1868–1943) in Vienna, who carried out his first operation with X-rays on a five-year-old girl to remove a hairy mole on her back in December 1896. There was little initial apprehension that any dangers might be involved in X-ray treatment. It was not long, however, before injuries and even deaths to practitioners and patients made it clear that X-ray technology was extremely dangerous if used without due precautions. Nevertheless by the time of the Second World War, X-rays were clearly here to stay (fig. 19.5).

The discovery of radioactivity at the end of the nineteenth century was another development in physics that was rapidly exploited for its medical uses. In 1896, the French physicist Henri Becquerel (1852–1908) found that the metal uranium emitted energetic rays of some sort. As a result, a Polish student studying in France, Maria Sklodowska (1867–1934)—later known as Marie Curie—chose the topic for her PhD dissertation. Along with her husband and fellow physicist Pierre Curie (1859–1906), she carried out extensive studies of the new radiation and in 1898 announced the discovery

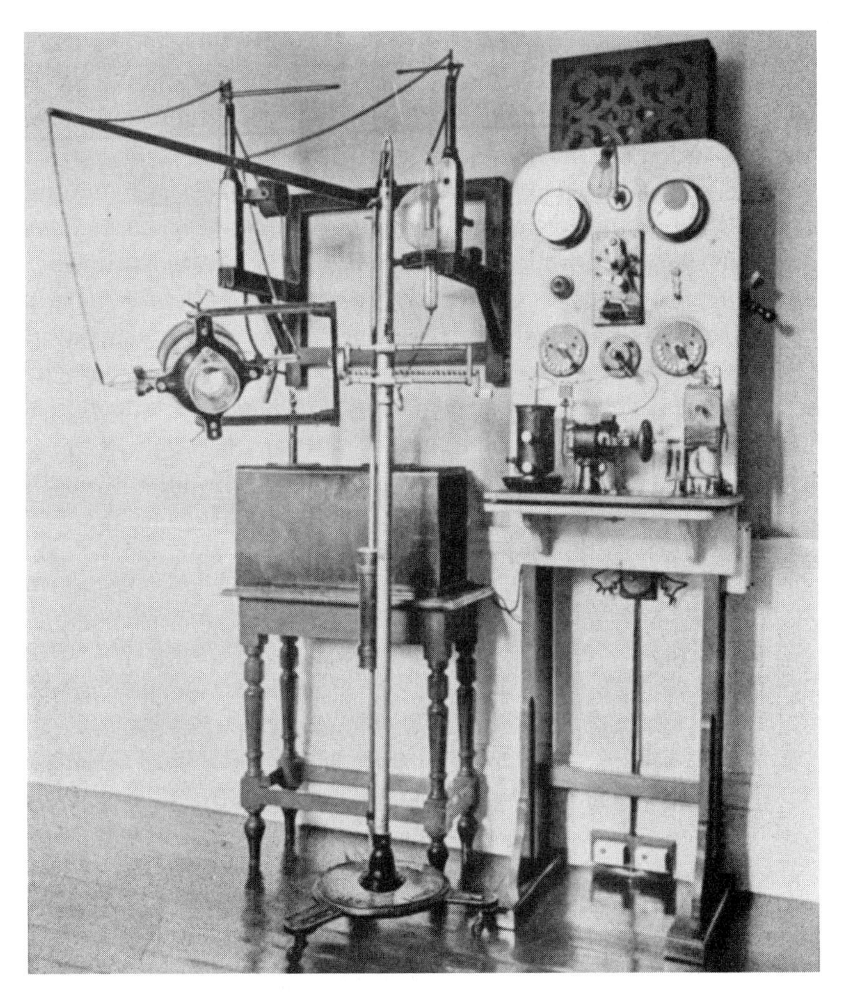

FIGURE 19.5 Early twentieth-century medical X-ray apparatus (Wellcome Medical Library, London).

of a new radioactive element—radium (see chap. 11, "Twentieth-Century Physics"). As was the case with X-rays, medical uses were soon found. As early as 1904 demonstrations showed that radium rays could apparently destroy diseased cells. Radium therapy rapidly became a popular fad. The telephone inventor Alexander Graham Bell suggested that a small vial of radium surgically inserted into the middle of a cancer would act to destroy it. Fringe medical practitioners took up radium and radioactivity with enthusiasm. Radioactive belts, radioactive toothpaste, and radioactive drinking water—marketed as "liquid sunshine"—were popular. Radioactive spas were also popular as were caves and mines where a large amount of radioactivity was present. Sufferers from "arthritis, sinusitis, migraines, eczema, asthma, hay fever, psoriasis, allergies, diabetes, and other ailments" were invited to cure their ills at places like the Merry Widow Health Mine or the Sunshine Radon Health Mine (Caufield 1989).

Radioactivity seemed to produce similar effects on the body to those of X-rays. Within orthodox medicine, radioactivity quickly joined X-rays as part of the range of treatments offered in hospital electrical departments. Again, just as with X-rays (and electricity a decade or so earlier), radium was used as a form of treatment for a large range of diseases, including heart trouble, cancers, impotence, and so forth. Early forms of treatment were usually a matter of strapping flat containers of radium against the affected part of body, injections of radium chloride to reach deeper injuries, or the insertion of small capsules into body cavities or directly into diseased tissue. Doctors and others were also increasingly concerned that radioactivity might itself be a health hazard. In 1928, the International Committee on X Ray and Radium Protection recommended that radium be well-shielded, handled with forceps, and kept in a lead safe. In 1934, the U.S. Advisory Committee on X Ray and Radium Protection recommended that limits should be established of safe exposure to radioactivity. Throughout the period, radium and other radioactive substances were routinely used in medicine. While there was increasing awareness of possible health risks, there was also a widespread perception that the quantities used in medicine were too small to cause any permanent damage. During the second half of the century, increasingly sophisticated technologies were developed to target radiation accurately at particular parts of the body and to regulate doses.

By the final few decades of the twentieth century, medical practitioners were exploiting developments in nuclear physics to provide a range of new ways of peering inside the human body (Kevles 1997). Computed tomography (or CT) scanning was first developed by the British engineer Godfrey Hounsfield (1919–) at EMI Laboratories in 1972. CT scanners used X-ray signals to construct images of the interior of the body in cross section. By 1974, clinical scanners were coming into use, although they did not become common until the 1980s. Early scanners usually only took images of the head, but later models were developed that could take images through the whole body. Hounsfield was awarded a Nobel Prize for his invention in 1979. Another technique developed during the 1970s was MRI (magnetic resonance imaging). The technique was initially developed by Raymond Damadian (1936–), working at the Downstate Medical Center in New York, making use of the fact that different atomic nuclei emit radio waves of predictable frequencies when exposed to a magnetic field. Damadian noted that tumorous cells emitted signals different from those emitted by healthy tissue and used this as the basis for a new technique for identifying cancers. Damadian and his fellow workers produced the first MRI scan of the human body in 1977. The key to both CT and MRI scanning, along with other techniques such as PET (positron emission tomography) was the development of powerful computers that could process information quickly and reliably to provide usable images.

Laennec's stethoscope at the beginning of the nineteenth century and Hounsfield's CT scanners at the end of the twentieth century had in common that they provided technological fixes to the problem of non-invasively looking inside the human body. Doctors traditionally used their own senses to try to visualize what they could not see. They listened to the sounds the body made. They manipulated limbs and pressed the flesh in an effort to find broken bones or internal swellings and contusions. Only after death could they literally look inside the body to see if their deductions were accurate. X-rays and the later computer-driven scanning technologies of the 1970s made it possible for doctors to see the body's internal machinery without first killing the patient. Similarly, what X-rays and radiation therapies offered doctors and patients was a way of dealing with the inside of the body without resorting to the surgeon's knife. Just as the antibiotic revolution placed commercial pharmaceutical companies at the heart of medical research, developments in physical medicine made medical diagnostics an important area of concern for electrical engineering and electronics companies. Electrical engineers were developing an interest in electrotherapy at the end of the nineteenth century. By the end of the twentieth century, companies such as IBM, Siemens, and Toshiba were at the forefront of medical research.

Conclusions

As we noted at the beginning of this chapter, science and medicine now appear very difficult to disentangle. Many people, if asked to identify a particular instance of the way in which science might be considered to have tangibly benefited humankind, would turn to modern medicine for their example. As the previous pages illustrate, this seemingly self-evident state of affairs did not come about inevitably or without much effort on the part of groups and individuals committed to new visions of how medicine might best be practiced. As part of this process, the cultural identity of doctors themselves underwent several transformations. Pasteur's and Koch's examples show how important the acquisition of secure institutional bases was for this process. The ways in which medical practitioners were trained and the ways in which they practiced in the eighteenth, nineteenth, and twentieth centuries were very different. An eighteenth-century gentleman physician would probably have some difficulty in recognizing the white-coated professional of today as a fellow practitioner of the same medical arts. During the course of this process, the ways in which doctors interacted with their patients and the kinds of therapies they offered them underwent dramatic change. During the twentieth century, scientific medicine became associated with big business and the state in a way that would have been considered unimaginable a few decades previously (Porter 1999). Even during the 1930s, the early researchers into penicillin in Britain were discouraged from patenting the processes they developed on the grounds that to do so would be "money grubbing."

The modern commercial context of scientific medicine remains a source of criticism. Pharmaceutical and other companies and their researchers are often accused of putting commercial interests before therapeutic concerns. The pharmaceutical industry usually responds by suggesting that without the profit motive there would be no incentive for them to invest in developing new life-saving drugs as they do. Another criticism frequently leveled at modern doctors is that by adopting a purely scientific perspective they have "medicalized" the human body. Michel Foucault's work on the "birth of the clinic" can, at least in part, be interpreted as coming from this tradition of criticism. The concern being expressed here is that by treating the body as an object—as a scientist might treat a piece of experimental apparatus—doctors are somehow dehumanizing it. From a historical perspective, debates like these can be seen as the modern equivalents of arguments concerning the proper practice of medicine and the appropriate cultural place of its practitioners, arguments that stretch back through history. Past debates about the relationship between science and medicine were couched in terms of the way in which using scientific tools might have an impact on the cultural image of the medical practitioner as often as they were about the extent to which science might provide tangible therapeutic benefits. Looking at the relationship historically, therefore, provides one way of putting contemporary concerns into perspective.

References and Further Reading

Brock, T. D. 1988. *Robert Koch: A Life in Medicine and Bacteriology.* Madison: University of Wisconsin Press.

Brock, William H. 1977. *Justus von Liebig: The Chemical Gatekeeper.* Cambridge: Cambridge University Press.

Burrows, E. H. 1986. *Pioneers and Early Years: A History of British Radiology*. Alderney: Colophon.

Bynum, William. 1994. *Science and the Practice of Medicine in the Nineteenth Century*. Cambridge: Cambridge University Press.

Bynum, William, and Roy Porter, eds. 1985. *William Hunter and the Eighteenth-Century Medical World*. Cambridge: Cambridge University Press.

Caufield, C. 1989. *Multiple Exposures: Chronicles of the Radiation Age*. Harmondsworth: Penguin.

Foucault, Michel. 1979. *The Birth of the Clinic*. Harmondsworth: Penguin.

French, Roger D. 1975. *Antivivisection and Medical Science in Victorian Society*. Princeton, NJ: Princeton University Press.

Geison, Gerald L. 1995. *The Private Science of Louis Pasteur*. Princeton, NJ: Princeton University Press.

Hall, T. S. 1975. *History of General Physiology*. 2 vols. Chicago: University of Chicago Press.

Holmes, F. L. 1974. *Claude Bernard and Animal Chemistry: The Emergence of a Scientist*. Cambridge, MA: Harvard University Press.

Jewson, N. 1976. "The Disappearance of the Sick Man from Medical Cosmology." *Sociology* 10:225–44.

Kevles, B. 1996. *Naked to the Bone: Medical Imaging in the Twentieth Century*. New Brunswick, NJ: Rutgers University Press.

Latour, Bruno. 1988. *The Pasteurization of France*. Cambridge, MA: Harvard University Press.

Lawrence, Christopher. 1985. "Incommunicable Knowledge." *Journal of Contemporary History* 20:503–20

———. 1994. *Medicine and the Making of Modern Britain, 1700–1920*. London: Routledge.

MacFarlane, G. 1984. *Alexander Fleming: The Man and the Myth*. London: Chatto & Windus.

Peterson, M. J. 1978. *The Medical Profession in Mid-Victorian London*. Berkeley: University of California Press.

Porter, Dorothy. 1999. *Health, Civilization and the State: A History of Public Health from Ancient to Modern Times*. London: Routledge.

Porter, Roy. 1997. *The Greatest Benefit to Mankind*. London: Harper Collins.

Porter, Roy, and D. Porter, eds. 1989. *Patient's Progress: Doctors and Doctoring in Eighteenth-Century England*. Cambridge: Cambridge University Press.

Shapin, S. 2000. "Descartes the Doctor: Rationalism and its Therapies." *British Journal for the History of Science* 33:131–54.

Spink, Wesley W. 1978. *Infectious Diseases: Prevention and Treatment in the Nineteenth and Twentieth Centuries*. Minneapolis: University of Minnesota Press.

Waddington, I. 1984. *The Medical Profession in the Industrial Revolution*. Dublin: Gill & Macmillan.

Weatherall, M. 1990. *In Search of a Cure: A History of the Pharmaceutical Industry*. Oxford: Oxford University Press.

Wilson, A. 1995. *The Making of Man-Midwifery*. London: UCL Press.

CHAPTER 20

SCIENCE AND WAR

DURING THE SCIENTIFIC REVOLUTION OF THE seventeenth century, Francis Bacon and others argued for the practical benefits that would result from the application of the new knowledge of nature. These appeals tended to focus on the benefits to industry and medicine, along with specialist applications such as navigational techniques. But from the start it was obvious that the same principle would apply to warfare and the arts of destruction—these, too, might be improved by the new sciences. Mathematics was already used for practical purposes in gunnery and the design of fortifications, and gunnery especially could benefit from a better theoretical understanding of projectile motion. By the nineteenth century, the involvement of science with industry was already beginning to include the design and manufacture of better explosives and guns, and there were suggestions of entirely new weapons such as poison gas. These trends were expanded by both sides in World War I, although to begin with, the successful interaction of science and the military was limited by lack of direct communications. These obstacles were largely overcome during World War II, when new inventions such as sonar (for detecting submarines) and radar played crucial roles. The application of a scientific way of thinking to complex practical problems led to operations research. But most obvious to succeeding generations, this war led to the creation of a new weapon with a destructive power so great that it potentially threatened the whole foundations of civilization: the atomic bomb. The Manhattan Project that generated the bomb started out from theoretical innovations in physics but led to the establishment of the first really large-scale integrated scientific-industrial-military research program. Continuing with the design of even bigger weapons based on nuclear fusion (the H-bomb) in the cold war, this area of interaction began to shape the environment within which a significant part of the scientific community would operate.

Some scientists feel very uncomfortable with the modern state of affairs. They know that the link between science and the military-industrial complex is seen by many as evidence that science itself is a harmful influence on our society. One avenue of escape is provided by the old argument that pure science generates impartial knowledge of nature—it is only applied science that can lead to harmful consequences, and then only when national emergencies focus efforts on military rather than peaceful applications. But modern historians of science are skeptical of such a separation between science and its applications. We know that, over the past several centuries, very few scientists have worked in total isolation from the world of applied science, especially as increasingly complex technical equipment became necessary to test hypotheses generated at a theoretical level. Many of the most innovative physicists of the nineteenth century, for example, were already concerned with practical questions generated by new industrial developments (see chap. 17, "Science and Technology"). Once that link was established, the involvement of scientists in the development of military technology became inevitable.

In some cases, the division between peaceful and warlike technologies is itself artificial. Better navigational techniques benefited all seafarers in the late eighteenth century, but it was Europe's navies that led the way—and the native peoples of many parts of the world would not have regarded the incursion of traders and colonists as a peaceful process. In the modern world, radar helps to make civil aviation safe but was applied first to detect military aircraft. New medicines such as penicillin and insecticides such as DDT were first developed under the pressure of war. Technologies developed to detect nuclear submarines provided information about the deep-sea bed that was crucial to the emergence of the modern theory of plate tectonics (see chap. 10, "Continental Drift"). There have been periods when scientists have openly rejected the call to do applied work for the military, but when their country or way of life seems threatened they do their patriotic duty like everyone else. Since the cold war ushered in an almost permanent state of anxiety about the safety of Western democracies, the possibility of stepping off the escalator of military development seems quite unrealistic—and Soviet scientists responded equally readily when their own country seemed under threat. Historians have to take it for granted that for most of the past century a significant amount of science has been done in collaboration with the military and must, therefore, explore the implications of this for the way science operates.

To simplify this question, in this chapter we will focus primarily on the direct application of science to military technologies. We begin with the first hesitant steps to use science to improve and eventually to design entirely new weapons, culminating with the somewhat fragmentary interactions with the military authorities in World War I. In the interwar years, there were efforts to intensify these connections even during the period when many hoped that war could be averted. Scientists then began to play a major role during World War II, providing the basis for new technologies such as sonar, radar, and the V-2 rockets that laid the foundation for later programs to design guided missiles. The project to build the atomic bomb will occupy a large proportion of the chapter, partly because it pioneered a new degree of intensity in the cooperation between the government, the military, industry, and the scientific community. But the bomb project also helps to focus our attention on the moral problems faced by scientists when they are asked to design weapons of mass destruction. The Allies raced for the bomb only to find out after the end of the war that their fears of a similar weapon being developed by Nazi Germany were unfounded. It has even been argued that German scientists actively avoided work that might have given Hitler the bomb. Then the American bombs were dropped on Hiroshima and Nagasaki, bringing home to everyone the horrors that would result from the widespread use of such weapons. Some scientists began to express doubts about participating in the arms race that accompanied the cold war with Soviet Russia—but others were anxious to help develop weapons that, they felt, were necessary to protect democracy. More disturbing still was the possibility that scientists were now actively suggesting new weapons so they could benefit from the resulting research funds. The moral and political dilemmas that face many scientists in the modern world became fully articulated.

The Chemists' War

It has been said that World War I was the chemists' war, while World War II was the physicists' war. This is an oversimplification, but it highlights the fact that much of the scientific effort that went into military applications between 1914 and 1918 was devoted to the production of better explosives and the first really new terror weapon—poison gas. In fact, neither side made effective use of its scientific expertise, and none of the weapons developed had a decisive effect on the outcome of the war. But at the very least, it had become apparent that the potential for the military application of science was considerable. The scientists themselves had been willing to offer their services when faced with a national emergency, and some quite senior figures had become directly involved with military research. The military establishment had been reluctant to take advice, however, and bridges between the two communities were only gradually and imperfectly built in the course of the war. In the end, perhaps the greatest legacy of this war was the creation of military research establishments that would go on to play a vital role in later conflicts (Hartcup 1988).

These hesitant steps built on a foundation that had been emerging over the previous century or more. Armies had included corps of engineers since the eighteenth century and were thus used to the applied sciences—what they were not prepared for was new initiatives coming from science and industry. The French revolutionary government had executed Lavoisier because he had collected taxes for the old regime, but France soon found that it did need chemists after all to suggest new means of producing saltpeter for gunpowder. In the course of the nineteenth century, new and more powerful explosives were developed, and some scientists even suggested the possibility of poison gas, although the military dismissed this as beneath its dignity. By the end of the century, however, the situation had begun to change. The inventor of dynamite, Alfred Nobel, was particularly effective in linking science with industry. He set up a research center in Berlin in 1897, with representatives of the armaments firm Krupps on the board of directors. In Britain, the Boer War in South Africa revealed alarming weaknesses in military equipment, and in 1900 the Ordnance Board set up an explosives committee headed by the eminent physicist Lord Rayleigh and containing among its members the chemist William Crookes. Crookes urged the use of TNT as a high explosive, although the British did not take this up until the start of World War I. Rayleigh also presided over an advisory committee on aeronautics to study the military use of the newly invented airplane.

Whatever limited preparations had been made, when war came in 1914 most European countries were slow to appreciate the potential for science to aid developments in military technology. Only in the following year did the British government, for instance, set up an advisory council of scientists, which soon became transformed into the Department of Scientific and Industrial Research under the direction of the physicist J. J. Thomson. Scientific teams were also set up in the Admiralty and the Ministry of Munitions. Even so, popular writers such as H. G. Wells continued to argue that the country's scientific expertise was being wasted. In 1916, a group of eminent scientists used the ineffectiveness of the military's handling of science to argue for a greater role for science education—as it was, most politicians and military officers were completely ignorant of science and hence could not appreciate its potential. The French government was somewhat more effective, setting up the Directory of Defense Inventions, which was linked to the universities. In Germany, the noted chemist Fritz Haber (who had invented a technique for "fixing" nitrogen to make fertilizers and explosives) placed his Institute for Physical and Electrochemistry at Dahlem in Berlin at the disposal of the military. It soon came completely under military control and in 1917 became the Kaiser Wilhelm Foundation for the Science of War Technology. In America, the founding of the National Research Council in 1916 was prompted by the increasing likelihood that the country would enter the war.

What did these various teams of scientists achieve? In some projects, quite a lot was done, although seldom without difficulties produced by the very different attitudes of scientists, industrialists, and the military. Chemists worked not only on new explosives but also on providing alternative means of manufacture when raw materials were in short supply. In Britain, J. J. Thomson and others worked on improvements in radio to aid military communications. A team from the Board of Invention and Research including Rayleigh, Ernest Rutherford, and W. H. Bragg helped to develop hydrophones for detecting submarines.

By far the most striking new initiative was the use of gas, and this was actively promoted by Fritz Haber as a weapon for the German army once it became clear that the conventional war had become bogged down in the trenches of the Western Front (Haber [1986]—a survey written by Haber's son). There was a Hague Convention forbidding the use of projectiles containing poisons, but Haber now suggested using chlorine from cylinders that would be released when the wind was in the right direction to carry it over the enemy trenches. With some reluctance, the army agreed to try out this idea. Haber's own institute provided the links with industry to set the program up, and a regiment was founded to deliver and operate the cylinders—it contained many young scientists who would go on to achieve eminence after the war. One hundred fifty tons of chlorine were released on 22 April 1915 on the Ypres salient, causing panic in the opposing French troops (although few fatal casualties). But the Germans gained little ground because the army was not ready to exploit the breakthrough. The British and the French responded much more rapidly than the Germans expected, and the rest of the war saw a succession of developments including the use of gas shells and the introduction of new chemicals such as mustard gas. Both sides also made advances in protecting against gas, various forms of masks being developed by teams of chemists and physiologists.

In the end, it was the Allies who made the most concerted use of their scientists—Haber always complained that despite his direct involvement with the army, the senior officers seldom took him seriously. The British set up a dedicated facility at Porton Down near Salisbury to work on chemical (and later biological) weapons. But it was the American Chemical Warfare Service that produced the most sustained scientific program in this area—by 1918 it included more university-trained scientists than all the other belligerents put together (Haber 1986, 107). Studies of poison gas continued, although neither side made use of it in the next war. The interwar years did, however, see the setting up of programs that developed weapons that were to have a far more substantial effect on the outcome of World War II.

World War II

Although most of the scientists recruited to help the military in the first war soon returned to civilian work, small numbers were retained on a permanent basis for defense research, especially by air forces and navies. There were now more applied scientists working in industry, including the armaments and aircraft industries. During the interwar years, many academic scientists looked down on their colleagues in industry and were reluctant to work on military research. A greater level of social awareness emerged in Britain during the 1930s when a prominent group of left-wing scientists began to criticize the extent to which applied science was being driven by military concerns. But the radicals were also aware of the growing menace of Nazi Germany, and when war came they, too, became willing to work on military research. The threat of bombing from the air became so apparent that the British government set up the Committee for the Scientific Survey of Air Defence in 1934 under Henry Tizard—this was to play a key role in the development of radar. But it was not all smooth sailing. The Marxist crystallographer J. D. Bernal led a movement to criticize the government's plans for civil defense, and it was only after the outbreak of war that he gained any influence on policy (Swann and Aprahamian 1999).

When the Nazis came to power in Germany they poured funding into a number of new weapons systems, including radar and long-range rockets. The Allies were warned of these new developments in the "Oslo report"—a paper smuggled to the British Embassy in Oslo in 1939 by H. F. Mayer, a German scientist whose sympathies were anti-Nazi. But in fact the more ambitious German programs had little effect—Hitler liked new military technology but had little sense of how to use it, and his regime consisted of a number of competing factions that often blocked each other's initiatives.

The year immediately preceding the outbreak of war in 1939 saw a rapid reinvigoration of scientific programs in Britain that effectively prepared the country for war. In America in 1940, Vannevar Bush of Massachusetts Institute of Technology persuaded President Roosevelt to set up a national defense research committee to coordinate scientific plans for war (Zachary 1999; more generally on science in World War II, see Hartcup [2000]; Johnson [1978]; Jones [1978]).

In the closing years of the first war, French scientists had proposed a technique for detecting submarines by reflecting sound waves off them underwater. The British continued this program, and although the system, known as asdic (later sonar) did not go into operation during this war, it was developed throughout the interwar years and was ready for use in the Battle of the Atlantic in World War II (Hackmann 1984). F. A. Lindemann was so confident in the effectiveness of asdic that he predicted the end of the submarine as a significant weapon (Hartcup 2000, 64–65). Events were to show just how wrong he was, since even with the new detection system British warships were unable to protect their convoys, and the country was nearly brought to its knees. A whole series of further developments in antisubmarine warfare were needed before the menace of the U-boats was defeated.

Perhaps the most important area of applied scientific research was the development of radar (Brown 1999; Buderi 1997; Price 1977). By the start of the war, both the British and the Germans had introduced radar systems for the detection of aircraft, although the British system was more efficiently applied. As noted above, the British had set up an aeronautical research committee in 1934, and one of its most important tasks was to work out a system for detecting incoming bombers. Scientists at the Radio Research Station showed that it was feasible to detect radio waves reflected from solid objects such as aircraft at considerable distances (curiously, the first calculations were done to disprove the idea of a "death ray" intended to destroy the aircraft). In the late 1930s, a large number of physicists from the Cavendish Laboratory at Cambridge were employed to develop the basis for what became known as the Chain Home radar stations. Working from large masts on the south coast of England, these played a vital role in the "battle of Britain" in 1940 when the German air force tried to gain control of British airspace as a prelude to invasion (fig. 20.1). The Oxford physicist F. A. Lindemann (later Lord Cherwell) encouraged work on other systems, including the detection of aircraft by infrared. Lindemann later became Winston Churchill's scientific adviser, and he and Tizard quarreled violently over the priority to be given to radar in the early years of the war.

FIGURE 20.1 Chain Home radar station on the south coast of Britain, 1940. These enormous towers were able to detect German aircraft heading toward Britain from German-occupied France early enough to give the Royal Air Force fighters a chance to intercept them.

The navy and the air force also wanted a system of short-range, high-accuracy radar, and this required the use of short wavelength (microwave) radio. There was no system available for generating microwaves at a useful power level until British physicists developed the cavity magnetron in 1940. An early model of this was flown to America in August of that year by a team led by Tizard, and soon microwave radars were in production on both sides of the Atlantic. They were used by night fighters to close in on enemy bombers but, more important, also by naval patrol aircraft searching for submarines (which had to spend part of their time on the surface to run their diesel engines). Thus radar joined sonar as a key weapon in the battle of the Atlantic.

The Atlantic battle also provided a classic example of the benefits of a scientific approach to management, increasingly known as operations research. The physicist P. M. S. Blackett, who worked on a number of weapons including magnetic mines, formed the Operations Research Section of the Royal Air Force's Coastal Command and conducted a systematic survey of the factors that influenced the fate of a convoy. Against the advice of naval experts, Blackett introduced larger convoys and was able to demonstrate that the larger the convoy, the smaller the proportion of its losses, and the use of larger convoys played a major role in easing the problem of supply. Operations research was also applied successfully in the management of the air offensive against Germany. By the end of the war, scientists from a wide range of backgrounds were employed in operations research, advising on issues such as the effectiveness of bombing and the best way to use the available forces in the invasion of Europe. They were not all physicists, either—one of the most influential British advisers in the later stages of the war and thereafter was the biologist Solly Zuckerman (see Peyton 2001; Zuckerman 1978, 1988).

The Germans used their applied scientists to develop a number of new weapons, but the confused state of the chain of command under Hitler (along with Hitler's own unstable temperament) often interfered with their introduction. The Germans had a good radar network but did not have a coordinated system for passing information on to their pilots. They also developed the jet engine, paralleling similar research led by Frank Whittle in Britain. In the later stages of the war, much attention was focused on the V weapons ("revenge weapons") designed to strike at long range. The V-1 was a pilotless aircraft driven by a pulse jet. Far more imaginative in terms of its future potential was the V-2, the world's first long-range rocket, developed by a team under the physicist Werner von Braun (Neufeld 1995). When used against Britain in the last year of the war, it was unstoppable, but by then it was too late for its impact to turn the tide against Germany. Von Braun and his team had solved a host of technical problems and were anxious to continue their work—like most rocket scientists of the time, they had they eyes on the exploration of outer space. At the end of the war, von Braun surrendered to the Americans and was soon leading the development of their program to develop rockets both for military ends and for space exploration. The Russians also scooped up a number of German experts and began employing them for the same purposes.

The Atomic Bomb

One question haunted the Allies throughout the war: Had the Germans begun to develop a bomb based on the energy released by radioactive elements (the atomic bomb)? The revolution in early twentieth-century physics had revealed the enormous power that was locked up in the atom (see chap. 11, "Twentieth-Century Physics"). Although most scientists were skeptical, there were occasional predictions that this power could be unleashed to give a bomb that might destroy a whole city. The first calculation that such a bomb might be feasible was made in 1940 by Jewish physicists who had fled the Nazi regime in Germany. But there were still eminent physicists left in Germany, most notably Werner Heisenberg, whose national loyalty might lead him to develop a bomb in wartime even if he disapproved of Hitler and his policies. It was the fear that Hitler might acquire such a superweapon that led the Allies to pour resources into what became the Manhattan Project to develop the bomb; unlike the V-2, the atomic bomb could have turned the tide in Germany's favor even at the last minute. In fact, German physicists had come nowhere near to developing a bomb, and their only nuclear reactor was virtually useless. When Heisenberg and his colleagues were interrogated after their capture by the Allies, it became clear that they had vastly overestimated the critical mass needed to start a chain reaction in uranium and had told the German military that the bomb could not be made. Controversy has continued ever since over the question of whether this overestimate was simple carelessness or a deliberate attempt to ensure that the Nazis did not get the bomb (Powers 1993; Rose 1998). A successful Broadway play, *Copenhagen,* was based on a notorious confrontation between Heisenberg and his mentor, the Danish atomic physicist Niels Bohr, in 1941, during which Heisenberg seems to have raised the issue of the bomb (Frayn 1998).

Unaware of the Germans' lack of interest in creating an atomic bomb and suffering from daily raids by conventional bombers, it was the British who made the first moves to explore the possibility of building a nuclear bomb (Gowing 1965). By 1939 it had become clear to Bohr and others that the only way to derive significant amounts of energy from the fission (breakup) of radioactive atoms was by starting a "chain reaction." Normally, the nuclei of such atoms fission spontaneously at a very slow rate, each liberating a small but significant amount of radiation. But some radioactive elements, most notably uranium 235 and the artificial element plutonium, also liberate neutrons, and these particles are capable of initiating fission if they collide with another nucleus. In small quantities of the radioactive element, the neutrons mostly escape before they can hit another nucleus, but if the quantity exceeds a "critical mass" the neutrons will begin to fission enough extra atoms to produce a cascade of further collisions—the chain reaction. In a nuclear reactor or "pile," the chain reaction is sustained at a level that will produce a constant amount of energy. But in an uncontrolled chain reaction, the whole mass of atoms will disintegrate in a fraction of a second, liberating a vast amount of energy in the form of an explosion. The simplest form of atomic bomb thus consists of a device to bring together two subcritical masses to create a critical mass, which will immediately explode. By 1940, a number of physicists had begun to think about this situation, and the central problem was: What is the critical mass? Heisenberg casually assumed it would be many tons, making a bomb impractical—but what if it were much less, say only a few kilograms?

The calculation was actually done in March 1940 by two German scientists, Otto Frisch and Rudolf Peierls, who had fled to escape the Nazis and were now working in England at the University of Liverpool. The answer was about five kilograms, certainly small enough to form a usable bomb—although there was as yet no way of extracting anything like that amount of fissionable material from natural sources. Most natural uranium consists of U-238, which cannot form a chain reaction; only 0.7% is the vital U-235, and to make a bomb some means of extracting the U-235 in quantity would have to be devised. But Frisch and Peierls' memorandum was sent to Henry Tizard, and a committee was soon set up to investigate the possibility of separating the isotopes and making a bomb. It was called the MAUD committee—Bohr had written of "Maud" in a telegram from Denmark, and it was thought to be a code word, although actually it was the name of a woman he knew in Britain. Its members included leading physicists: G. P. Thomson, James Chadwick, Mark Oliphant, and P. M. S. Blackett. Work began at Oxford on devising a process of isotope separation by gaseous diffusion, eventually under the cover name of Tube Alloys.

Blackett and other members of the committee felt that, with the imminent threat of invasion, the actual production would be best done in the United States. Oliphant visited America in August 1941 to discuss radar research, but he was also instructed to convey to the Americans the importance now being attached to the bomb project by the British. So far, the Americans had been inactive, although in 1939 Albert Einstein, prompted by the Hungarian physicist Leo Szilard, had written to President Roosevelt warning of the dangers. Now Oliphant gained the attention of Ernest Lawrence, who convinced the administration's key scientific advisers Vannevar Bush and J. B. Conant that the project was likely to be successful. On 6 December 1941 (the day before the Japanese attack on Pearl Harbor), Roosevelt approved funds for research, and by the summer of the following year, pilot plants for production were being planned. Work also began on the design of the bomb itself (Hoddeson et al. 1993).

As yet no chain reaction had actually been observed, and the theory was not confirmed until December 1942, when Enrico Fermi built a reactor in the basement of a football field at the University of Chicago and initiated a controlled chain reaction. One function of the reactor was that it would convert uranium-238 into plutonium, another potential source of fissionable material for a bomb. In fact, the construction of reactors to make plutonium offered a better way to make fissionable material because it could easily be extracted by chemical means, while the separation of U-235 and U-238 involved very delicate physical processes using gaseous diffusion or electromagnetic techniques. Plans went ahead on both fronts, with the aim of making bombs with both U-235 and plutonium. Brigadier General Leslie Groves was put in charge of what became known as the Manhattan Project. Groves was highly experienced in managing large projects, and his organization skills were vital—yet he was not a scientist and was disliked by many of the scientists recruited for the project, who found his military approach uncongenial. He was also anti-British and for a while British scientists were excluded from the project, although this situation later changed, and even Bohr joined the project after he escaped from occupied Denmark.

The scale of the project became truly enormous—the plants built at Oak Ridge, Tennessee, to extract U-235 and at Hanford, Washington, to make plutonium both used more hydroelectric power than a large city (fig. 20.2; Hughes 2002). The technical skills of the scientists and engineers who designed the equipment were taxed to the limits. Meanwhile, design of the bomb itself began at Los Alamos, New Mexico, under J. Robert Oppenheimer. Oppenheimer was a leading figure in the American physics community, which had now developed to the stage where it was on equal terms with the long-established European traditions (Goodchild 1980; Kevles 1995). He then faced a new challenge, in which his abilities as an inspired leader would be put to more practical ends. Significantly, although the Manhattan Project as a whole was organized by Groves and the military, the scientific teams working on technical problems were all civilians and were led by scientists. This meant that they were not simply taking orders from the military and were free to think about the consequences of what they were doing. Eventually this freedom would allow major debates to emerge over the morality of working on the bomb, but in the short term, the perceived threat from Nazi Germany encouraged most scientists to throw themselves into the work.

Although a brilliant physicist, Oppenheimer knew that in this new environment where practical results were all that mattered, the scientists' traditional individualism would not work. He found it necessary to adopt a quasi-military style of management that required the whole team to focus on the immediate goal but still left room for individual creativity in the solution of the problems. Oppenheimer also became skilful at working with government and military committees, emerging as a new kind of scientific leader, as much at home in the corridors of power as in the laboratory. In a

FIGURE 20.2 The Alpha-1 racetrack, Y-12 Plant in 1944. (U.S. Corps of Engineers, Manhattan Engineering District, Oak Ridge, Tennessee. Photograph by James E. Westcott). The Alpha-1 racetrack was used in the separation of uranium isotopes. This device gives some impression of the scale at which big science began to operate when the resources of the military-industrial complex were thrown behind it. The wiring used 6,000 tons of silver obtained from the U.S. treasury.

sense, the Manhattan Project was changing the way science was done, requiring leading scientists to engage in much closer cooperation with military and industrial interests. Oppenheimer realized that scientists would have to learn to work in these new ways if they were to have any influence over what was being done with their work.

Meanwhile, technical problems were emerging that required even closer cooperation between the theoretical physicists and the engineers. These problems demanded new theoretical concepts for their solution, and the theories could not be tested without building the hardware for the bomb. Far from seeing applied science as a chore to be done reluctantly under pressure of war, the physicists often found themselves fascinated by the theoretical innovations they were forced to make to solve problems generated by practical applications. The original design for a bomb was based on a

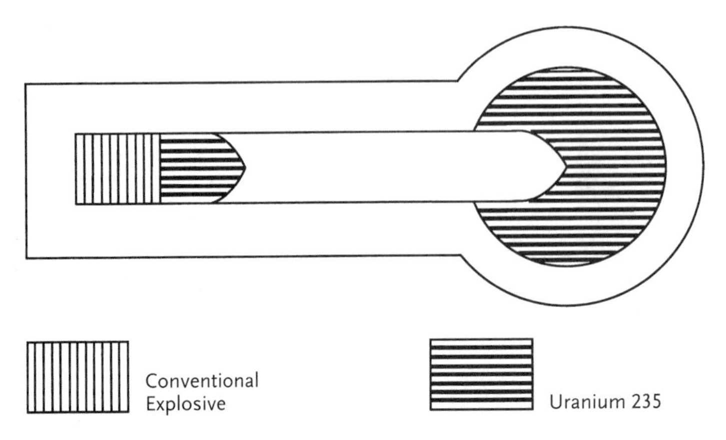

FIGURE 20.3 Diagram to show the "gun" method of exploding a uranium-235 bomb. A charge of conventional explosive fires the small slug of uranium down the barrel of the gun into the larger body on the right, raising it above the critical mass and allowing the chain reaction to begin.

"gun" that fired a slug of U-235 down a barrel to smash into a target of the same material (fig. 20.3). The combined mass was above the critical point and would instantly undergo an uncontrolled chain reaction. But in the spring of 1944, tests with plutonium showed that the gun method would not work with this element because it had such a high spontaneous fission rate that each subcritical mass would begin to fragment even before the two pieces had come together. This would disrupt the fissionable material before it could be combined in a small enough region for an effective chain reaction to take place. A whole new type of bomb had to be designed using an "implosion" method in which a slightly subcritical mass is compressed by a carefully shaped sphere of conventional explosive to achieve a critical state. British physicists (including the German refugee Peierls), now back on the project, did much of the work on this new design. But the proposal was so radical that science advisers such as J. B. Conant doubted that it would work. This was why the plutonium bomb was tested in the desert at Alamogardo, New Mexico, on 16 July 1945. It yielded an explosion equivalent to 20,000 tons of TNT, even more than the scientists had predicted (figs. 20.4 and 20.5). On witnessing the explosion, Oppenheimer famously quoted a line from the Hindu epic the *Bhagavad-Gita:* "I am become death, the destroyer of worlds." Another physicist, Kenneth Bainbridge, made a more down-to-earth comment: "Well, now we're all sons of bitches" (quoted in Schweber 2000, 3).

The actual use of the bombs followed quickly to end the war with Japan

FIGURE 20.4 Explosion of the first atomic bomb.

(Germany had already surrendered). On 6 August, the B-29 bomber *Enola Gay* obliterated the city of Hiroshima with a "Little Boy" uranium bomb. Three days later a "Fat Man" plutonium bomb was dropped on Nagasaki. Controversy has raged over the actual motivations for using the bombs. The official position was that they forced a rapid Japanese surrender, thereby saving hundreds of thousands of American soldiers who might have died in an invasion. But this was certainly an overestimate, and suspicion has

FIGURE 20.5 J. Robert Oppenheimer and General Groves at the Trinity site after the explosion of the first atomic bomb (Popperfoto/Retrofile.com). Oppenheimer was a brilliant physicist, but in the new world of big science he had to learn to cooperate with figures in authority within the military and big business.

lingered that the new American president, Harry Truman, used the bombs to gain extra leverage over the Russians in postwar negotiations (Alperowitz 1996; Giovannitti and Freud 1965; Walker 1996).

More relevant to our own theme is the question of how the scientists themselves felt about their involvement in the creation of so devastating a weapon. There can be little doubt that the initiative to create the bomb came from those scientists who realized that it might be possible to exploit nuclear fission in this way. Had the scientists not promoted the idea, the project would not have begun—this was what actually happened in Germany. But with the fear that the Nazis might explore the same possibility, there seems to have been little reluctance among British and, later, American scientists to push ahead. It was a brutal war anyway, and cities were already being obliterated by conventional bombing. The crunch came when Germany collapsed, leaving Japan (which had only a small nuclear program) as the only target. At this point, some scientists did begin to argue that the bomb should not be used or, at least, should be dropped in a remote location in Japan first, as a warning. Leo Szilard, who had originally encouraged Einstein to write to Roosevelt about the possibility of a nuclear weapon, then emerged as a leading critic of the military's policy to use the bombs. He pressured the Committee of Social and Political Implications under the physicist James Franck to issue a report arguing for a demonstration first (reprinted in Giovannitti and Freud, 111–15). But many scientists refused to endorse Szilard's proposals, some because they accepted the argument for saving American lives, others because they were still so deeply involved in the last-minute technical problems that they had no time to step back and rethink their position. Oppenheimer himself accepted the view that it would save American lives and seems to have done little to encourage debate at Los Alamos—although after the war he became a leading critic of the decision to build the even more powerful hydrogen bomb.

Science and the Cold War

In the postwar era, international tensions continued with the Soviets replacing the Nazis as the perceived threat to Western democracies. Once the muted hostilities of the cold war fell into place, it was easy for scientists on both sides to revive the old argument that involvement in military research was justified. Only a few influential figures stood out against the trend, and they faced the risk of being ostracized for disloyalty. But there were other reasons to keep up the involvement with what was now becoming known as the military-industrial complex. It was only under the threat from external powers that governments were likely to invest the huge sums of money that were needed for research in areas of "big science," where even the testing of theories required the building of vastly expensive equipment. The temptation for scientists to involve themselves with, perhaps even to encourage, projects with military applications was thus immense—it often seemed the only way of getting the funding to do research at this level. The atomic bomb project had also required an interpenetration between pure and applied science that made it difficult to distinguish between theoretical innovation and practical application. Many areas of science thus remained wedded to the military-industrial complex, and scientists would sometimes initiate projects with military implications so they could obtain funding for research they wanted to do anyway (Mendelsohn, Smith, and Weingart 1988).

The Soviets were quick to respond to the threat of the American atomic bomb (Holloway 1975). Before the war, their physicists had done good research in this area, despite government indifference. The environmental scientist V. I. Vernadskii had encouraged the search for uranium as a raw material in the hope that it could be used for peaceful purposes. During the war, Soviet officials got some information about British and American nuclear projects from spies, but when it became clear that the Germans were not involved, Stalin lost interest. His henchman Beria even suspected that stories about the Manhattan Project had been planted to encourage the Soviets to waste money in this area. Once it became clear that the Americans had the bomb, however, Stalin soon decided that it was a major threat to Soviet influence in the world, if not an actual threat that might be used in war, and a crash program was begun to build a bomb. Soviet scientists cooperated because they shared Stalin's feeling that the Americans should not be allowed to wield this power on their own. Partly as a result of information transmitted by spies, their progress was rapid, and to the consternation of the Americans they exploded their first bomb in October 1949. In the course of the 1950s, the world moved into a state of nuclear stalemate, as both sides acquired enough weapons to eliminate the other completely.

The British, too, felt left out of the nuclear club. They had initiated this area of research and had played an important role in the Manhattan Project. In the postwar era they had lost much of their international influence and saw the development of an independent nuclear deterrent as a way to preserve at least a semblance of their old position in the world. They went on to build bombs of their own, and the aircraft to deliver them, but as the superpowers moved into the age of intercontinental missiles and nuclear submarines, their status as a second-rank power became more apparent.

Even so, the cold war led to Britain's scientists benefiting more than those of any other European country from the funding made available for military research (Bud and Gummett 1999). That scientists actively promoted new military projects was confirmed later by the government's scientific adviser, Solly Zuckerman: "Our 'experts' would then inform and persuade their civil service and military colleagues—not a difficult task—and the idea would then find its way upwards until as often as not it reached Ministers" (Zuckerman 1988, 390). All too often, the resources needed to put the project into operation were beyond those available to a second-rank power—although the research had been done before operational constraints became apparent.

In America, the explosion of the first Soviet atomic bomb threw another debate into sharp relief. It was apparent to physicists that there was another, yet more powerful, bomb that could be made by fusing the atoms of hydrogen together, in effect duplicating the power source of the sun itself. This would only be possible using the immense temperatures and pressures reached in the explosion of an atomic bomb, so the hydrogen bomb would require an atomic bomb as a detonator. The architect of the program to build this "superbomb" was the physicist Edward Teller (York 1976). As a Hungarian Jew by origin, Teller had relatives in Europe living under Soviet occupation. He was acutely conscious of the threat posed by the Soviets' determination to impose their system on the world and saw the retention of American superiority in the arms race as essential. He had begun working on the physics of the fusion bomb at Los Alamos and lobbied relentlessly for support within the military and the government. News of the first Soviet atomic bomb added a new urgency to his campaign. In October 1949 the General Advisory Committee of the Atomic Energy Commission, chaired by Oppenheimer, recommended the development of improved atomic bombs but rejected Teller's arguments for the superbomb. Teller saw this as tantamount to surrender and began to use all his contacts with government to undermine Oppenheimer's position. Oppenheimer was vulnerable because he had had contacts with left-wing organizations as a young man, and this was the era of the anti-Communist witch hunts led by Senator Joseph McCarthy. After a lengthy investigation, Oppenheimer's security clearance was revoked in 1954, and he was evicted from the whole atomic energy program. J. B. Conant, who shared Oppenheimer's reservations about the H-bomb project, was also marginalized.

In 1949 the Atomic Energy Commission had supported Teller and his fellow "hawks" and rejected the advice of Oppenheimer's committee. In the following year President Truman, under the advice of the National Security

Council, authorized the development of the hydrogen bomb. The key technical problem was overcome with the invention of the Teller-Ulam device at Los Alamos. The first bomb was exploded at Eniwetok Atoll in the Pacific late in 1952, yielding the equivalent of 10 million tons of TNT—one thousand times the power of the bomb that had destroyed Hiroshima. The American lead was short-lived, however: the Soviets solved the technical problems in a different way and exploded their own first hydrogen bomb late in 1955. The possibility that nuclear weapons might destroy civilization, if not all life on earth, was now all too real and had a powerful effect on the public (Boyer 1994). Many scientists felt uncomfortable with Teller's hawkish strategy, which had given America only a temporary superiority and had ratcheted up the arms race to a new level of danger. Oppenheimer had become a somewhat isolated figure, even within the scientific community, although many were stirred by his assertion that the freedom necessary for scientific enquiry required an equivalent degree of freedom in society as a whole. Resistance to the unrestrained use of science to develop new weapons came more effectively from the German émigré Hans Bethe at Cornell University, who would eventually receive the Nobel Prize for having worked out the theory of nuclear fusion within stars (Schweber 2000). Although he had worked on the nuclear weapons project, Bethe became increasingly concerned about the implications of a nuclear war and played an important role as an adviser to the American team that negotiated the test-ban treaty of 1963.

The development of more powerful nuclear weapons was not, of course, the only scientific contribution to the arms race. Von Braun and his teams built on the achievements of the V-2 to found a rocket program that made possible a new delivery system, the intercontinental ballistic missile, but also laid the foundations of the American space program. The latter was, in fact, stirred into action by the rivalry of the cold war and the Russians' early achievements in this area, most notably the launching of the *Sputnik* satellite in October 1957. Soon the missiles were being launched from nuclear-powered submarines that could stay submerged for months in the hope of escaping detection. Navies wanted new methods of locating those submarines and demanded a better knowledge of the deep-sea bed where they might be hiding—one spin-off from this was better information about the sea floor that provided crucial evidence for the theory of plate tectonics. Studies of how the radiation from atomic bombs might increase the mutation rate in humans and other species represented an important source of funding for biologists (Beatty 1991). The interaction between science and the military thus began to flourish in many different ways, and the flow of information has not always been one way. What starts as applied science in one area sometimes provides evidence for new insights in an entirely different area.

Conclusions

The twentieth century saw a massive expansion in the relationship between science and the military. The early phases were tentative in nature: patriotic scientists suggested ways of improving weapons (or devising new ones) under pressure of national emergency, often to be greeted with hostility or derision by the military authorities. World War I saw the emergence of the first attempts to streamline the interaction, though none of the new weapons turned out to be decisive. During the interwar years, several nations built on these early efforts and began the integrated programs linking scientists, industry, and the military that generated genuinely new systems such as radar, capable of transforming the way navies and air forces (especially) would fight. World War II laid the foundations for scientists' involvement with the military-industrial conflict during the cold war. As a consequence of these developments, theoretical science acquired a new degree of involvement with industry, the military, and the government. The line between pure and applied science became increasingly blurred, especially in those areas where enormous amounts of funding were needed for equipment. Scientists also realized that technical problems could sometimes generate fascinating theoretical issues. Leading scientists now managed large projects absorbing vast amounts of industrial and government money and needed the managerial skills necessary to interact with those who provided the funds.

The emergence of a close relationship between science and the military had been delayed by the mutual suspicion inevitable between two professions with such different origins. But once that relationship was established, it is hardly surprising that scientists should be attracted by the funding it made available—especially if it allowed them to work on projects in which they were genuinely interested. By the 1950s, 90% of the funding provided for research in physics at American universities came from the Atomic Energy Commission, much of it for work on military projects (Hoch 1988, 95; see also Forman 1987). Small wonder that many scientists were willing to slant their research in this direction and to acquire the managerial skills needed to interact with the world of government and industry. More serious, for those concerned with the moral consequences of the relationship, was the temptation to promote the development of new weapons systems simply because this would open up the government's coffers to fund new areas of research. Teller almost certainly wanted the H-bomb because he feared the threat from the Soviet Union—but the more recent proposal of the Star Wars missile defense system has raised suspicions that the weapons designers have moved into the driving seat. Those scientists who actually work in defense industries are controlled by engineers and managers with commercial priorities.

After World War II, there were some efforts in the West to reestablish the ideal of pure science carried out solely to gain knowledge, partly because the Soviet system encouraged the rival view that scientists, like everyone, else, should work for the common good (invariably identified with the state). The leading American science adviser Vannevar Bush wrote a report in 1945 titled "Science: The Endless Frontier" in an attempt to re-create the image of the disinterested search for an understanding of nature. A firm foundation in pure research was necessary to ensure that technological spin-offs would subsequently emerge. This is still the orthodox notion of science promoted by many academic scientists, but it fails to acknowledge the extent to which much apparently pure research is now done with finance provided by industry and the military. Those scientists who most effectively confronted the moral dilemmas posed by the new situation were not those who retreated into isolationism, but those who accepted the engagement with the practical world and argued that scientists must use their influence to control the ways in which their work was exploited. This might involve active campaigning against the temptation to promote a new military technology just because it offered opportunities for research, but it might also involve constructive engagement with the military and political realities, as with Bethe's contribution to the signing of a treaty that would limit, at least, the dangers from the testing of nuclear weapons.

References and Further Reading

Alperowitz, Gar. 1996. *The Decision to Use the Atomic Bomb.* London: Fontana.

Beatty, John. 1991. "Genetics in the Atomic Age: The Atomic Bomb Casualty Commission, 1947–1956." In *The Expansion of American Biology,* edited by Keith R. Benson, Jane Maienschein, and Ronald Rainger. New Brunswick, NJ: Rutgers University Press, 284–324.

Boyer, Paul. 1994. *By the Bomb's Early Light: American Thought and Culture at the Dawn of the Atomic Age.* New ed. Chapel Hill: University of North Carolina Press.

Brown, L. 1999. *A Radar History of World War II.* Philadelphia: Institute of Physics.

Bud, Robert, and Phillip Gummett, eds. 1999. *Cold War, Hot Science: Applied Re-*

search in Britain's Defence Laboratories, 1945–1990. Amsterdam: Harwood Academic Publishers.
Buderi, Robert. 1997. *The Invention That Changed the World: The Story of Radar from War to Peace.* Boston: Little, Brown.
Forman, Paul. 1987. "Behind Quantum Electronics: National Security as a Basis for Physical Research in the United States, 1940–1960." *Historical Studies in the Physical and Biological Sciences* 18:149–229.
Frayn, Michael. 1998. *Copenhagen.* London: Methuen Drama.
Giovannitti, Len, and Fred Freud. 1965. *The Decision to Drop the Bomb.* New York: Coward-McCann.
Goodchild, Peter. 1980. *J. Robert Oppenheimer, "Shatterer of Worlds."* London: BBC.
Gowing, Margaret. 1965. *Britain and Atomic Energy, 1939–1945.* London: Methuen; New York: St. Martin's Press.
Haber, L. F. 1986. *The Poisonous Cloud: Chemical Warfare in the First World War.* Oxford: Clarendon Press.
Hackmann, Willem. 1984. *Seek and Strike: Sonar Anti-Submarine Warfare and the Royal Navy, 1914–1954.* London: HMSO.
Hartcup, Guy. 1988. *The War of Invention: Scientific Developments, 1914–1918.* London: Brassey's Defence Publishers.
———. 2000. *The Effects of Science on the Second World War.* London: Palgrave.
Hoch, Paul K. 1988. "The Crystallization of a Strategic Alliance: The American Physics Elite and the Military in the 1940s." In *Science, Technology and the Military,* edited by Everett Mendelsohn, Merritt Roe Smith, and Peter Weingart. Dordrecht: Kluwer, 1:87–116.
Hoddeson, Lillian, Paul W. Henrickson, Roger A. Meade, and Catherine Westfall. 1993. *Critical Assembly: A Technical History of Los Alamos during the Oppenheimer Years, 1943–1945.* Cambridge: Cambridge University Press.
Hogan, Michael J., ed. 1996. *Hiroshima in History and Memory.* Cambridge: Cambridge University Press.
Holloway, David. 1975. *Stalin and the Bomb: The Soviet Union and Atomic Energy, 1939–1956.* New Haven, CT: Yale University Press.
Hughes, Jeff. 2002. *Manhattan Project: Big Science and the Atom Bomb.* Cambridge: Icon Books.
Johnson, Brian. 1978. *The Secret War.* London: BBC.
Jones, R. V. 1978. *Most Secret War.* London: Hamish Hamilton.
Kevles, Daniel. 1995. *The Physicists: The History of a Scientific Community in America.* New ed. Cambridge, MA: Harvard University Press.
Mendelsohn, Everett, Merritt Roe Smith, and Peter Weingart, eds. 1988. *Science, Technology and the Military.* 2 vols. Dordrecht: Kluwer.
Neufeld, Michael J. 1995. *The Rocket and the Reich: Peenemünde and the Coming of the Ballistic Missile Era.* New York: Free Press.
Peyton, John. 2001. *Solly Zuckerman: A Scientist out of the Ordinary.* London: John Murray.
Powers, Thomas. 1993. *Heisenberg's War: The Secret History of the German Bomb.* London: Jonanthan Cape.
Price, Alfred. 1977. *Instruments of Darkness: The History of Electronic Warfare.* New ed. London: Macdonalds & Jane's.
Rose, Paul Lawrence. 1998. *Heisenberg and the Nazi Atomic Bomb Project: A Study in German Culture.* Berkeley: University of California Press.

Schweber, Sylvan S. 2000. *In the Shadow of the Bomb: Bethe, Oppenheimer, and the Moral Responsibility of the Scientist.* Princeton, NJ: Princeton University Press.

Swann, Brenda, and Francis Aprahamian, eds. 1999. *J. D. Bernal: A Life in Science and Politics.* London: Verso.

Walker, J. Samuel. 1996. "The Decision to Use the Bomb: A Historiographical Update." In *Hiroshima in History and Memory,* edited by Michael J. Hogan. Cambridge: Cambridge University Press, 11–37.

York, Herbert F. 1976. *The Advisers: Oppenheimer, Teller, and the Superbomb.* San Francisco: W. H. Freeman.

Zachary, G. Pascal. 1999. *Endless Frontier: Vannevar Bush, Engineer of the American Century.* Cambridge, MA: MIT Press.

Zuckerman, Solly. 1978. *From Apes to Warlords: The Autobiography (1904–1946).* London: Hamish Hamilton.

———. 1988. *Monkeys, Men and Missiles: An Autobiography, 1946–1988.* Reprint. New York: Norton.

CHAPTER 21

SCIENCE AND GENDER

THE RELATIONSHIP BETWEEN SCIENCE AND GENDER has been one of constant controversy for the last half-century and more. Science is frequently taken to be the ideal of objective enquiry, untainted by the class, political and religious conviction, race, or gender of its practitioners. As we have already seen, many developments in the history, philosophy, and sociology of science have in recent decades made this image of science as the ultimate in value-free knowledge increasingly hard to maintain. Few critiques of scientific objectivity have caused more controversy than those mounted by feminist scholars. Feminists have pointed to a number of problems with the traditional picture of objective scientific enquiry. A number of key texts published during the 1960s and 1970s, for example, have accused science of being a fundamentally masculine activity, some even suggesting that there are fundamental differences in the ways in which men and women interact with the natural world. Others have pointed to the fact that science has historically been an overwhelmingly male activity in terms of its practitioners. Others have pointed the finger at historians of science, too, accusing them as well as scientists of simply disregarding women's contribution to scientific endeavor. In this chapter, we will take a look at some of the key issues raised by feminist scholars and the arguments they put forward concerning the fundamentally gendered nature of scientific activity.

Commentators such as Evelyn Fox-Keller and Carolyn Merchant have suggested that the so-called Scientific Revolution of the sixteenth and seventeenth centuries brought about a transformation in the ways in which Europeans interacted with the natural world (see chap. 2, "The Scientific Revolution"). In particular, they associate the Scientific Revolution with the increasing dominance of a distinctively male way of looking at nature. Broadly speaking, they argue that before the Renaissance natural philosophers emphasized the importance of living in harmony with the world around them. The predominant image of nature was of the Earth Mother. With the rise of the New Science, however, nature increasingly came to be regarded as a resource to be exploited. Natural philosophers increasingly described their activities in terms of the exposure and penetration of a passive and female nature. Women were increasingly marginalized in the search for knowledge. Natural philosophers and scientists were (and still are) predominantly men. Some feminist scholars have suggested that the contributions of women to the study of science have been systematically written out of history. They argue for the importance of studying distinctively female ways of understanding nature by recovering the careers and lives of otherwise forgotten women scientists. By reassessing women's contributions to the sciences and encouraging more women to take up scientific careers, they hope to decisively change the practice of science and its relationship with nature.

Feminist historians of science have argued that women's bodies themselves increasingly became the subjects of scientific inquiry in the aftermath of the Scientific Revolution. The historian Thomas Laqueur, for example, argues that during this period a shift occurred from regarding male and female bodies as essentially similar to regarding them as fundamentally different (Laqueur 1990). While male bodies were considered normal, female bodies were increasingly regarded as pathological and hence more and more subject to medical and scientific intervention. Other historians have charted the ways in which eighteenth-century anatomists represented female skeletons as having smaller skulls (and therefore smaller brains) than their male counterparts. By the nineteenth century, doctors and scientists increasingly regarded women's bodies as being in need of careful medical regulation. While men's bodies were supposed to be securely under the control of their minds, women's minds were widely regarded as being under the control of their bodies, particularly of their reproductive organs. As a result, women were viewed as being inherently mentally and intellectually inferior to men. Claims like these were used later in the century to argue against women's education and against women's participation in the political process. Opponents of women's emancipation (like advocates of white European racial superiority) could argue that science showed that women (like non-Europeans) were physically and mentally unfit for a university education or for anything other than a subservient and domestic existence.

Is science inherently sexist then? Some feminist scholars argue that science as it has developed from the early modern period onward represents a fundamentally male perspective on nature. They argue that science plays a major role (if not the key role) in sustaining an essentially exploitative relationship between humans and the rest of the natural world. Furthermore, science and scientists are guilty of having systematically denigrated other, essentially female, ways of knowing that would encourage a more nurturing and ecologically friendly relationship with nature. There are other ways in which science might be regarded as inherently sexist, too. Science remains by and large an overwhelmingly male activity. It is certainly the case that the practice of natural philosophy and science in previous centuries has been an almost entirely male preserve. The few women who were able to engage in scientific pursuits were usually relegated to the margins of science. This might be taken as evidence of systematic discrimination against women by men of science. Then again it might be taken as evidence that science is the result of fundamentally male ways of thinking and that as a result few women find it an attractive pursuit. There are several ways of looking at these issues, and in this chapter we will only be able to provide a brief overview.

Mastering Nature

Some feminist historians of science have viewed the Scientific Revolution of the sixteenth and seventeenth centuries in a very different light from the way in which it is conventionally portrayed. Traditionally, at least, the Scientific Revolution has been widely regarded as the dawn of a new age of enlightenment. According to this view, the emergence of the New Science heralded the victory of experience over authority. The rise of the experimental method and the systematic application of human reason to understanding the laws of nature were seen as having decisively broken with the old scholastic Aristotelian philosophy. From this point of view, the Scientific Revolution was unquestionably progressive and essentially benevolent. It was fundamentally a good thing. As we have already seen, a new generation of historians of science has cast some doubt on this rosy traditional picture of unproblematic scientific progress (see chap. 2, "The Scientific Revolution"). Historians and philosophers of science are now far less convinced that there is any such thing as a unique scientific method. Historians of science are now more inclined to see the emergence of the New Science in the particular context of early modern European culture rather than regarding it as the inevitable outcome of the application of a universal human reason. Some feminist historians of science have suggested, in addition, that the Scientific Revolution was in theory and practice an overwhelmingly male and sexist enterprise.

In an influential account of the emergence of modern science published in 1980, the feminist environmental historian Carolyn Merchant suggested that the Scientific Revolution overturned traditional ideas about living in harmony with nature in favor of ecological exploitation that also sanctioned the subjugation of women (Merchant 1980). She pointed to the "age-old association" between women and nature and argued that the Scientific Revolution was responsible for bringing about a new mechanistic worldview that was directly responsible for the exploitation of both women and nature. Traditional philosophies of nature had regarded it as essentially feminine. The earth was a nurturing mother who provided for the needs and wants of humankind. This image of the earth as a mother carried with it strong ethical constraints against the exploitation of natural resources. For humankind to pillage the earth of its resources would be the moral equivalent of a child turning on its mother. From this perspective, traditional philosophies of nature advocated living in harmony with nature rather than seeking to exploit it. Along with the image of nature as a mother came the view that the cosmos should be regarded as an organic unity. The dominant metaphor for the universe was that it was a living body (fig. 21.1).

Merchant and others, such as Evelyn Fox-Keller, have argued that a key outcome of the Scientific Revolution was to overturn this traditional metaphor of the universe as a living female being and replace it with the image of the universe as a machine (Merchant 1980; Fox-Keller 1985). Where premodern Europeans had regarded the cosmos as being alive, the instigators of the Scientific Revolution argued that it was best understood as an inanimate collection of mechanical parts. The clock was their favorite metaphor for the operations of nature. The Greek philosopher Plato in his *Timaeus,* for example, explicitly described the universe as a living being with a female soul. His Neoplatonist Renaissance successors such as the English alchemist Robert Fludd similarly portrayed the world soul as a woman. Images like this expressly supported the idea that the universe itself was a living (female) being. Promoters of the New Science, including René Descartes, conversely , viewed nature in explicitly mechanical terms. Nature was a soulless machine set in motion by God. Even animals had no souls, according to Descartes. Other seventeenth-century natural philosophers such as the Englishman Francis Bacon or the Anglo-Irish Robert Boyle

FIGURE 21.1 The female soul of the world, illustrated in Robert Fludd, *Utriusque cosmi maioris scilicet et minoris metaphysica* (1617).

viewed nature in much the same light. What feminist historians of science have suggested is that the increased dominance of the machine metaphor brought about a radical change in the way in which Europeans saw themselves in relationship to nature. Nature was no longer a nurturing mother but a resource to be exploited.

In fact, some feminist historians have suggested that the increasingly pervasive metaphor describing the New Science's relationship with nature was that of rape. Insofar as the Scientific Revolution's instigators still regarded nature as female, they described their relationship with it in terms of domination and penetration. Francis Bacon described the process of experimentation as "the inquisition of nature" and suggested "that there are still laid up in the womb of nature many secrets of excellent use." The purpose of the New Science was to unveil nature, lay bare her secrets, and penetrate her mysteries (Merchant 1980). Fox-Keller similarly draws attention to Bacon's use of language in this context and the way in which he portrayed the experimental method in terms of forcing a feminine nature to submit to masculine power and authority (Fox-Keller 1985). Bacon's account of the scientific utopia of Solomon's House in the *New Atlantis* has little room for women's knowledge. Natural philosophy was increasingly characterized as an inherently masculine activity in which there was little if any role for women. Feminist historians of science draw parallels between the increasing masculinization of science with the rise of the mechanical philosophy and the increasing economic marginalization of women and attacks on women's cultural place through institutions such as witchcraft trials.

From this perspective, the Scientific Revolution is viewed as being intimately linked to the rise of capitalism and the beginnings of industrialization (see chap. 17, "Science and Technology"). Modern science is characterized as a philosophy that at minimum justifies widespread environmental destruction and the systematic overexploitation of natural resources. Merchant argues that organic views of nature as a nurturing mother figure at the very least acted as a brake on excessive environmental abuse. She points out that ancient authors such as the Roman writer Pliny explicitly drew on the Mother Earth metaphor in warning against excessive mining and deforestation, suggesting, for example, that earthquakes were expressions of the earth's displeasure at such pillage of her treasures. Destroying the sanctity of the earth's body through overeager exploitation of resources was regarded as an expression of avarice, selfishness, and lust. By destroying traditional accounts of the cosmos as an organic unity and describing it as a soulless machine, the mechanical philosophy sanctioned widespread attacks on the environment. Natural philosophers such as Bacon were explicit in their claim that "knowledge is power" and that the purpose of natural philosophy was to make nature's resources available for man's economic benefit. Natural philosophy in general—and the mechanical philosophy in particular—might be regarded in this way as philosophical and ideological justification for unlimited commercial and industrial expansion.

These arguments concerning the relationship between science as an expression of male power and as both a tool and a justification for environmental exploitation are an indication of the increasing links between the feminist and environmental movements during the second half of the twentieth century. Carolyn Merchant, for example, was quite explicitly trying to foster the growth of radical ecofeminism through her writings. Merchant and others saw their accounts of science both as efforts to place what they perceived as modern science's masculine perspective in historical context and as attempts to revive a more holistic and feminist view of humanity's relationship to the natural world. Much of what they have to say about early modern natural philosophy as a fundamentally masculine and antifeminine activity is difficult to argue with. The worldview of seventeenth-century natural philosophers was unquestionably and overwhelmingly male-oriented. Whether this makes natural philosophy any more of a gendered activity than any other during the early modern period is, however, rather more questionable. Their claims about the more organic and female-oriented philosophies of ancient and medieval writers are a little more difficult to accept at face value. More or less organic and mechanistic views of nature alike have been put forward by different thinkers throughout history. There seems little evidence that more organically inclined philosophers, like Plato, for example, were notably more woman-friendly than their mechanistic counterparts.

Scientific Heroines

While some feminist historians of science seek to demonstrate the essentially male-gendered nature of scientific activity, others try to demonstrate that women have in the past made a number of important and influential contributions to scientific knowledge. The main purpose of such studies is often twofold. On the one hand, some feminist historians attempt to show how male scientists (and historians of science) have systematically discriminated against women, belittling or ignoring women scientists' achievements. On the other hand, many efforts to recover the lost histories of female contributors to the sciences are frankly celebratory in nature. Their aim is simply to celebrate women's contribution and to offer female role models to aspiring women scientists (Alic 1986). Some also try to offer the example of past women scientists as case studies of the ways in which women approach the study of nature differently from men (Fox-Keller 1983). In this way, they hope to demonstrate that women's participation in the sciences might change the nature of scientific knowledge itself. At the very least, looking at women's contributions to the development of the sciences helps shift the focus away from the traditional view of science as the outcome of successive heroic discoveries and insights by great men. It helps demonstrate the extent to which a range of alternative views as to what science is, how it should be practiced, and by whom have always been with us (Abir-Am and Outram 1987).

One woman cited by Carolyn Merchant among others as an important example of the ways in which women's approach to the study of the natural world might differ from that of the prevailing male ethos is the early modern natural philosopher Anne Conway (Merchant 1980). Born into a well-to-do and politically influential family (her father had been Speaker of the House of Commons), Conway carried out an extensive correspondence with the Cambridge Platonist Henry More, who had been one of her brothers' tutors. In letters to More, she embarked on a philosophical critique of Cartesian dualism. She also corresponded with the Hanoverian philosopher Gottfried Wilhelm von Leibniz, who was later a particularly vociferous critic of Newtonian natural philosophy. Leibniz probably derived the term "monad," used in his philosophical attacks on dualism, from her writings. In later life, Conway became a Quaker—a dangerously independent move in seventeenth-century England (see chap. 15, "Science and Religion"). She died young, and her only complete philosophical work, *The Principles of the Most Ancient and Modern Philosophy,* was published posthumously in 1690. Conway's philosophical acumen was widely admired. More claimed that he had "scarce ever met with any Person, Man or Woman, of better Natural parts than Lady Conway." Feminist scholars frequently cite her Platonism and her opposition to Descartes's philosophical dualism and materialism as indicative of a distinctively feminine opposition to the prevailing trend toward the mechanical philosophy in early modern intellectual circles.

Like Conway, the English philosopher Margaret Cavendish was also an opponent of materialism. Coming from a family of Royalists during the reign of Charles I and the English Civil War, Margaret was one of the queen's ladies in waiting and fled with her mistress to Paris after the Royalists' defeat. There she married William Cavendish, a prominent Royalist and eminent natural philosopher. During her exile in France and after her return to England, Cavendish published widely on a number of topics, including natural philosophy. This in itself was highly unusual for a seventeenth-century woman. In 1667 Cavendish was granted permission to attend a meeting of the Royal Society to witness experiments carried out by Robert Boyle. Fellowship of the Royal Society was, of course, limited to men only, and there was considerable debate over whether a woman, however eminent, should be allowed to attend a meeting at all (see chap. 2, "The Scientific Revolution"). In a utopian tract published in 1666, *The Description of a New World Called the Blazing World,* Cavendish described an ideal scientific academy headed by a woman (herself) in which knowledge of nature was acquired through the help of anthropomorphic animal helpers. In writings like her *Observations upon Experimental Philosophy* (1666) and her *Grounds of Natural Philosophy* (1668), she argued for a view of nature as self-knowing and disputed some of Robert Boyle's claims concerning the role of experiment in natural philosophy.

The nineteenth-century woman about whom some of the most grandiose claims concerning her scientific contributions are made is undoubtedly Ada Lovelace (Stein 1985). She is often hailed as the "first computer programmer." Lovelace was the daughter of the English romantic poet Lord Byron and his wife, Anne Isabella. Her parents separated soon after her birth and she never met her father. Ada was privately educated by, among others, the Cambridge mathematician William Frend and Augustus de Morgan, the first professor of mathematics at the University of London. Ada moved socially in philosophical circles and was acquainted with a number of eminent scientists including Michael Faraday and Charles Babbage. In 1843 she translated for Babbage a description of his analytic engine by the Italian engineer L. F. Menebrea, including her own notes in which she described, among other things, a possible method for programming the analytic engine to tabulate Bernoulli numbers. This is the basis for her claim to fame as the first computer programmer or even, more recently, the "first computer hacker." Despite the utter anachronism of the description in an age more than a century before the invention of the first electronic computer, Lovelace does provide a good example of the kind of role some women did play in the early nineteenth-century scientific community (Toole 1992). She had a social status that allowed her to move easily in philosophical circles. She had the leisure and the inclination to become well-informed in natural philosophy, and her views and opinions were clearly taken seriously by her male scientific interlocutors. What she did not have as a woman was a systematic scientific education or the opportunity to join scientific societies and become a recognized contributor to science in her own right.

It was only toward the end of the nineteenth century that women started to get access to university-level scientific education in any numbers, though it is worth noting that until the middle of the century relatively few male scientists received formal university training in their subjects either. One of the first women to make a significant impact on the increasingly professional world of late nineteenth-century physics was Marie Curie, born in Poland as Maria Skodlowska. Educated in Paris at the Sorbonne, she had become interested in investigating the mysterious new form of radiation noticed by the French physicist Henri Becquerel in samples of uranium salts. Along with her husband Pierre Curie, Marie isolated two new radioactive substances—polonium and radium. In 1903 she and her husband were awarded the Nobel Prize for their research. Marie Curie was the first woman to be awarded the honor. Following her husband's death, Curie continued her research as the leading authority in the new field of radioactivity that she had played a key role in establishing. Curie became a real power in the world of physics not only by continuing to make important contributions but by becoming director of her own laboratory and building links between science and industry, too (fig. 21.2). Despite her stature however, she certainly found more obstacles on her path to success than would have been the case for a male scientist. Her career was almost destroyed, for example, when she was suspected of having an affair with fellow physicist Paul Langevin (Curie 1938; Quinn 1995).

The example of Rosalind Franklin is often used to illustrate graphically the difficulties and prejudices faced by women scientists in gaining recognition for their work (Maddox 2002). Franklin studied natural sciences at Newnham College, Cambridge, and gained her degree there in 1941. She completed a PhD in physical chemistry in 1945 before going to work in the Laboratoire Central des Services Chimiques de l'Etat in Paris, where she became familiar with the latest techniques in X-ray crystallography. Working at King's College, London, during the early 1950s, she was the first to produce clear X-ray pictures of DNA, which proved crucial in helping Francis Crick and James Watson discover the double helix structure of the DNA molecule. Her contribution to the discovery was consistently underregarded by her male colleagues, and she often found herself excluded from the informal gatherings where they discussed their work. Her pioneering X-ray photographs of DNA were shown to Crick and Watson without her

FIGURE 21.2 Marie Curie at work in her laboratory (image courtesy of the American Institute of Physics, College Park, MD).

permission (see chap. 8, "Genetics"). Franklin died of ovarian cancer in 1958, at the age of thirty-seven, four years before Crick and Watson, along with her King's College colleague Maurice Wilkins, were awarded a Nobel Prize for the discovery. James Watson, in his bestselling *The Double Helix,* describing the discovery of the structure of DNA, characterized Franklin as a frustrated and obstructive bluestocking, largely dismissing the role of her photographs in clarifying DNA's structure (Watson 1968).

Franklin's case is a good example of the difficulties faced by women scientists in a male-dominated professional world. Her fellow X-ray crystallographer, Dorothy Crowfoot Hodgkin, has been used as an example of the ways in which a woman scientist might try to forge a distinctive female career in a male-dominated scientific world. Dorothy Crowfoot studied chemistry at Oxford before moving to Cambridge to work with the Irish X-ray crystallographer and Marxist, J. D. Bernal. Like her mentor she was a socialist and a pacifist and was actively involved in groups such as the Association of Scientific Workers and the Cambridge Scientists' Anti-War Group. She married Thomas Hodgkin, a lecturer to the Workers' Education Association, in 1937. Hodgkin carried out her scientific research using X-ray crystallography to help work out the structure of medicinally valuable molecules such as insulin, vitamin B_{12}, and penicillin. Her explicit aim in this work was to put scientific knowledge to humanitarian use. She was awarded the Nobel Prize in Chemistry for her crystallography work in 1964. In keeping with her socialist ideals, Hodgkin also regarded science as a cooperative, rather than individualist activity. As a laboratory director, she encouraged openness and the sharing of ideas rather than competition. Traits such as these have been regarded as indications of a distinctively female approach to science (Hudson 1991).

Individual female scientists' careers have, as we can see, been used in a number of different ways by historians of science. They have been used to demonstrate that women have indeed made important contributions to scientific inquiry. They have been used to show the extent to which women have been marginalized and denigrated in their scientific efforts. They have been used to show how some women scientists have practiced a distinctively female science. Historians of science have also started looking more generally at the ways in which women have supported and sustained scientific activity. During the eighteenth and nineteenth centuries, wives and sisters to men of science frequently played important roles as helpers and assistants. The French chemist Lavoisier's wife took an active part in his experimental research, and the Anglo-German astronomer William Herschel was often assisted by his sister, Caroline Herschel. Nineteenth-century women such as Jane Marcet or Mary Somerville played important roles as scientific popularizers, writing scientific books for wide audiences (Neeley 2001). In addition, women were often an important component of the audience for science during the eighteenth and nineteenth centuries (see chap. 16, "Popular Science"). They also regularly played important roles in furthering alternative sciences, such as mesmerism and phrenology (Winter 1998). This is increasingly the way that historians of science now look at the role of women in science. Rather than trying to fit women into the traditional picture of a succession of great men making great discoveries, they look at the changing place of women within scientific culture.

Defining the Body

Much recent historical attention has been devoted to the ways science has been used in the past as a way of defining gender characteristics. Feminist historians often argue not only that science itself is a predominantly (or even essentially) male activity but that the ways in which science has in the past described and defined women and women's bodies are inherently sexist as well. According to this perspective, women's bodies have in the past been characterized as inherently inferior to those of men, and that inferiority has been assumed to have consequences for women's mental capacities and their place in society. Women's bodies—their reproductive organs in particular—have been regarded as making them particularly susceptible to mental or nervous disorders. Women have been characterized as being less capable than men of abstract reasoning (and hence as less capable of being good scientists). During the nineteenth century, these kinds of arguments were frequently put forward in opposition to women's education. New theories in physics, for instance, the conservation of energy, and in the life sciences, such as Darwin's theory of natural selection, were used to explain the inherent inferiority of women, both mentally and physically. In much the same way that scientific racism was used to justify the subjugation of non-Europeans, these kinds of scientific theories were used to justify the social subordination of women to men.

The historian and anthropologist Thomas Laqueur has suggested that the modern view of male and female bodies as being inherently and essentially distinct is of comparatively recent origin (Laqueur 1990). From the time of the ancient Greeks through to the early modern period, the physical difference between men and women was often characterized as one of degree rather than kind. Women's bodies were simply regarded as being less perfect versions of male bodies. Women's reproductive organs were viewed as being inverted male reproductive organs. The ovaries were regarded as equivalent to testicles, for example. The uterus was an inverted scrotum. The vagina was an inverted penis. According to the Greek philosopher Aristotle, the primary difference between male and female bodies lay in the amount of heat each possessed. Male bodies were hotter than female bodies, and it was this that resulted in male genitalia being pushed outward while female genitalia remained within the body. Up until the sixteenth or seventeenth centuries, popular tales of young women being transformed into men as a result of a sudden shock causing their reproductive organs to fall outward were widely circulated and taken at face value by natural philosophers and medical men. Increasingly, from the seventeenth century onward, however, men and women's bodies were regarded as being anatomically distinctive. The one-sex model of gender was being replaced by a two-sex model.

The historian of science Londa Schiebinger suggests that by the end of the eighteenth century anatomists were increasingly taking the view that the physical difference between men and women entailed much more than a difference in the location and function of the reproductive organs. It encompassed the entire body. She quotes an early nineteenth-century commentator as observing that the "entire life takes on a feminine or masculine character" (Schiebinger 1989). By the middle of the eighteenth century, a new generation of anatomists were drawing illustrations of the details of the human body—the skeleton in particular—that showed men and women as being anatomically distinctive at all levels. Male skeletons were typically drawn with longer legs than their female counterparts. Female skeletons were represented as having a broader and stronger pelvic girdle as befitted their function as childbearers. Female skulls were also often drawn as being proportionally smaller with regard to the rest of the body than were male skulls, as a mark of men's superior intellectual capacities. In his 1829 *Anatomy of the Bones of the Human Body,* the Edinburgh anatomist John Barclay drew the male skeleton compared with a horse, accentuating the male's structural strength and robustness. The female skeleton, by way of contrast, was drawn compared with an ostrich, emphasizing the large pelvis, the elegant neck, and the comparatively small skull (figs.21.3, 21.4).

By the nineteenth century, women were increasingly being represented as particularly prone to nervous and mental disorders resulting from their physical constitution. As several historians have pointed out, male bodies were typically regarded as normal, while female bodies were regarded as pathological and hence in need of constant medical and scientific intervention (Moscucci 1991). Women were viewed as being particularly susceptible to hysteria as a result of the disturbing action of the reproductive organs on the brain. In fact the term "hysteria" derives from the Greek word for "uterus." Mid-nineteenth-century experts on women's nervous diseases, such as the Edinburgh professor Thomas Laycock, argued that disorders of the female reproductive organs stimulated a reflex action in the brain, bringing about mental instability. As a result of such disorders, the "gentle, truthful, and self-denying woman" became "cunning, quarrel-

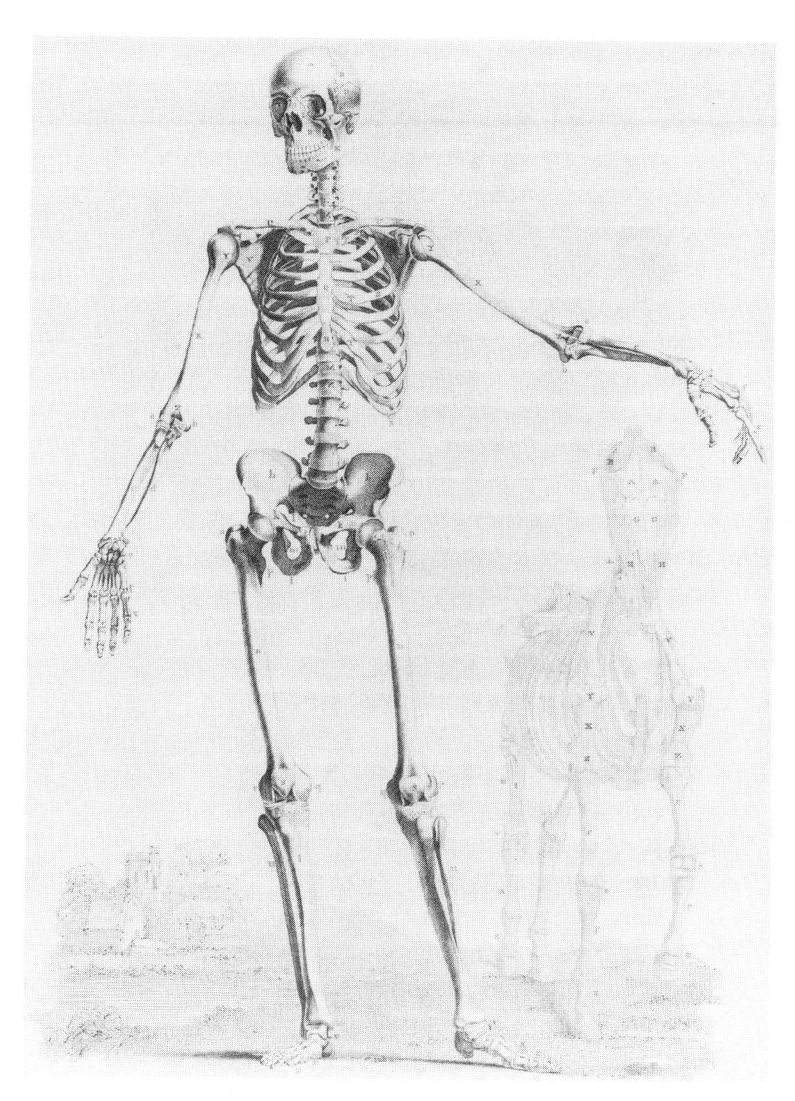

FIGURE 21.3 The human male skeleton's robust and masculine qualities demonstrated by comparison with the skeleton of a horse, from John Barclay, *The Anatomy of the Bones of the Human Body* (1829).

some, selfish; piety has degenerated into hypocrisy, or even vice, and there is no regard for appearances, or for the feelings of others." Victorian ideals of femininity were represented as scientifically established norms of female behavior (Showalter 1987). Deviations from that ideal were therefore often regarded as evidence of mental illness. Experts such as Laycock or Henry Maudsley in midcentury or such as Jean-Martin Charcot or Sigmund Freud

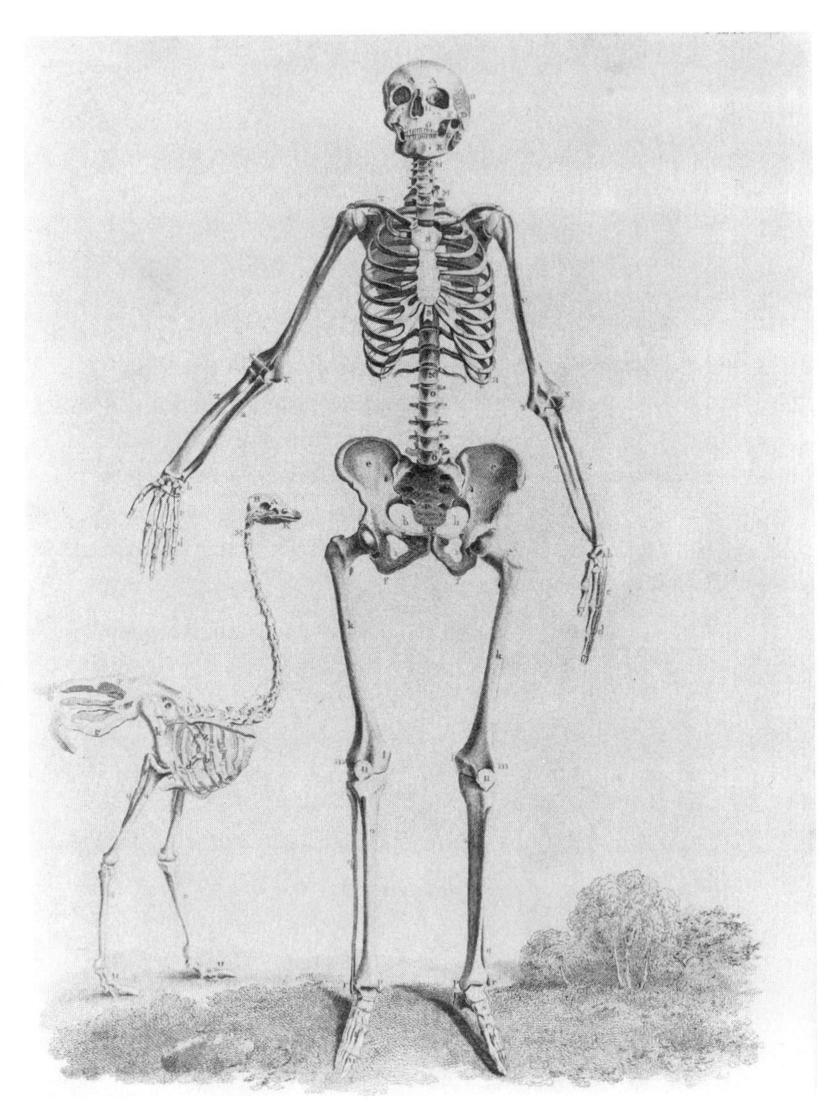

FIGURE 21.4 The human female skeleton's fragile femininity emphasized by comparison with the skeleton of an ostrich, from John Barclay, *The Anatomy of the Bones of the Human Body* (1829).

later in the century represented themselves as having a properly scientific knowledge of the workings of the female mind and body that allowed them to keep women's natural tendency toward mental deviance under control (Masson 1986).

As new scientific theories came along, they were often quickly adopted to provide an explanation for women's intellectual and physical inferiority. The mid-nineteenth-century physical theory of the conservation of energy is a case in point. Many nineteenth-century doctors and scientists broadly subscribed to the view that the human body contained only a finite amount of nervous force and that using too much of that energy for one purpose would lead to their not being enough available for other functions. The theory of the conservation of energy provided a powerful new rationale for this widespread assumption. It demonstrated, among other things, the dangers of women's education. If women became too educated, too much of the body's finite resources of nervous energy would be used up by the brain with the result that not enough would be available for other uses, such as reproduction, for example. The conservation of energy could therefore be used as an argument suggesting that allowing women access to university education would lead to sterility. It also suggested why most women were incapable of taking advantage of such an education. Quite simply, so much of their available store of nervous energy was used up in maintaining their reproductive organs that there was comparatively less available for intellectual activity. The very physics of women's bodies seemed to suggest that they were better suited for domestic life rather than public or professional activity (Russett 1989).

In much the same way, Darwin's theories of evolution by natural selection were used to show how women's place in society was determined by nature rather than by any social constraints. According to this view, those physical and mental characteristics that Victorians regarded as typically male or female were simply the result of evolution by natural selection. In particular, Darwin argued that the differences between men and women were largely due to the process of sexual selection. Men competed with each other to gain access to the most sexually attractive women. The result was that only the strongest and most resourceful bred successfully. Women in these circumstances were selected for simply on the basis of sexual attractiveness rather than any other qualities such as physical strength or intellectual capacity. Darwin's view was that the end result of natural and sexual selection was "man attaining to a higher eminence, in whatever he takes up, than women—whether requiring deep thought, reason or imagination, or merely the use of the senses and hands." Views like these concerning the evolutionary adaptation of men and women for particular roles in society were also developed by others, such as Darwin's friend and ally T. H. Huxley (see chap. 6, "The Darwinian Revolution' and chap. 18, "Biology and Ideology"). Late nineteenth and early twentieth-century anthropologists argued in a similar way concerning the ways in which men and women were adapted for particular social roles in different cultures (Richards 1989).

Examples such as these, feminist historians of science argue, show how science has been used to provide powerful support for the social subordination of women. Science from this perspective can be seen to have reinforced, if not actually created, social attitudes prejudicial to women's place in society. Arguments like these have often been put forward as examples of the way in which misogynist male scientists have allowed their prejudices to distort their scientific objectivity. According to this view, it is not so much science itself but individual male scientists who are responsible for promulgating stereotypical views of female inferiority. Similar arguments have been made concerning scientific racism. This view takes for granted that science is essentially objective and untainted by the culture in which it takes place. It assumes that there are "good" and "bad" ways of doing science and that sexist science, like racist science, is simply bad science. Others argue that science itself is inherently sexist and that it is therefore unsurprising that it produces views of women that reinforce male prejudices. From this perspective there is no such thing as good science. If we take the view that science is always the product of particular cultural circumstances, however, we might be less surprised to note the ways in which it often reflects the values of those particular cultures in which it was produced.

Is Science Sexist?

The most radical feminist perspective on science is that science itself, or at least science as it is currently practiced, is an inherently sexist activity. This argument is usually expressed in one of two forms. Some commentators point to substantial gender imbalance in the makeup of the scientific community both now and historically. They argue that this is an indication of institutional sexism within the scientific community, discouraging women from participating in scientific activity. Some such commentators argue for the need to introduce particular measures to make science appear more attractive to women. This is one motive for the trend we discussed earlier of attempting to recover the role of women in the past as significant contributors of new discoveries and insights. Some historians hope that such figures could be put forward as role models for potential women scientists. More radical feminist critics of science, however, see the gender imbalance as evidence of a deeper problem. According to this view, women tend to be underrepresented within the scientific community because science is the outcome of overtly male, fundamentally sexist ways of thinking and interacting with the world around us. From this perspective, the gender imbalance is far more than just a correctable historical trend: it is built into the very fabric of science itself (Harding 1986).

To a large extent, this argument is built on the claim, which we discussed at the beginning of this chapter, that modern science emerged from a view of nature as a female body waiting to be violated. These radical feminists critics point to the prevalence of metaphors of penetration, rape, and violation in early modern accounts of the scientific method—the experimental method, in particular—and draw the conclusion that they are an indication of something fundamental about the way science, then and now, regards the world. They take the view that metaphors like these are integral to the scientific worldview—that they are at the very heart of scientific inquiry. Furthermore, radical feminist critics argue that the kind of thinking that they say lies at the heart of science is fundamentally male. From this perspective it is hardly surprising that women tend not to become scientists. To become scientists, they would have to start thinking like men.

At the core of many of these feminist critiques of science lies the view that modern science maintains a fundamentally exploitative and destructive relationship with the natural world. This is what Harding has in mind, for example, with her characterization of science as maintaining "a conception of nature as separate and in need of control." Again, feminist critics argue that this is a characteristically male way of thinking. Men typically regard themselves as apart from nature and therefore as needing to be able to control it, while women typically regard themselves as part of nature and therefore as being in harmony with it. The critic of science Brian Easlea, in his influential book *Science and Sexual Oppression,* argues not only that science is inextricably linked to male oppression of women, but that it is more widely associated with (male) Western oppression of non-European cultures and environmental destruction. Easlea suggests that "when the potential that science offers and continues to offer for improving the life of all humankind is measured against the oppression and destructive reality that has so often characterized post-sixteenth-century science, then there can be little doubt that scientific practice has been overwhelmingly irrational" (Easlea 1981). He suggests that the only way of redeeming science is to overturn the dominant male perspective on nature and social relations at its foundation.

As an alternative to male science, many feminist critics offer the possibility of a science based on essentially female ways of knowing. They argue that rather than acquiescing to the dominant male perspective, women should develop their own feminist science. At their most radical, such commentators argue that far from encouraging more women to take up careers in science, feminists should actively try to dissuade women from engaging in a fundamentally misogynistic enterprise. This feminist science would be built on essentially female characteristics promoting harmony with nature. According to this view, just as male science is based on fundamentally male ways of thinking, a feminist science would be based on fundamentally female ways of thinking. Such a science would be intuitive rather than rational, for example, practical rather than abstract, cooperative rather than competitive, or nurturing rather than exploitative. Ironically, perhaps, some of these feminist critics of science appear to agree with their misogynist Victorian forebears that men and women do indeed think in radically different ways. In fact, they often appear to agree as to just what those differences are. The contrast, of course, is that feminist critics celebrate these essentially female ways of knowing as being superior to male perspectives on the world, while Victorian thinkers denigrated them.

Some feminist critics of science, however, have turned to postmodernism as offering a solution to the problems of masculine science. Rather than attempting to replace male scientific objectivity with an opposing and supposedly more inclusive female objectivity, feminists such as Donna Haraway advocate an acceptance of the fact that there are an indefinite number of ways of interacting with and making sense of the natural world. She suggests that all these different ways of knowing should be recognized as being equally valid (Haraway 1991). The model she offers is one of conversation. Haraway argues that rather than seeing the world as something passive to be mapped out and manipulated, scientists should regard nature as having its own agency and interact with it on those terms. Rather than adopting the traditionally masculine view of scientific objectivity as the "view from nowhere" she suggests that scientists and others should recognize and celebrate the fact that all knowledge is "situated." Haraway suggests that "Situated knowledges require that the object of knowledge be pictured as an actor or agent, not a screen or a ground or a resource, never finally as slave to a master that closes off the dialectic in his unique agency and authorship of 'objective' knowledge" (1991, 188). What she means by this is that postmodern perspectives can be taken to suggest that scientists should visualize themselves as being on the same level as the rest of the natural world (rather than somehow above or outside it) as they try to understand it.

Conclusions

As we have seen, feminist accounts of science operate on a number of levels. Some feminist historians of science have argued that science was imbued from its very beginnings with masculine, if not outright misogynist, implications. They argue that science adopted a view of nature as female, passive, and available for domination and control. Other historians of science have attempted to recover the contribution made by women in the past to scientific development. They argue that women's contribution to the sciences has been unjustly neglected and try to find scientific heroines to match scientific heroes such as Newton or Einstein. Others try to recover the wider role of women in the sciences as audiences, helpers, or popularizers. Most successfully, perhaps, some feminist historians of science have demonstrated how particular scientific theories and practices have been used in the past to sustain prevailing beliefs about the proper and subordinate place of women in society. Science was invoked to demonstrate that women's subordination was the result of nature rather than of culture. Some feminist historians have expressed this view in terms of male scientists' deliberate distortion of the evidence to support their misogynist beliefs, producing "bad science" along the way. Others have recognized that such "distortions" are the product of particular historical circumstances rather than being the result of a deliberate conspiracy.

Some feminist accounts of science are, as we have already hinted, inclined toward essentialism. In other words, they take it for granted that science has an "essence"—an unchanging core of defining features that remain static through its history. As historians, philosophers, and sociologists of science increasingly move toward the view that science is best understood as a patchwork of often competing activities, attitudes, ideas, practices, theories, and worldviews in a constant process of change, it becomes more difficult to accept that science is an inherently masculine institution or that there are inherently male and female ways of knowing. Not all the different feminist perspectives on science outlined here are mutually coherent. It is difficult to reconcile the view of some feminists that science is inherently sexist, for example, with the efforts by others to demonstrate women's scientific achievements as a way of providing role models for budding female scientists. According to the first view, after all, presumably there should be no good female role models in science. Feminist historians of science have nevertheless played a crucial role in producing a more balanced and nuanced account of scientific activity and its social relations. Very few historians of science would now deny that science has indeed in the past played a central and deleterious role in sustaining social inequality. It is also clear that scientific institutions have often been, in modern terms, institutionally sexist, discouraging and excluding women from equal participation in scientific activity. Feminists have certainly been successful in demonstrating that if gender discrimination exists in any society, then since science is a cultural activity, the science that society produces will reflect that discrimination, too.

References and Further Reading

Abir-Am, Pnina, and Dorinda Outram, eds. 1987. *Uneasy Careers and Intimate Lives: Women in Science, 1787–1979.* New Brunswick, NJ: Rutgers University Press.

Alic, M. 1986. *Hypatia's Heritage: A History of Women in Science from Antiquity to the Late Nineteenth Century.* London: Virago.

Curie, E. 1938. *Madame Curie.* London: Heinemann.

Easlea, Brian. 1981. *Science and Sexual Oppression: Patriarchy's Confrontation with Women and Nature.* London: Weidenfeld & Nicolson.

Fox-Keller, E. 1983. *A Feeling for the Organism: The Life and Times of Barbara McClintock.* San Francisco: W. H. Freeman.

———. 1985. *Reflections on Gender and Science.* New Haven, CT: Yale University Press.

Haraway, Donna. 1991. *Simians, Cyborgs, and Women: The Reinvention of Nature.* London: Routledge.

Harding, S. 1986. *The Science Question in Feminism.* Ithaca, NY: Cornell University Press.

———. 1991. *Whose Science? Whose Knowledge?* Ithaca, NY: Cornell University Press.

Hudson, G. 1991. "Unfathering the Thinkable: Gender, Science and Pacifism in the 1930s." In *Science and Sensibility: Gender and Scientific Enquiry, 1780–1945,* edited by M. Benjamin. Oxford: Blackwell.

Laqueur, Thomas. 1990. *Making Sex: Body and Gender from the Greeks to Freud.* Cambridge, MA: Harvard University Press.

Maddox, Brenda. 2002. *Rosalind Franklin.* London: HarperCollins.

Masson, J. M. 1986. *A Dark Science: Women, Sexuality and Psychiatry in the Nineteenth Century.* New York: Farrar, Straus and Giroux.

Merchant, Carolyn. 1982. *The Death of Nature: Women, Ecology and the Scientific Revolution.* San Francisco: W. H. Freeman.

Moscucci, O. 1991. *The Science of Woman: Gynaecology and Gender in England, 1800–1929.* Oxford: Oxford University Press.

Neeley, K. A. 2001. *Mary Somerville.* Cambridge: Cambridge University Press.

Quinn, Susan. 1995. *Marie Curie: A Life.* New York: Simon & Schuster.

Richards, Evelleen. 1989. "Huxley and Women's Place in Science: The 'Woman Question' and the Control of Victorian Anthropology." In *History, Humanity and Evolution,* edited by James R. Moore. Cambridge: Cambridge University Press, 253–84.

Russett, Cynthia E. 1989. *Sexual Science: The Victorian Construction of Womanhood.* Cambridge, MA: Harvard University Press.

Schiebinger, L. 1989. *The Mind Has No Sex? Women in the Origins of Modern Science.* Cambridge, MA: Harvard University Press.
Showalter, Elaine. 1986. *The Female Malady: Women, Madness and English Culture, 1830–1980.* New York: Pantheon.
Stein, D. 1985. *Ada: A Life and a Legacy.* Cambridge, MA: Harvard University Press.
Toole, B. A. 1992. *Ada, Enchantress of Numbers.* San Francisco: Strawberry Press.
Watson, James D. 1968. *The Double Helix.* New York: Atheneum.
Winter, Alison. 1998. *Mesmerized: Powers of Mind in Victorian Britain.* Chicago: University of Chicago Press.

CHAPTER 22

EPILOGUE

If anyone has read the chapters of this book consecutively, they should be aware by now that there is little possibility of bringing the whole to a neat conclusion. It is not our purpose to present the rise of modern science as the triumph of a coherent methodology and worldview with clearly defined consequences for the way we think and live our lives. On the contrary, by ending with the topics of science and war and science and gender we have revealed the diversity of interests and effects that the modern historian of science has to work with in trying to assess the nature of science and its interactions with society. The first part of our study revealed the many different directions in which science has developed and the variety of methodologies and theories that have emerged in these different areas. There is no single scientific method because a nuclear physicist simply does not ask the same kind of questions as an evolutionary biologist, let alone use the same techniques to arrive at answers. Nor does the theoretical framework of the physicist intersect with that of the biologist, except via a collection of intermediate areas, each of which, from biochemistry to geology, has its own problems and techniques.

The ideology of modern science sees an overarching unity in the commitment to the use of rational argument and objective evidence to decide between competing hypotheses. The objectivity of scientific knowledge is, according to this interpretation, guaranteed by the fact that it actually works when applied in practice. If we can predict how nature behaves well enough to control it through technology, then we must be getting ever closer to a true picture of how it works. There is clearly something in this argument but not enough to support the claim that science can build a single, unified, and permanently valid model of the world. The demand for objective testing certainly imposes limits that prevent scientists making up theories out of thin air, but it does not guarantee that there is only a single model that will provide accurate predictions. This is confirmed by the fact that scientific theories change through time, with the later theories providing better or wider predictions on the basis of foundations quite different to those accepted before. History suggests that the scientists' commitment to objectivity is all too often limited by constraints on their freedom to conceptualize how nature might work, some of them obvious, others so subtle they go unrecognized except in hindsight.

The second part of our survey has explored these constraints and influences to drive home the message that science can only appeal to objective evidence within a framework that is defined by its social environment. The involvement of scientists with the military-industrial complex is the most obvious illustration of the fact that the direction of research is to some extent determined by who is paying for it. There may still be room for pure intellectual curiosity, but that incentive will get far more results if it is applied in an area where funding is available. In some high-tech areas, progress is virtually impossible unless industry or governments can be persuaded to foot the bill. But even when socially privileged scientists were free to engage in research out of pure curiosity, the vision of nature within which they developed their theories was shaped by influences derived from religion, philosophy, and political ideology. The ubiquity of these external influences has persuaded almost all modern historians that it is impossible to identify lines of pure, objective research that remain uncontaminated by all subjective factors. Supporters of even the successful theories had their broader agendas, explicit or implicit, as the case may be. By showing that "good" science (i.e., science that became incorporated into the body of orthodox knowledge) was often influenced by scientists' religious and political opinions, we make it impossible to dismiss everything else as "bad" science distorted by values and opinions.

By ending with the theme of gender, we raise perhaps the most difficult problem of all for those who would defend the traditional ideal of scientific objectivity. Some scholars argue that the gradual exclusion of women from science has resulted in the emergence of attitudes toward nature that reflect a harsher and more masculine viewpoint, thereby marginalizing theories that look for a more interactive and holistic view. If this is so, then scientists need to face up to the possibility that ideas that they take for granted, and thus assume are morally neutral, are in fact the reflection of values so deeply buried that they are virtually impossible to identify and challenge. From this perspective it seems far more plausible that social and political values reflecting race and social class may have influenced—perhaps quite unconsciously—the kind of theories that scientists explored in the past. Darwinism and social Darwinism cannot be separated into valid science and invalid rhetoric, nor can genetics and eugenics. This does not mean that the theory of natural selection and the concept of the gene must be abandoned as figments of the ideologues' imagination. But it does mean that the inspiration for ideas that worked in the field and laboratory may have been derived from sources that—when viewed from a later era—were less than purely objective. The methodologies by which these ideas were tested could also be open to manipulation in ways that made the possibility of falsification less obvious at the time. If this is the lesson of history, it should be learned by everyone engaged in modern debates about science and its implications, scientists and nonscientists alike.

INDEX

HAVERFORD COLLEGE
3 1795 00540 6325